Environmental Geomechanics

Environmental Geomechanics

Contributors

Evgenii Sharkov and Valentina Svalova et al.

AURIS
Reference

www.aurisreference.com

Environmental Geomechanics

Contributors: Evgenii Sharkov and Valentina Svalova et al.

Published by Auris Reference Limited

www.aurisreference.com

United Kingdom

Environmental Geomechanics

ISBN: 978-1-78154-836-3

British Library Cataloguing in Publication Data
A CIP record for this book is available from the British Library

Printed in the United Kingdom

Exclusively distributed by CBS Publishers & Distributors Pvt. Ltd.

Sales & Distribution Rights only for India, Pakistan, Bangladesh, Sri Lanka, Nepal and Bhutan. This book is not to be sold outside these territories.

Contents

List of Abbreviations

ACF	Auto-Correlation Function
AWCP	Acoustic Water Column Profilers
CBM	Coal Bed Methane
CFE	Comisión Federal de Electricidad
CRP	Conservation Reserve Program
DEM	Digital Elevation Model
DITF	Drilling-Induced Tensile Fracture
DNRA	Dissimilatory Nitrate Reduction to Ammonium
ECD	Electron Capture Detector
FEM	Finite Element Method
GAM	Generalized Additive Model
GAMM	Generalized Additive Mixed Model
GLM	Generalized Linear Model
GLMM	Generalized Linear Mixed Model
LOT	Leak-Off Tests
LPB	La Popa Basin
MC	Multilayered Crust
MIMS	Membrane Inlet Mass Spectrometry
NOR	Nitric Oxide Reduction
PAL	Passive Aquatic Listener
REV	Representative Elementary Volume
SDR	Seaward-Dipping Reflector
SRV	Stimulated Reservoir Volume
TEB	Trans-Eurasian Belt

List of Contributors

E.V. Sharkov
Institute of Geology of Ore Deposits, Petrography, Mineralogy and Geochemistry RAS, Moscow, Russia

V.B. Svalova
Institute of Geoecology RAS, Moscow, Russia

B.A. Dyachkov
East-Kazakhstan State Technical University, Kazakhstan

M.A. Mizernaya
East-Kazakhstan State Technical University, Kazakhstan

Nina Maiorova
East-Kazakhstan State Technical University, Kazakhstan

Zinaida Chernenko
East-Kazakhstan State Technical University, Kazakhstan

Victor Maiorov
East-Kazakhstan State Technical University, Kazakhstan

O.N. Kuzmina
East-Kazakhstan State Technical University, Kazakhstan

Vsevolod Yutsis
Universidad Autónoma de Nuevo León, Facultad de Ciencias de la Tierra, Linares, N.L., Mexico

Antonio Tamez Ponce
Posgrado en Ciencias Geológicas, Facultad de Ciencias de la Tierra, Universidad Autónoma de Nuevo León, Linares, N.L., Mexico

Konstantin Krivosheya
Universidad Autónoma de Nuevo León, Facultad de Ciencias de la Tierra, Linares, N.L., Mexico

Feng Gui
Baker Hughes, Perth, Australia

Khalil Rahman
Baker Hughes, Perth, Australia

Daniel Moos
Baker Hughes, Menlo Park, USA

George Vassilellis
Gaffney, Cline &Associates, Houston, USA

Chao Li
Gaffney, Cline &Associates, Houston, USA

Qing Liu
Baker Hughes, Beijing, China

Fuxiang Zhang
PetroChina Tarim Oil Company, Korla, China

Jianxin Peng
PetroChina Tarim Oil Company, Korla, China

Xuefang Yuan
PetroChina Tarim Oil Company, Korla, China

Guoqing Zou
PetroChina Tarim Oil Company, Korla, China

Nima Gholizadeh Doonechaly
School of Petroleum Engineering, University of New South Wales, Sydney, Australia

Sheik S. Rahman
School of Petroleum Engineering, University of New South Wales, Sydney, Australia

Andrei Kotousov
School of Mechanical Engineering, the University of Adelaide, South Australia, Australia

Víctor Manuel Arellano
Instituto de Investigaciones Eléctricas, Gerencia de Geotermia, Cuernavaca, México

Rosa María Barragán
Instituto de Investigaciones Eléctricas, Gerencia de Geotermia, Cuernavaca, México

Miguel Ramírez
Comisión Federal de Electricidad, Gerencia de Proyectos Geotermoeléctricos, Morelia, México

Siomara López
Instituto de Investigaciones Eléctricas, Gerencia de Geotermia, Cuernavaca, México

Alfonso Aragón
Instituto de Investigaciones Eléctricas, Gerencia de Geotermia, Cuernavaca, México

Adriana Paredes
Instituto de Investigaciones Eléctricas, Gerencia de Geotermia, Cuernavaca, México

Emigdio Casimiro
Comisión Federal de Electricidad, Residencia Los Azufres, Campamento Agua Fría, México

Lisette Reyes
Comisión Federal de Electricidad, Residencia Los Azufres, Campamento Agua Fría, México

Marian Petre
Institute of Geology of Ore Deposits, Petrography, Mineralogy and Geochemistry (IGEM), Russia

Sharkov
Russian Academy of Sciences, Moscow, Russia

Charles B. Moss
Food and Resource Economics Department, University of Florida, Gainesville, FL, USA

Andrew Schmitz
Food and Resource Economics Department, University of Florida, Gainesville, FL, USA

Yugui Yang
State Key Laboratory for Geomechanics and Deep Underground Engineering, China University of Mining and Technology, Xuzhou, Jiangsu 221008, China
School of Mechanics and Civil Engineering, China University of Mining and Technology, Jiangsu 221116, China

Feng Gao
School of Mechanics and Civil Engineering, China University of Mining and Technology, Jiangsu 221116, China

Yuanming Lai
State Key Laboratory of Frozen Soil Engineering, Cold and Arid Regions Environmental and Engineering Research Institute, Chinese Academy of Sciences, Lanzhou 730000, China

Jatta Saarenheimo
Department of Biological and Environmental Science, University of Jyväskylä, 40014, Jyväskylä, Finland

Antti J. Rissanen
Department of Biological and Environmental Science, University of Jyväskylä, 40014, Jyväskylä, Finland

Lauri Arvola
Lammi Biological Station, University of Helsinki, 16900, Lammi, Finland

Hannu Nykänen
Department of Biological and Environmental Science, University of Jyväskylä, 40014, Jyväskylä, Finland

Moritz F. Lehmann
Department for Environmental Science, University of Basel, CH-4058, Basel, Switzerland

Marja Tiirola
Department of Biological and Environmental Science, University of Jyväskylä, 40014, Jyväskylä, Finland

Jennifer L. Miksis-Olds
Applied Research Laboratory, The Pennsylvania State University, State College, Pennsylvania, United States of America

Laura E. Madden
Applied Research Laboratory, The Pennsylvania State University, State College, Pennsylvania, United States of America

Preface

Environmental Geomechanics covers a range of topics that are of increasing importance in engineering practice: natural hazards, pollution, and environmental protection through good practice. Transport of contaminants and other substances may occur in the fluids, e.g. water, water vapour and air, filling the pores of geomaterials as happens in waste disposal problems or durability problems. Mass transport also takes place in reservoir engineering problems, where the fluids involved are oil, water, and gas. First chapter deals with "alive" tectonomagmatic processes, which are reflected in neotectonics, topography, geophysical fields, and the present-day magmatic activity. Geotectonic position and metallogeny of the greater Altai geological structures in the system of the Central-Asian mobile belt is proposed in second chapter. Third chapter presents an approach on geophysical modeling of the surroundings of La Popa Basin, Ne Mexico, with gravity and magnetic data. Fourth chapter provides a comprehensive geomechanical study to optimize stimulation for a fractured tight gas reservoir in the northwest Tarim Basin. In fifth chapter, an innovative analytical approach based on the distributed dislocation technique is developed to simulate the roughness induced opening of fractures in the presence of compressive and shear stresses as well as fluid pressure inside the fracture. This provides fundamental basis for computation of aperture distribution for all parts of the fracture which can then be used in the next step of modeling fluid flow inside the fracture as a function of time. It also allows formulation of change in aperture due to thermal stresses. The objective of sixth chapter was to investigate the exploitation-related processes through the analysis of geochemical and production data of 39 wells. The goal of seventh chapter is to show that the present-day tectonomagmatic activity within the TEB can be interpreted that a new ocean has begun to open here. Eighth chapter examines the changes in land use between 1947 and 2007 focusing on the possibility that commercial uses generate significant environmental benefits. In ninth chapter, it is recognized experimentally that the compressibility of warm and ice-rich frozen soil is remarkable under loading, which will cause a significant deformation and affect the stability of infrastructure constructed in cold region. In tenth chapter, we studied potential links between environmental factors, nitrous oxide (N_2O) accumulation, and genetic indicators of nitrite and N_2O reducing bacteria in 12 boreal lakes. In last chapter, we identify specific environmental parameters, including components of the ambient background sound that are predictive of ice seal presence in the Bering Sea.

Chapter 1

GEOLOGICAL-GEOMECHANICAL SIMULATION OF THE LATE CENOZOIC GEODYNAMICS IN THE ALPINE-MEDITERRANEAN MOBILE BELT

E.V. Sharkov[1] and V.B. Svalova[2]

[1]Institute of Geology of Ore Deposits, Petrography, Mineralogy and Geochemistry RAS, Moscow, Russia

[2]Institute of Geoecology RAS, Moscow, Russia

INTRODUCTION

Alpine-Mediterranean Mobile Belt, which currently ongoing to develop, is one of the best sites for studying of geodynamic mechanisms for formation of such regions. In this case we are dealing with "alive" tectonomagmatic processes, which are reflected in neotectonics, topography, geophysical fields, and the present-day magmatic activity. Last circumstance allows independent control of petrological processes in the underlying mantle of the belt. Comparison of geological-geophysical data available with the results of mechanic and mathematic simulation allows us to establish the relationships of all these processes and the character of their manifestation on the Earth's surface. This is the purpose of our study.

GEOLOGY AND PETROLOGY OF ALPINE BELT

Alpine-Mediterranean Mobile Belt (Alpine Belt) represents the western part of the huge Alpine-Himalayan collision zone, which appeared in the late Cretaceous-early Paleogene after closure of the Tethys Ocean. The suture of this neotectonic zone is traced by chain of late Cenozoic andesite-latite volcanism, stretching across Eurasia from Mediterranean to the Indonesian Island Arc and back-arc seas of the Western Pacific as well as areas of continental rifting and areas of intraplate basaltic volcanism.

The most complicated structure of this belt is in its west, in the Alpine segment (Fig. 1), where there is a whole system of mountain ridges, andesite-latite volcanic arcs and back-arc basins with thinned crust of intermediate to oceanic-type (Alboran, Tyrrhenian, Aegean Sea, and Pannonian Basin) occurs. Despite the differences in the morphology of these structures, they have several common features: along their periphery volcanic arcs and fold-thrust belts which form arc-shaped mountain ridges are developed. Among their thrust slices are often observed deep-water sedimentary rocks of Tethys, ophiolitic complexes, and sometimes blocks of the lower crust and upper mantle. In general, the situation in many aspects is similar to that which takes place on the active margins of continents and oceans. Such structures are characterized mainly for the West Mediterranean, while for the Eastern Mediterranean, as well as for the Black and Caspian seas typical passive margin. For this reason, we divide the Alpine Belt into two segments: the eastern, or the Aegean-Caucasian, and western, or proper Alpine, which will be considered separately.

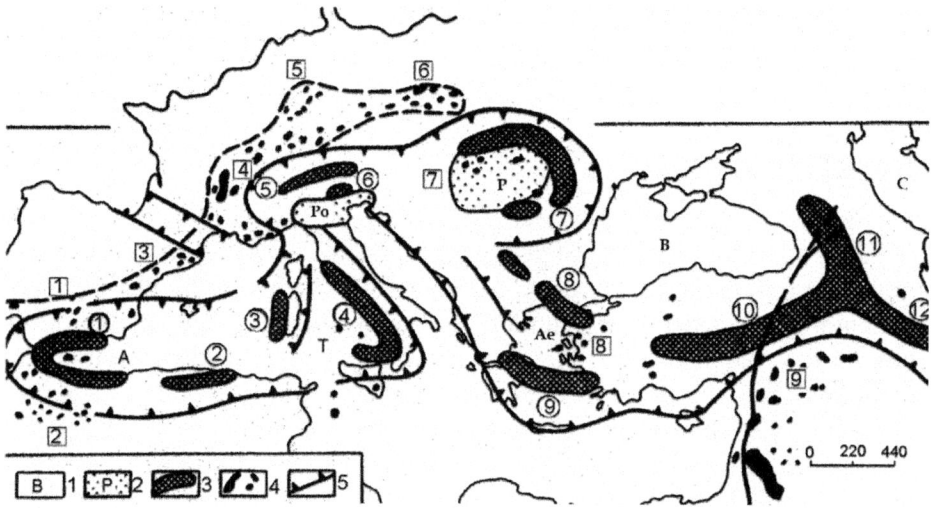

Figure 1. Development of the Late Cenozoic igneous rocks within the Alpine Belt1 – back-arc seas (A – Alboran, T – Tyrrhenian; Ae – Aegian) and "downfall" seas (B –Black, C – Caspian); 2 – back-arc sedimentary basins (P- Pannonian, Po – Po valley); 3 – Late Cenozoic andesite-latite volcanic arcs (in circles): 1 – Alboran, 2 – Cabil-Tell, 3 – Sardinian, 4 – South-Italian, 5 – Drava-Insubrian, 6 – Evganey, 7 – Carpatian, 8 – Balkanian, 9 – Aegian, 10-12 – Anatolia-Elbursian (10- Anatolia-Caucasian, 11 – zone of the Modern Caucasus volcanism, 12 – Caucasus-Elbursian); 4 – areas of flood basaltic volcanism (in square): 1 – South Spain and Portugal, 2 – Atlas, 3 – Eastern Spain, 4 – Central France massif, 5 – Rhine graben, 6 – Czech-Silesian, 7 – Pannonian, 8 – Western Turkey, 9 – northern Arabia; 5 – suture zones of major thrust structures interaction of a superplume head with mobile continental lithosphere. Good example of

such situation is the TEB (Sharkov, this book), where processes of collision continue now. The main feature of this belt is wide spread of the Late Cenozoic-derived volcanism, which has displayed practically coeval conditions on all its length, presuming existence of a superplume (or asthenospheric rise) beneath it. The belt has the most complicated structure within the Alpine segment, where a system of andesite-latite volcanic arcs and back-arc basin, bordering by nappe-folded mountain ridges were observed. In front of these ridges in Western Europe, north-west Africa and Arabia, coeval rift systems and flood basaltic volcanism often occur

Aegean-Caucasian Segment

Caucasian part of the segment is located to the north of the major suture zone traced by ophiolites of Cyprus, Syria, southeastern Turkey, Zagros, etc. (Periarabian ophiolitic arc). Late Cenozoic Anatolian-Elbursian andesite-latite volcanic arc is developed at its rear. It, in turn, is formed by two arcs – Anatolian-Caucasian and Caucasian-Elbursian, touching in the transverse (Transcaucasian) zone of the latest volcanism (Fig. 1). The Black and Caspian seas with oceanic crust, cut Pre-Pliocene structure of the Caucasus and the Kopet Dag; they are filled with Mesocenozoic sediments of 12-15 km thick. The nature of these deposits, until the late Miocene was generally similar to developed within the Caucasus and the Kopet Dag (Zonenshain, Le Pichon, 1986).

The Caucasus is located in the zone of the Arabian syntax (Burtman, 1989), where Arabian plate is subducted beneath Eurasian. Specific analog of deep-see trench is represented by the Mesopotamian trough here, which, beginning from the Eocene, is experiencing an active submergence, due to thick molasse accumulated (Ponikarov et al, 1969). The northern part of the Arabian plate began to rise above sea level in the late Oligocene and early Miocene, about 26-25 Ma when there began the development of basaltic volcanism (Sharkov, 2000) and the Red Sea rift opened up in southwest, separating Arabian plate from Africa. Rate of ascending movements sharply increased to the Miocene-Pliocene boundary, when Gulf of Aqaba opened approximately 5 Ma and Arabian plate began quickly shifted to the north along a large Levant Fault (Dead Sea transform) (Kopp, Leonov, 2000; Prilepin et al, 2001; Sharkov, this book). However, this displacement is hardly manifested in the Greater Caucasus shift and, moreover, GPS data indicate that the width of the central Caucasus is not decreasing but increasing (Shevchenko et al, 1999)

Specific structure occurred at the northern side of the Black Sea. Judging from the geological and geophysical data along the profile of Tuapse-Armavir, the Black Sea microplate is separated from the Eurasian by narrow subvertical zone of strong positive gravitational anomalies (Shempelev and et al, 2001). It concerned to large blocks of deformed and metamorphic rocks, close on the density to crust-mantle mixture. This zone can be traced to depth of 60-70 km and the Moho is not established here. The northern blocks move upward along their separating steeply-dipping faults, ensuring the existence of mountain relief of the Western Caucasus.

Formation of the Black Sea began, apparently, in the early Cretaceous, but significant deepening of the basin occurred at the Oligocene-Early Miocene boundary (Zonenshain, Le Pichon, 1986, Nikishin et al, 2001), followed in the Miocene by filling of the deep-water depressions by sediments and a gradual shallowing of the basin (Kazmin et al, 2000). Since the Pliocene-Quarternary new significant deepening of the Black Sea basin has occurred (Nikishin, Karataev, 2000), which occurred almost simultaneously with the uplift of the Caucasus and Crimea, which in the Oligocene-early Miocene were not expressed in the relief (Neotectonics…, 2000; Kostenko, Panina, 2001). The close sequence of events took place in the South Caspian Basin, which is a similar structure (Zonenshain, Le Pichon, 1986; Grachev, 2000).

Aegean part of the segment is characterized by island arc associated with subduction of the oceanic East-Mediterranean plate beneath the Eurasian (Papazachos et al., 1995). There are all typical structures of the active zones of transition from continent to ocean here: the deep-water Hellenic trough, Aegean volcanic arc and back-arc Aegean Sea with basaltic magmatism in its periphery (in the west of Asia Minor, near Izmir).

Numerous deformations of extension in subhorizontal submeridional direction (strike-slip and normal faults, nappe-thrusts, grabens, etc.) are known in the Aegean basin and in adjacent parts of Greece and Turkey (Prilepin et al, 2001). At that for Balkan Mountains in the north are characteristic north-vergentes imbricate nappes and thrusts whereas for Hellenides-Aegides-Taurides they are south-vergent. Judging from the seismic data, the stress state of the subhorizontal N-S stretching is characteristic of only for the upper 50-60 km of the Aegean basin lithosphere, leading to its expansion. At greater depths within the mantle beneath the basin both in south and north are fixed compression conditions. The regions of extension and compression are in direct contact by subhorizontal section at depth of 70-80 km; probably, it is the boundary of lithosphere and plastic material of extended plume head.

Aegean volcanic arc is a Pleistocene in age, but the development of the Aegean Sea, started earlier, about 12 Ma (Evsyukov, 1998). Apparently, earlier the main subduction zone located north, and its residues were survived in Dinarides and western Asia Minor (Western Anatolian: Fig. 1).

Descent of the Eastern Mediterranean (Ionian Sea, underwater Medina Ridge, Levant Basin) began in the late Miocene (Evsyukov, 1999).

Approximately at the middle Pliocene (about 3-3.5 Ma) processes reinforced on the east: judging by the results of the 5th Cruise of R/V "Akademik Strakhov", Sinai plate descended here beneath the sea level to depths of 2-2.5 km (Geological..., 1994). Fragments of this plate are preserved in the form of Eratosthenes Seamount and smaller rises.

The northern part of the Eastern Mediterranean is separated from the southern part by zone of powerful deformations, which runs along the base of the Cyprus arc (Zverev, 2002). As in the case of the northern side of the Black Sea, large sub-vertical faults and a sharp increase in the boundary velocities up to 7-7.6 km/sec is recorded here. In the northern zone is observed uplift of blocks of basement, represented by ophiolites of Cyprus. This deformation zone begins at the northern end of the Arabian plate, and following along Periarabian ophiolite belt, comes to Cyprus and then, continuing to the west, joins the Aegean subduction zone (see Figure 1).

The situation on the eastern passive margin of the Mediterranean Sea also looks like northern passive margin of the Black Sea. System of subparallel mountain ranges (Lebanon, Anti-Lebanon, Jabal al-Ansari, Amanus, etc.), separated by a system of steeply dipping faults, occur along the coast here. The largest of the latter is the aforementioned Levant (Dead Sea) Transform Fault. Parallel to it, already under water, there is a zone of Pelizium faults bounding the east Levant deep depression (Khair, Tsokas, 1999). This depression has the oceanic crust, overlapping by thick (approximately 10 km) sequence of Phanerozoic sediments, and in this respect no different from the Black and Caspian seas.

Origin of within-plate Palmirides folded-thrusted zone of deformed platform cover is related to development of Levant Fault. According to Kopp and Leonov (2000), the formation of this structure was caused by braking of the western edge of the Arabian plate in the zone above the bending S-like curve of this fault at its motion to the north during the Neogene-Quaternary. Palmirides were formed under compression of the crust by approximately 20-25 km, compensating the northern Arabian plate movement that started in the Middle-Late Miocene and continues today.

Areas of intraplate moderate alkaline basaltic volcanism are widely developed in the northern Arabian plate, indicating the presence of a mantle plume here. Judging from the isotopic dating, basalt eruption began at the Eocene-Miocene boundary, about 25-26 Ma, and almost without interruption continued until historic times. The most powerful eruption occurred in the late Miocene-early Pliocene and Late Pliocene-Quaternary (Sharkov, 2000; Lustrino, Sharkov, 2006).

The Caucasus-Aegean segment in geophysical terms is characterized by two strong positive isostatic anomalies, one of which is confined to the area of the Aegean Sea, and the second – to the Transcaucasian zone of modern volcanism on the north of Eurasian-Arabian syntax. It likely evidence about uncompensated excess of mass beneath these structures, presumably associated with the kinematics of the mantle plume ascending and extending of its head. This is supported by seismic tomography data (Gök et al., 2003) and consistent with the wide development of the Neogene-Quaternary platobasaltic volcanism in the north Arabian plate (Trifonov et al., 2011) and more rare – in Transcaucasia. In this regard, attention is drawn to the isotopic characteristics of lavas of Mount Elbrus, for which was establish the impurity plume material increases with time (Chernyshev et al., 2002) Another indication of the existence of the plume under the South Caucasus is a found of a mantle helium in the Lake Van in the north-eastern Turkey (Kipfer et al., 1994).

However, under the Eastern Mediterranean and the Caspian Sea, conversely, are a strong minimum isostatic anomalies, indicating the mass deficit beneath them, which is probably due to the presence beneath them descending flows in the mantle.

Structure and Development of the Western (Alpine) Segment

Judging on geological data, the current structure of the Alpine segment was formed mainly in the late Cenozoic, largely on the continental crust of the African plate. Remnants of this plate commonly observed along the northern coast of the Mediterranean Sea, and formed south of Spain (Betic Cordillera), Balearic Islands, Corsica and Sicily and the Apennines, the southern part of the Alps, large parts of the Balkans and Asia Minor (Ricou, 1986).

For the Alpine segment, in contrast to the Caucasian, is characteristic of complex configurations of the major structures related to thrusting the African plate beneath the Eurasia. Andesite-latite volcanic arcs, which associated with compression zones, represented by ridges, partly bordering back-arc basins with thinned crust and with often well expressed basaltic volcanism (Fig. 1). The formation of these subduction-related arcs occurs mainly in middle-late Miocene and Pliocene, and South-Italian arc is active till now. The feature of these subduction zones is that they involved material continental crust (Pino, Helmberger, 1997: Morales et al., 1999; Marson et al., 1995). For these arcs is characteristic their distinct migration in space – Alboran arc has moved to westward (Morales et al., 1995), Carpathian – to the east (Royden, 1989), and the Tyrrhenian – to the southeast direction (Rehault al., 1987).

Back-arc basins of Alpine segment (Tyrrhenian and Alboran seas, Pannonian Basin), became formed around the same time (Marotta et al.,

1995; Duggen et al., 2004). Originally the back-arc seas were developed as continental rifts, which submerged under the sea level approximately at the boundary of the Miocene and Pliocene. At the same time under the sea level began sink the South Balearic Basin, which became faster deeper in the late Miocene (Trifonov et al., 1999).

It is noteworthy that the region of the Alpine orogen, including all Western Mediterranean, is surrounded by a broad band of Late Cenozoic basaltic volcanism associated with rift structures of Central and Western Europe (grabens Rona, Rhine, Hessen, Polabian, etc.) as well as numerous basaltic plateaus, stretching to west from the French Massif Central via south of Spain to Portugal. It further extends beneath waters of the Atlantic Ocean (seamounts Amper, Josephine, and others), as well as on the islands of Madeira and Canary, boardering Alboran arc from the west. Powerful basaltic volcanism of Atlas occurs to the south of the arc. In the southwestern part of the Alpine segment basaltic volcanoes of islands Sicili (including Etna), Pantelleria, Lemos and seamount volcanoes on the Tunisia Threshold are occurred.

Together with platobasalts of Syria and southern Turkey, they form anorogenic circum-Mediterranean magmatism with common source – so-called Common Magmatic Reservoir (Lustrino, Wilson, 2007). It is, obviously, evidence about existence beneath the region a present-day mantle superplume; Alpine orogen with complicate combination of mountain ranges and basins is located in its inner part (Sharkov, this book). Earlier this entire basaltic volcanism was considered as the final, which appeared after the cessation of collision, but recent studies have shown that it is the beginning of a new distructive phase of the Europe development (Grachev, 2003).

The Alpine segment in geophysical terms represents a region with decreasing overall thickness of the earth's crust due to moving away of high-velocity layer of lower crust, characteristic of the East European Craton, and the high density of heat flow (Gize, Pavlenkova, 1988). The compression zones, represented by ridges, are characterized by deep roots (till 200 km in the Alps: Laubscher, 1988); part of lithoplastines, especially adjacent to the back-arc basins, has a steep attitude (Ricou, et al., 1986). In cases where the butt-end parts of lithoplastines come to the surface, within these contours are observed blocks of lower crust and even upper-mantle rocks lifted from 70-120 km, as it observed in Western and Eastern Alps and the Gibraltar arc (Magmatic... 1988; Harley, Carswell, 1995).

At the same time, oceanic crust, composes the floor of the newly formed seas of the Western Mediterranean, appeared due to thinning and rupture of continental lithosphere of the African plate. The latter survived along the periphery of these seas, in particular, the Tyrrhenian, where seismic data from the periphery to its center set reduction of crustal thickness from 20-30 to 6-8 km to the extent to complete disappearance of the "granite" layer (Royden et al., 1986; Marson et al., 1995). The development of this basin began from continental rifting in the middle-upper Miocene, and the open of the sea has occurred at the boundary of the Late Miocene and Pliocene and continues today (Bartole, 1995; Storti, 1996). In the Pannonian Basin reducing crust has the same trend - it is reduced from 30 to 18-20 km mainly at the expense of the "basaltic" layer (Nikolaev, 1988; Horváth et al., 2006). Develop-ment of the basin started in the Middle Miocene, 11-10.5 Ma, simultaneously with the appearance of basaltic volcanism here (Neotectonics..., 2000).

Attention is drawn to another feature of the back-arc basins of Alpine segment - they, as well as to the Aegean Sea, are associated with the maxima of isostatic anomalies of average (Alboran and Tyrrhenian seas) and large (Pannonian) intensity (Artemiev, 1971; Sparkman et al., 1993; Artemieva et al., 2006) (Fig. 2). As in the case of the Aegean-Caucasian segment, it may indicate uncompensated excess of mass beneath these structures associated with ascending of mantle plumes. These facts, along with materials on magmatism, show an essential deep mantle roots of observed geological processes here, as evidenced by extensive manifestations within the Alpine Belt epicenters of intermediate-focus earthquakes of the depths of 100 to 500 km (Tyrrhenian Sea, the Carpathians, Caucasus, etc.: Gize, Pavlenkova, 1988).

Thus, the formation of the major geological structures of the Alpine Belt began approximately at the boundary the Oligocene and Miocene and proceeded almost synchronously on all territory. At the first stage of development gentle uplift and subsidence of the relief of the region solid surface was dominated and began the formation of cavities of back-arc basins in the western Mediterranean, deepening of the Black and Caspian seas has occurred, platobasalt eruptions began in the north Arabian plate. An the second stage, which started in the late Miocene-early Pliocene, intense of tectonic activity increased sharply, began to form mountain ranges, as well as sharply increased basaltic volcanism along the periphery of the Alpine belt. All major features of the structure of the region were formed at that time (Trifonov et al, 1999), sharp deepening of the Black and Caspian seas, and the Eastern Mediterranean occurred; uplift of Caucasus and Crimea has begun as well as appearance of the Dead-Sea (Levant) Fault and Palmirides fold-thrusting structure. All of these processes has got impulse in the Pliocene-Quaternary, when finalized tendencies inherent in the current stage of the Alpine Belt development.

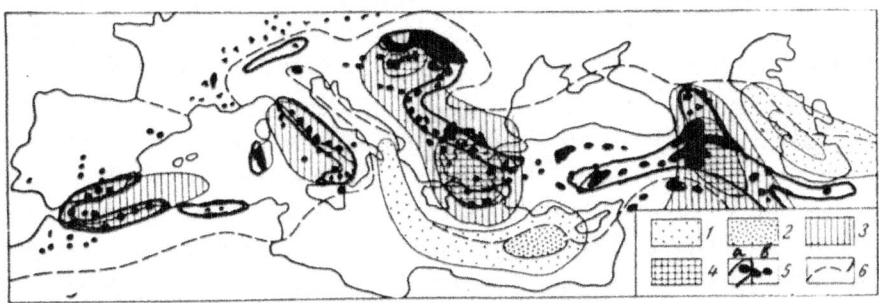

Figure 2. Distribution of major regional isostatic anomalies and areas of Cenozoic volcanism in the Alpine Belt. After M. Artemiev (1971) with corrections.1- regional lows of average intensity; 2 - of High intensity; 3 - regional highs of average intensity; 4 – of high intense; 5 - volcanic rocks (a - calc-alkaline series, b - basalts); 6 - boundaries of the Alpine Belt. Some depressions of the Alpine belt (Tyrrhenian, Aegian, Alboran, Pannonian) are characterized by positive anomalies, which evidence about excess of mass beneath them. Probably, they represent the present-day plume heads, which support basaltic volcanism and lead to onwards displacement of andesitic volcanic arcs in time. According to geophysical data, the plumes joint together in common layer at the depth 200-250 km, forming of circum-Mediterranean common magmatic reservoirs (Lustrino & Wilson, 2007). This rise starts in the North Atlantic, near Azores and extends to the east to Western Europe

MECHANICAL-MATHEMATICAL SIMULATIONS OF DEEP-SEATED PROCESSES IN THE ALPINE MOBILE BELT

Almost simultaneous occurrence of all these tectonomagmatic processes on the vast territory assumes that in this place we are faced with combination of present-day continental collision zone and a mantle superplume (Sharkov, this book). The relief of the superplume head is complicated by numerous protuberances (local plumes), controlling the position of modern depressions in the Alpine Belt and are caused extended zones on the general context of compression. The presence of such superplume under Alpine Belt also supported by seismic tomography data (Anderson, Dzevonsky, 1984; Sparkman et al., 1993; Hearn et al., 1999). This uplift starts in the Eastern Atlantic, extending eastward into parts of Western and Central Europe (Hoerne et al., 1995).

Surface of the superplume head is highly variable, apparently reflecting the development of gravitational Rayleigh-Taylor instability at the boundary of the rigid lithosphere and the heated the superplume head. Judging by the foregoing isostatic anomalies, beneath the back-arc depressions of the Alpine orogen (Tyrrhenian, Aegean, and Alboran seas, Pannonian Basin, etc.) an excess of mass is occurred, obviously connected with the existence beneath them heads of local plumes (protuberances). Extending of these heads led to displacement of subduction zones and their andesite-latite volcanic arcs (Harangi et al., 2006). Judging from the observations in the Aegean region, the thickness of an extended plume head does not exceed 40-50 km, and its spreading leads to appearance of field strong subhorizontal strength in the lithosphere in its front (see above). These plumes at depths of 200-250 km are merged into a single asthenospere layer, corresponding, apparently, to body of Alpine superplume head, which is the major source of geodynamic activity in the region.

The exception to this general rule is the North-Arabian-Transcaucasian plume, where while there were no basin, but there is a clear shift of the Anatolian-Elbursian subduction zone to the north. Perhaps this is due to the spread of the plume head to the north and its relative youth. Obviously this is due to the current increase in the width of the Central Caucasus, to which attention was drawn above.

In contrast to these structures, the Eastern Mediterranean, as well as the Black and Caspian seas are characterized by negative isostatic anomalies, which evidence of downward mantle flows beneath them ("cool plumes") located between ascended "hot plumes" (Sharkov, this book). Unlike the back-arc seas, all of them have typical passive margins and significant thickness of Meso-Cenozoic sediments. They look like "downfall", which cut-off earlier geological structures of the continent. Origin of such "cool plumes" obviously linked with appearance of excess of mantle material between extended plume heads.

Apparently, in these basins survived oceanic crust of Tethys. Judging from the northern sides of the Black Sea and Eastern Mediterranean, belts of strong positive gravity anomalies occurred along the sides of these basins. They formed by blocks of high-density rocks, separated by subvertical faults, receding into the mantle to the depths 60-70 km. These belts are reminiscent of similar zone of strong magnetic and gravitational anomalies developed along the passive margins of the Atlantic Ocean and are known as "seaward-dipping reflectors» (SDR) (Bogdanov, 2001; Larsen, 2002).

Very likely that such structures appear along the boundary between simultaneously active ascending (plumes) and descending currents in the mantle, due to rocks here are undergone by powerful deformation and metamorphism. As seen from the presented data, blocks, adjacent to the descending currents in the mantle are penetrated into its, and adjacent to the plume - ascended upstairs. Obviously, to the same circumstance, i.e., with the ascending of mantle plumes through the thickness of the lithosphere, can be related processes of exhumation of deep-seated rocks, large slices of which, as shown above, are observed in the mountain ridges on the periphery of plumes.

Thus, we can assume that the situation within the Alpine Belt defined by the presence beneath it the large (approximately 2000 x 5000 km) superplume head with a complex relief of surface. At the sites of the large uprising of this relief are usually placed back-arc basins now. The highest elevation of the superplume material is observed near the Tyrrhenian Sea, which probably can be interpreted as a modern center of activity of the whole zone. It is here very thin lithosphere (up to 30 km) and the maximum heat flux (above 3 E.T.P.) occurred. On the other hand, the most powerful isostatic anomalies are observed in areas of the Pannonian Basin and the Aegean Sea, which are reminiscent of the Tyrrhenian Sea at the early stages of its development, whereas the Caucasus syntax can be the very beginning of the process. Apparently, this implies that the center of activity will shift in the future.

However, it is possible to describe some of the characteristics of these structures, in particular areas of back-arc spreading over regional uplifts of the superplume surface relief even now. They exactly are the centers of deep-seated activity, which are very largely determined all the tectono-magmatic processes. For this analysis, we used the general model of high-viscosity incompressible fluid, the parameters of which vary from layer to layer (Zanemonets et al, 1974).

Development of Structured Above Regional Uplifts Astheno-sphere (Plume Heads): Mechanical-Mathematical Simulations

As it was shown above, depth of asthenosphere (superplume) surface under Alpine Belt changes from 30 km in the centre of Tyrrhenian sea up to 70-100 km in depressions of East Mediterranean, strongly changing on lateral. The characteristic size of depressions is 500-1000 km and more, distance between them is 1000-1500 km.

Hence we have characteristic parameters of a problem: $h_3 \sim 10$ km - thickness of sedimentary cover, $h_2 \sim 100$ km - thickness of lithosphere, $L \sim 1000$ km - horizontal scale, $= h_3/L = 10^{-2}$-small parameter.

Decomposing velocities and pressure in line on $\sqrt{\varepsilon}$ and considering the boundaries between layers as material ones, it is possible to receive in zero approximation the equation of a day surface ζ_3 and a surface of the basement ζ_2 in dependence of dynamics of mantle plume $U_0, W_0|_{\zeta_1}$:

$$\begin{cases} \dfrac{\partial^2 \zeta_3}{\partial X^2} = \beta \left[h_2 \dfrac{\partial U_0}{\partial X} - W_0 \right] \\[2ex] S \dfrac{\partial \zeta_2}{\partial t} + U_0 \dfrac{\partial \zeta_2}{\partial X} + \alpha \left[h_2 \dfrac{\partial U_0}{\partial X} - W_0 \right] = 0 \end{cases}$$

$$\alpha = \frac{\left(h_3 \right)^3}{\left(h_3 \right)^3 + \dfrac{\mu_3}{\mu_2} \left(h_2 \right)^3}, \quad \beta = \frac{1}{\dfrac{\rho_3}{3} \left[\dfrac{\left(h_3 \right)^3}{\mu_3} + \dfrac{\left(h_2 \right)^3}{\mu_2} \right]}$$

(1)

$S = \dfrac{L}{u_1 t_0}$ - Strukhal Number, u_1 - characteristic velocity of the lithosphere matter, t_0 - characteristic time of the processes, $_i$ - viscosity, i - density.

Let us set a field of velocities and morphology of boundary ζ_1 as:

$$U_0 = a \, \text{th} \, kX, \quad \zeta_1(X,t) = -\gamma \, \text{sh}^2 \, kX - (h_2 + h_3) + \frac{D}{S} t$$

(2)

where k, a characterize intensity of rifting: k-in the centre of structure, a - far from the centre: - allows to vary the form of rising plume; D - velocity of the plume rise $\quad D = S \dfrac{\partial \zeta_1}{\partial t}$ (fig. 4a).

The given field of velocities qualitatively enough well reflects the basic features of a considered class of movements: rise of plume, rifting above it and lowering of substance on sufficient distance from the centre. Quantitative conformity at comparison with the available geological-geophysical data is achieved with the help of a variation of coefficients in a modelling field of velocities and their change during considered process at preservation of the general structure of movements. From the decision of system (I) we shall receive for enough big t:

$$\begin{cases} \zeta_2 = -h_3 - \alpha \gamma \, \text{sh}^2 \, kX + \alpha h_2 \ln(\text{ch} \, kX) + \alpha(D - h_2 ak)\dfrac{t}{S} \\[3ex] \zeta_3 = \beta \left[\dfrac{h_2 a}{k} \ln(\text{ch} \, kX) + \dfrac{\gamma a}{(2k)^2} \, \text{ch} \, 2kX - \left(\dfrac{\gamma a + D}{2} \right) X^2 \right] + C_1(t) \end{cases}$$

C1 (t) is determined from balance of mass.

The analysis of the received equations shows, that there is a critical depth of rise of mantle plume $h_2 = 2$, when the characteristic form of the lithosphere layers changes. If $h_2 > 2$, there is a deflection of a surface of the basement in the centre of rifting, that really takes place in considered back-arc seas. If $h_2 < 2$ (depth of plume is insignificant) or velocity of its rise is essential ($D > h_2 ak$)

then the swelling of the basement surface corresponds to swelling and rise of a surface of the mantle plume (fig. 3).

When on periphery of basin there are the conditions interfering free rifting of the lithosphere of region, for example, caused by collision of the Arabian-African and Eurasian plates, the field of velocities on the bottom boundary of layers can be modelled as:

$$U_0 = \frac{\text{th } X}{\text{ch}^2 X}, \quad \zeta_1 = -\text{sh}^2 X - (h_2 + h_3) \tag{3}$$

For the greater presentation of result the coefficients in a modelling problem are omitted.

Then:

$$\begin{cases} \zeta_3 = -\dfrac{\beta h_2}{2}\dfrac{1}{\text{ch}^2 X} + \dfrac{\beta}{2} X^2 - \beta \ln(\text{ch} X) + C(t) \\ \zeta_2 \cong -h_3 - \dfrac{\alpha h_2}{S} t - \alpha(1-2h_2)\text{sh}^2 X + \alpha(1-2h_2)(\text{sh} X)^{\frac{2(h_2-1)}{1-2h_2}} \exp\left[\dfrac{2-3h_2}{1-2h_2}\left(\dfrac{t}{S} - \text{sh}^2 X\right)\right] \end{cases}$$

C (t) it is determined from balance of mass.

Now there are two critical depths of the asthenosphere upwelling, when cross-section of layers changes its structure. At $h_2 > 2/3$ in the centre of structure the deflection is formed. At $1/2 < h_2 < 2/3$ the surface of the basement is inclined, and at $h_2 < 1/2$ it reflects the morphology of plume in the centre of rifting and forms concavity of the basement on periphery of basin (fig. 3b).

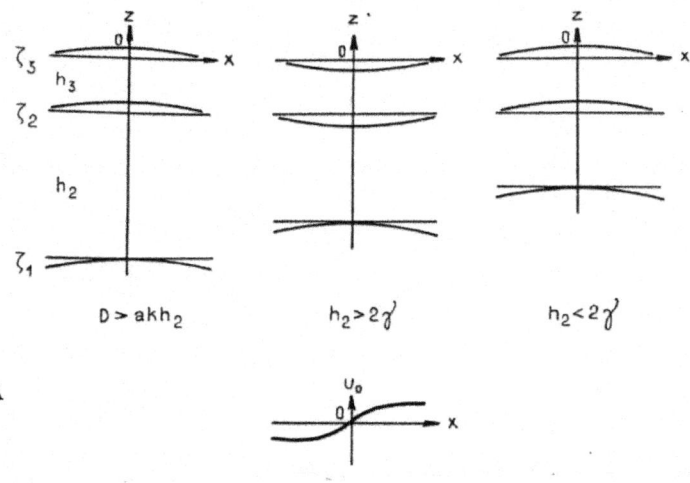

A

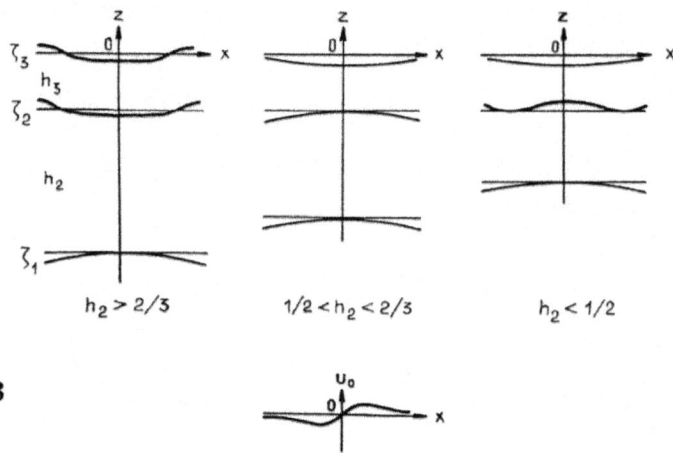

Figure 3. Characteristic cross-sections of layers of earth crust and subjacent lithosphere above ascending mantle plume: a – without lateral limitations ($U_0 = $ th kX), b – with lateral limitations ($U_0 = $ th kX / ch^2 kX; D = 0)

The first type of velocity (2) can simulate the early stages of structures development, and the second type (3) - Alboran, Tyrrhenian and Aegean seas, as well as the Pannonian depression. The second type of activity, according to geological data, often accompanied by a extending of the local plume heads. Under conditions of the collision zone, at the boundary of the extended plume head and limiting its blocks of the continental lithosphere arise powerful strengths. It lead to formation of zones of deformations and metamorphism, which can later develop into zone of downward flowage of the material (subduction zone), to which involves an excess of crustal material, appeared as a result of displacing of the material. It is often activated already existed subduction zones also, as evidenced by the data on the Western Mediterranean (Morales et al., 1999). Change of regime of upwarping to structure of deepwater basin is confirmed by a number of geological factors: the change in the regime of sedimentation and the removal of terrigenous sediments, wedgeout of layers of sedimentary cover, changing the direction of flow paleorivers, evolution of paleodepths of basins, etc. (The crust..., 1982).

An example of the interaction of ascended plumes with the earth's crust in the absence of side limitations, apparently are intracontinental rifts such as Baikal Rift. As Grachev (1987) shown, during pre-rift stage of their development an overall rise of the territory occurs, and the on the rift stage itself – its descent with the formation of extensional structures – grabens and sedimentary basins. In contrast to the of collision zones, deep-water basins with oceanic crust are not formed. The case is usually limited to thinning of the crust and the relatively small submergence of the crystalline basement for some kilometers. Only in exceptional cases, when a powerful inflow of plume material may be breaking the crust with formation of structures such as the Red Sea.

Thus, the proposed model seems to adequately describe the mechanism of formation of geological structures associated with the plume tectonics. From this it follows that the formation of depressions over the ascending mantle plumes depends on the geodynamics of the deep-seated layers. Determinants are mechanical processes that reflect the general direction of movements from formation of upwarping to formation of basin with unidirectional motion (plume ascending). At that formation of deep basins do not require much stretching of the layers. Morphology of the deep-seated boundaries determined by the shape of the plume, the rate of its ascensing and the intensity of moving apart a material over it, i.e., effective viscosity of the lithosphere layers. For sufficiently large gradients of plume heads surface that ascending in areas of collision, above them are formed deep-water depressions such as the Tyrrhenian Sea.

Development of Volcanic Arc – Backarc Systems

The principal components of such systems a subduction zone accompanied by andesite-dacite magmatism and a newly formed back-arc basin with transitional to oceanic crust, formed originally on the continental crust (Bogatikov et al., 2009). Both structures generally seem to have been initiated and developed about the same time, possibly implying a common reason for their formation. However, the nature of back-arc basins and their role in geodynamic processes were largely overlooked until recently, even though these settings are likely to have played the most crucial role here. High heat flows, positive isostatic anomalies, extensive basaltic magmatism, and the presence of basalt-hosted mantle xenoliths suggest that the back-arc basins have developed above the asthenoshere rises or extended plume heads. This is in good agreement with seismic tomography results, which revealed that the back-arc basins are underlain by a hot mantle as deep as 400 km (Anderson et al., 2002). Some of Mediterranean basins (Alboran, Tyrrhenian, Aegean, and Pannonian) are also marked with large positive isostatic anomalies mostly confined to back-arc basins (Fig. 2). This suggests uncompensated excess mass which can be related to mantle plumes upwelling beneath these structures. Like the Pacific back-arc basins, Mediterranean's back-arc basins are also characterized by extensive basaltic magmatism and intermediate- to deep-focus earthquakes of 100-500 km in depth. During stretching of the relatively soft oceanic upper mantle (asthenosphere), its edges exert mechanical pressure on the continental block. The rotational motions of the continental block cause some part of the asthenospheric material to move beneath the softened base of the latter and then begin to ascend as an independent "asthenospheric" plume (Sharkov, Svalova, 2005). In our case, the role of asthenosphere plays material of superplume head and protuberances on its surface (local plumes). Above a certain depth, after reaching the buoyancy level, the plume head begins to spread out laterally to form an extensional zone, such as a continental rift zone developed above the plume, as was the case for the initial stages of the development of the Sea of Japan and Sea of Okhotsk (Lelikov, Emel'yanova, 2007). However, unlike the ordinary rift zones, where the situation is symmetric on either side of the spreading axis, here it is sharply asymmetric: on one side is the massive cold continental lithosphere, and on the other side is a less dense oceanic plate. Under such conditions, the spreading may have occurred in a different way, mainly oceanward, to the side of mechanical downdragging. Accordingly, the continental material transported by the spreading plume head is expected to move in the same direction, where both material flows, migrating from the ocean and continental side consequently, will eventually collide.

We consider the sequence of events on the basis of the mechanical-mathematical model for a multilayer viscous incompressible fluid, describing the dynamics of "granite" and "basalt" layers of the earth's crust, lithosphere, and asthenosphere (Sharkov, Svalova, 1991; 2005). At the initial stages, when deflection of the layer boundaries from their original position is still insignificant, the base of the lithosphere always dives into the asthenosphere. In other words, during the early stages of structural evolution in zones where interaction between the plates is the most active, the lithosphere sinks into the less dense asthenospheric material to form a subduction zone. The calculations imply that the most dense rocks of the ancient continental lithosphere, its mantle and lower crust, made up of garnet granulites, which are much denser than rocks of oceanic crust, were first to begin descending. Much sialic material of the upper continental crust (its granitic layer) from the back-arc region, which was sandwiched between two subsiding plates (oceanic and lower continental crust), may have also been involved into the overall motion. Having a subduction rate of 7-10 cm/year, this motion causes rocks of this layer to be sucked in the subduction zone. As a result, a MORB-type oceanic lithosphere in a back-arc setting will develop when the old continental lithosphere is partially or fully removed to be involved in this subduction zone and buried in the deep mantle (Fig. 4).

As the subducting plate (slab) sinks, its rocks transform to high-dense amphibolites, garnet granulites, and eclogites as well as relatively light sedimentary rocks, volcanics, and gneisses (including granite-gneisses).

Meanwhile the latter, being metamorphosed under ultrahigh pressure and moderate temperature ($P > 2.8-4$, possibly, up to 8.5 GPa, $T = 600-900°C$), often retain their structural and textural characteristics in a metastable state and are thus easily recognizable (Ernst, 2001).

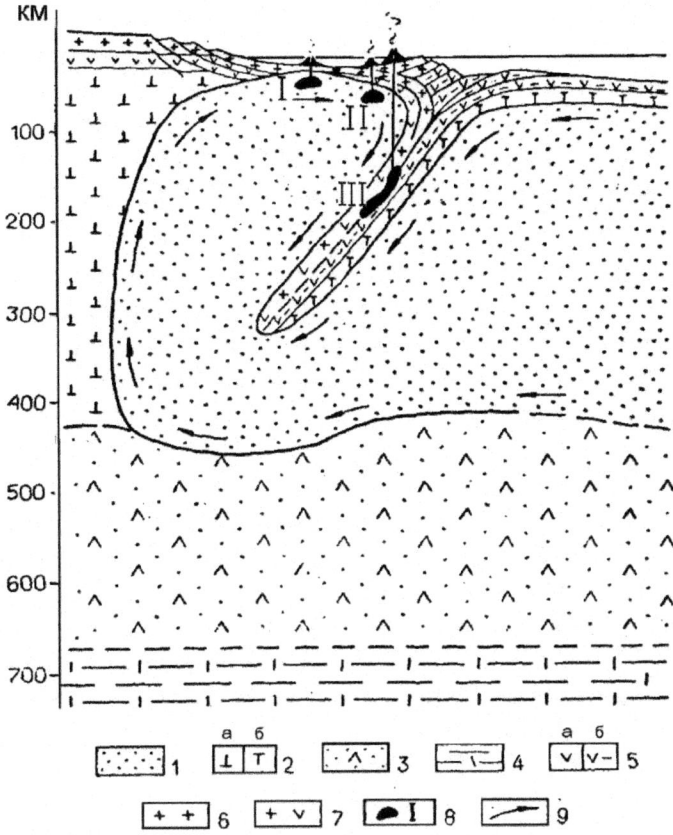

Figure 4. Schematic view showing the evolution of back-arc spreading1) Young soft upper mantle (asthenospheric) beneath oceans; 2) lithospheric mantle beneath: (*a*) continent, (*b*) ocean; 3) upper mantle beneath the 410 km discontinuity; 4) lower mantle; 5) lower crust of (*a*) continent, (*b*) ocean; 6) sialic continental crust; 7) mixture of the sialic and basite crustal materials within a subduction zone; 8) magma chambers (I - tholeiite series; II - boninite series; III – calc-alkaline series); 9) flow direction of the soft oceanic upper mantle.

Thus, the development of Late Cenozoic volcanic arcs and their adjacent back-arc basins, both in the Pacific and Mediterranean, have evolved in similar way. Their principal component is a back-arc mantle (asthenospheric rise or plume head) that possibly spreads out laterally towards the less dense lithosphere. At the place where the mantle flows collide, new subduction zones and back-arc basins at their rear have been formed. This results in the gradual deepening of the back-arc sea, thinning of its parental continental crust, and formation of transitional and oceanic type crust. At early stages, these systems might have looked like active continental margins, which then evolved in a complex arc-backarc system. Through continuous involvement in subduction processes, crustal material from the back-arc region is removed from the tectonosphere and stored in the "slab cemetery," revealed in the mantle by seismic tomography (Karason, van der Hilst, 2000). Only a minor portion of crustal materials is returned to the surface in form of subduction-related magmatism.

So, asymptotic models provide an opportunity to explore the main features of the formation and development of geological structures. Ascending of mantle plumes determines the depth of the deep-water basins initiation. The lithospheric material above them (the uppermost mantle and crust) is moved apart to make room for an oceanic crust. As already mentioned, the resulted excess of this lithospheric material is involved in subduction zones with formation of volcanic arc-backarc basin system. The process of subduction depends on the difference in density of the lithospheric and plume material, and can only contribute to spreading, freeing up space for moved apart crust from the back-arc basin. However, existence of mantle plume itself, which occurred against the background zone of lithospheric plate collision, is the result of collision of deep-seated mantle flows, contributing to pumping and ascending of local protuberances in the form of plumes with the formation structures of extension over its extended heads. Relationship between the area of collision of mantle flows at the depth and the zone of collision of lithospheric plates, along with relative velocities and ratios of densities, determines the dippig of subduction slab also.

Thus, complex processes in the zone of lithospheric plates collision are interrelated and interdependent. The relative rapidity of the observed processes evidence about defining role of mechanical motions in the formation of structures, forced by the influence of thermal factors. The Alpine Belt is an area of elevated heat flows due to removal of abyssal heat by ascending of plumes caused by deep-seated (up to the core) mantle activity and complex interaction with the thinned lithosphere. Analysis of times of formation of the Mediterranean's deep-water basins shows that the process of activation occurred from the periphery to the center, which was caused by compression and ascending of superplume material between lithospheric plates.

It is possible that a complex dynamic pattern of interaction between a plume and litho-sphere is not limited to the examples above. As seen in Fig. 1, volcanic arcs are lenticular in shape and are distributed in space fairly chaotic, which differ them from the Western Pacific island arcs or active margins of the both Americas. However, as it know, active continental margin occurred along the northern margin of northern Tethys in this place in the Mesozoic, where zone of subduction existed and there were a powerful eruptions of lavas of calc-alkaline series (Khain, 1984). In this regard, it has been suggested that this ancient subduction zone during early-middle Miocene split into two fragments, one of them roll back, forming the Tyrrhenian Sea and the modern Calabrian arc, and another rolled to west, forming Alboran Sea and Betik Reef arc (Lonengran, White, 1997).

Continuing this logic, one might think that one more fragment rolled to the north-east, forming the Carpathian arc. In this case the Anatolian-Caucasus-Elbursian arc may represent the eastern fragment of this old subduction zone. As already stated, under the central part of this arc the northern end of a mantle plume occurs with which probably connected the bend in place of Transcaucasian structure, where the area of modern magmatism of the Caucasus occurs. From this perspective, we can assume that in the near future two composing arcs (Anatolian-Caucasian and Caucasian-Elbursian) under the effect of the plume extending will be separated and will exist independently, like the most of Late Cenozoic arc of Alpine Belt. From this obviously implies that extended plume heads under conditions of large collision zone can impact not to only ancient lithosphere, but also to shift and even break off into individual pieces of existing subduction zone.

CONCLUSIONS

- The Late Cenozoic Alpine-Mediterranean Mobile Belt (Alpine Belt) has appeared under condition of collision of lithospheric plates above superplume head. It's surface is complicated by number of protuberances (local plumes), which are a cause of emergence of extending zones on the background of the overall structure of compression. Geological situation in the belt is considered with complex interaction of converging lithospheric plates with plastic plume material.

- It is shown that two types of depression with a predominance of the oceanic crust occur within the belt: (1) newly-formed back-arc basins above extended heads of local plumes (Western Mediterranean, Aegean Sea, Pannonian Basin), and (2) fragments of ancient oceanic crust of the Tethys, which has descended under the influence of downward movements in the mantle between plumes (Eastern Mediterranean,

Black and Caspian seas). The second type is characterized by basins with passive margins, along which are developed steep deep faults; these areas in their structure resembles the structure characteristic of Atlantic passive margins.

- The exception is the region of the North-Arabian - Caucasus plume without depression above it. This plume, apparently bends to the north the surface of the subduction zone, ensuring the existence of Anatolian-Elbursian andesite-latite volcanic arc to form a transverse area of modern volcanism of the Caucasus.

- Geodynamics of the Alpine Belt has developed from the periphery of this structure to its center, in other words to the central part of the superplume head. Maximum geodynamic activity is now in the Tyrrhenian Sea, where thickness of the lithosphere is minimal; in the future it will probably be moved into the region of the Aegean, Pannonian, and Caucasus, where the most powerful positive isostatic anomalies occur.

- Asymptotic models provide an opportunity to explore the major features of formation and development of geological structures. Ascending of mantle plumes determines the depth of initiation of deep-water basins. The lithospheric material above them (the uppermost mantle and crust) is moved apart to make room for an oceanic crust. The resulted excess of this lithospheric material is involved in subduction zones with formation of volcanic arc-backarc basin system.

- The process of subduction depends on the difference in density of the lithospheric and plume material. It can only contribute to back-arc spreading, freeing up space for moving apart ancient crust of the back-arc basin which involved in subduction and further buried in the deep mantle. Existence here of the mantle plume itself is a result of collision of deep-seated mantle flows, contributing to pumping and ascending of local protuberances in the form of plumes with the formation of extensional structures over their extended heads. Relationship between the area of collision of mantle flows at the depth and the zone of collision of lithospheric plates, along with relative velocities and ratios of densities, determines the dip of subduction slab also.

- Complex processes in the zone of lithospheric plates collision are interrelated and interdependent. The relative rapidity of the observed processes evidence about defining role of mechanical motions in the formation of structures, forced by the influence of thermal factors. The Alpine Belt is an area of elevated heat flows due to removal of abyssal heat by ascending of plumes caused by deep-seated mantle activity (up to the core) in the complex interaction with the thinned lithosphere.

Analysis of times of formation of the Mediterranean's deep-water basins shows that the process of activation occurred from the periphery to the center, which was caused by compression and ascending of superplume material between lithospheric plates.

REFERENCES

1. M. E. Artemiev, 1971 Some peculiarities of deep-seated structure of depressions of Mediterranean type: evidence from data on isostatic gravity anomalies. Bull. Soc. of the Nature Investigators of Moscow. Geol. Dept., 4: 5-10 (in Russian)

2. I. M. Artemieva, H. Thybo, M. K. Kaban, 2006 Deep Europe today: geophysical synthesis of the upper mantle structure and lithospheric processes over 3.5 Ga. In: European Lithosphere Dynamics. D.G. Gee and R.A. Stephenson (eds). Geol. Soc. London Mem. 32 1142 .

3. R. Bartole, 1995 The North Tyrrhenian- Northern Apennines Post-Collisional System- Constraints for a Geodynamic Model. Terra Nova, 7(1), 7-30.

4. O. A. Bogatikov, E. V. Sharkov, A. V. Vesselovsky, V. B. Mehscheryakova, 2009 Synchronism of development of Cenozoic volcanic arcs and back-arc basins: Reasons and consequences. Doklady Earth Sciences, 427A (6), 907 EOF911 EOF .

5. N. A. Bogdanov, 2001 Continental margins: general problems of structure and tectonic evolution. In: Fundamental problems of general geotectonics. Yu.M. Puscharovsky (Ed.). Moscow, Nauchnyi Mir Publ., 231249 .

6. V. S. Burtman, 1989 Geodynamics of Arabian syntax. Geotectonics, № 2 6775 .

7. I. V. Chernyshev, V. A. Lebedev, S. N. Bubnov, et al. 2002 Isotopic geochronology of the Quaternary volcanoes eruptions of the Great Caucasus. Geochim. Intern., 40 11511166 .

8. S. Duggen, K. Hoernle, P. van den, C. Bogaard, Harris, 2004 Magmatic evolution of the Alboran region: The role of subduction in forming the western Mediterranean and causing the Messinian Salinity Crisis. Earth and Planetary Science Letters, 218 91108 .

9. V. A. Krasheninnikov, J. K. Hall, 1994 Geological structure of the Eastern Mediterranean (Eds.). Jerusalem, 1994.

10. R. Girbacea, W. Frisch, 1998 Slab in the wrong place: Lower lithospheric mantle delamination in the last stage of the eastern Carpatian subduction

retreat. Geology, 26 (7). 615618 .

11. P. Gize, N. I. Pavlenkova, 1988 Structural maps of the Earth's crust of Europe. Physics of the Earth, N 10 314 .

12. V. S. Gobarenko, S. B. Nikolova, T. B. Yanovskaya, 1986 Structure of the upper mantle of South-Eastern Europe, Asia Minor and Easterm Mediterranean based on data of discrepancy time of race P-wave. Physics of the Earth, № 8 1523 .

13. R. Gök, E. Sandvol, N. Türkelli, D. Seber, M. Barazangi, 2003 Sn attenuation in the Anatolian and Iranian plateau and surrounding. Geophys, Research Letters, 30 (24), 8042, doi: 10.1029/2003GL018020.

14. A. F. Grachev, 2003 Final volcanism of Europe and it's geodynamic nature. Physics of the Earth, N 5 1146 .

15. A. F. Grachev, 1987 Rift zones of the Earth. Moscow, Nedra Publ., 247 p. (in Russian)

16. N. L. Dobretsov, A. G. Kirdyashkin, A. A. Kirdyashkin, 2001 Deep Geodynamics. GEO Publ., Novosibirsk (in Russian).

17. Y. D. Evsyukov, 1998 Origin and development of morphostructuree of northern Aegian Sea. Oceanology, 38 (2), 286-292.

18. Y. D. Evsyukov, 1999 Origin and main stages of development of Medina Ridge (Central Mediterranean, Doklady Earth Sciences, 366 (3), 364-368.

19. Earth crust and history of development of the Mediterranean Sea. 1982 Moscow, Nauka Publ., 207 p. (in Russian).

20. S. Harangi, H. Downes, I. Seghedi, 2006 Tetriary-Quaternary subduction-related magmatism in the Alpine-Mediterranean region // In: European lithosphere dynamics. Edited by D.G. Gee and R.A. Stephenson. Geol. Soc. Mem. 32. London, Geol.Soc. London Publ., 167190 .

21. S. L. Harley, D. A. Carswell, 1995 Ultradeep crustal metamorphism: A prospective view. Journal of Geophysical Research, 100 (B5), 8367-8380.

22. T. M. Hearn, 1999 Uppermost mantle velocities and anisotropy beneath Europe. Jour. Geophys. Res., 104 (B7), 15,123-15,139.

23. K. Hoernle, Y. S. Zhang, D. Graham, 1995 Seismic and geochemical evidence for large-scale mantle upwelling beneath the eastern Atlantic and western and central Europe. Nature, 374 (6517), 34-39.

24. V. G. Kazmin, A. A. Shreider, I. Finetti, et al. 2000 Early stages of the Black Sea development according seismic data. Geotectonics, № 1 4660 .

25. V. E. Khain, 1984 Regional Geotectonics. Alpine-Mediterranean Belt.

Moscow, Nedra Publ., 344 p.

26. K. Khair, G. N. Tsokas, 1999 Nature of the Levantine (eastern Mediterranean) crust from multiple-source Werner deconvolution of Bouguer gravity anomalies. Jour. Geophys. Res., 104 (B11), 12469-12478.

27. R. Kipfer, W. Aeschbachhertig, H. Baur, et al. 1994 Injection of Mantle Type Helium into Lake Van (Turkey)- The Clue for Quantifying Deep Water Renewal. Earth and Planetary Science Letters, 125 (1-4), 357-370.

28. M. L. Kopp, Y. G. Leonov, 2000 Tectonics. In: Outline of geology of Syria. Yu.G. Leonov (Ed.). Moscow, Nauka Publ., 7-104 (in Russian).

29. N. P. Kostenko, L. V. Panina, 2001 Some pecularities of neotectonic structure and development of Caucasus on conerosional stage. In: Tectonic of the Neogea: Common and regional aspects. Contributions of Tectonic meeting, 1 Moscow, 248-252 (in Russian).

30. H. S. Lasen, 2002 Investigations of rifted margins. In: Achievements and Opportunities of Scientific Ocean Drilling. Spec. Issue of Joides Journal, 28 (1), 85-90.

31. H. Laubscher, 1988 Material balance in Alpine orogeny. Geol. Soc. Amer. Bull.,100 13131328 .

32. E. P. Lelikov, T. A. Emel'yanova, 2007 Volcanogenic complexes of The Okhotsk and Japan seas (comparative analysis). Marine geology, 42 (2), 294-303

33. L. Lonergan, N. White, 1997 Origin of the Betic-Rif mountains belt. Tectonics, 16 (3), 504-522.

34. rocks. Magmatic, . Vol, rocks. Ultramafic, 1988 E.E. Laz'ko and E.V. Sharkov (Eds.). Moscow, Nauka Publ., 500 p. (in Russian)

35. A. M. Marotta, R. Sabadini, 1995 The Style of the Tyrrhenian Subduction. Geophysical Research Letters, 22 (7), 747-750

36. I. Marson, G. F. Panza, P. Suhadolc, 1995 Crust and upper mantle models along the active Tyrrhenian rim. Terra Nova, 7 (3), 348-357.

37. J. Morales, I. Serrano, A. Jaboloy, et al. 1999 Active continental subduction beneath the Betic Cordillera and the Alboran Sea. Geology, 27 (8), 735-738.

38. A. F. Grachev, 2000 Neotectonics, geodynamics and seismicity of the Northern Eurasian, 2000, (ed.). PROBEL, Moscow, 487 p. (in Russian).

39. A. M. Nikishin, M. V. Korotaev, 2000 History of the Black Sea basin formation. In: General Problems of Tectonics. Tectonics of Russia. Moscow, GEOS, 360-363 (in Russian).

40. V. G. Nikolaev, 1986 Pannonian Basin. Moscow, Nauka Publ., 120 c. (in Russian)

41. C. B. Papazachos, P. M. Hatzidimitriou, D. G. Panagiotopoulos, G. N. Tsokas, 1995 Tomography of the crust and upper mantle in southeast Europe. Jour. Geophys. Res., 100 (B7), 12405-12422.

42. N. A. Pino, D. V. Helmberger, 1997 Upper mantle compressional velocity structure beneath the west Mediterranean Basin. Jour. Geophys. Res., 102 (B2), 2953-2967.

43. Ponikarov. V. G. В.П., V. V. Kazmin, et. Kozlov, al, 1969 Syria. Leningrad, Nedra Publ., 216 p.

44. M. T. Prilepin, T. V. Guseva, A. A. Lukk, V. I. Shaevcheko, 2001 Modern geodesic measuring and major geodynamic conceptions. In: Tectonic of the Neogea: Common and regional aspects. Contributions of Tectonic meeting, 1 Moscow, 135-138 (in Russian).

45. Rehault. E. Т.Г., A. Moussat, Fabri, 1987 Structural evolution of Tyrrhenian back-arc basin. Marine Geology, 74 (1-2), 123-150.

46. L. E. Ricou, J. Dercourt, J. Geyssant, et al. 1986 Geological constrain on the Alpine evolution of the Mediterranean Tethys. Tectonophysics, 123(1-4), 83-122

47. L. H. Royden, 1989 Late Cenozoic tectonics of the Pannonian Basin System// Tectonics, 1989. 8 N 1. 5161 .

48. E. V. Sharkov, 2000 Mesozoic and Cenozoic basaltic magmatism. In: Outline of geology of Syria. Yu.G. Leonov (Ed.). Moscow, Nauka Publ., 177-200 (in Russian).

49. E. V. Sharkov, V. B. Svalova, 1991 Possibility that continental lithosphere is involved in subduction in back-arc spreading. Intern.Geol. Review, 33 (12), 1184-1198.

50. E. V. Sharkov, V. B. Svalova, 2005 Late Cenozoic geodynamic of Alpine Folded Belt in connection to formation of withincontinental seas (petrologo-geomechanical aspects). Proc. of Higher Educational Establishments. Geology and Exploration, № 1, 3-11 (in Russian)

51. A. G. Shempelev, N. I. Prutsky, I. S. Feldman, S. U. Kuchmasov, 2001 Geologo-geophysical model along profile Tuapse-Armavir. In: Tectonic of the Neogea: Common and regional aspects. Contributions of Tectonic meeting, 2 Moscow, 316-320 (in Russian).

52. V. I. Shevchenko, T. V. Guseva, A. A. Lukk, et al. 1999 Present-day geodynamics of the Caucasus (results of GPS measurements and seismological data. Physics of the Earth, № 9 318 .

53. W. Spakman, S. van der Lee, R. van der Hilst, 1993 Travel-time tomography of the European-Mediterranean mantle down to 1400 km. Phys. Earth Planet Inter., 79 374 .

54. F. Storti, 1996 Tectonics of the Punta Bianca promontory: Insights for the evolution of the Northern Apennines Northern Tyrrhenian Sea basin. Tectonics, 14 (4), 832-847.

55. V. G. Trifonov, A. E. Dodonov, E. V. Sharkov, et al. 2011 New data on the Late Cenozoic basaltic volcanism in Syria, applied to its origin. Journal of Volcanology and Geothermal Res., 199. 177192 .

56. V. B. Zanemonents, V. D. Kotelkin, V. P. Myasnikov, 1974 About dynamics of lithospheric movements. Physics of the Earth, № 4 4354 .

57. L. R. Zonenshain, X. Le Pichon, 1986 Deep basins of the Black Sea and Caspian Sea as remnants of Mesozoic back-arc basin. Tectonophysics, 123 (1-4), 181-212.

58. S. M. Zverev, 2002 Features of structure of sedimentary sequence and basement in frontal zone of the Cyprian arc (Eastern Mediterranean). Oceanology, 42 3 416428 .

Chapter 2

GEOTECTONIC POSITION AND METALLOGENY OF THE GREATER ALTAI GEOLOGICAL STRUCTURES IN THE SYSTEM OF THE CENTRAL-ASIAN MOBILE BELT

B.A. Dyachkov, M.A. Mizernaya, Nina Maiorova, Zinaida Chernenko, Victor Maiorov and O.N. Kuzmina

East-Kazakhstan State Technical University, Kazakhstan

INTRODUCTION

According to the geotectonic position the Greater Altai geological structures are located in the Central Asian (or Kazakhstan-Okhotskij) global mobile belt of a sub lateral direction embracing geo-structures of Kazakhstan, Northern China and Mongolia. Within its border the structures of the Greater Altai are located on the north-western flank of Altai-Alashan curved shape mobile zone that envelop the Siberian platform from the south-west and the south.

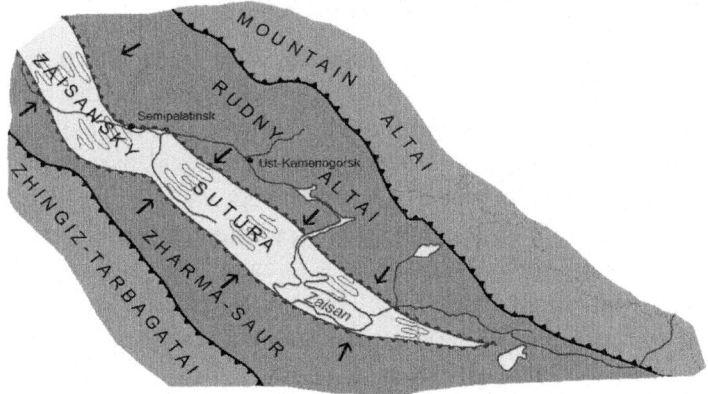

Figure 1. Main geological structures of Great Altai.

The territory comprises the Rudny Altai, Kalba-Narym zone, West Kalba and Zharma-Saur which are considered in the Irtysh-Zaisan geosynclinal fold systems. The board structures of the north-east are the Rudny Altai caledonides (Charyshskij, Holzunskij-Chuiskij-Sitsikheskij and Tsunhu-Chinkheskij structural zones), and the caledonides of Chingiz-Tarbagatai fold system are on the south-western part (Fig.1). On the south-eastern flank of the Greater Altai structures is narrowing sharply due to the thrust from the Djungarian array and West-Siberian plates. The north-western flank can be observed up to Russia and is mostly blocked by loose cover of the Kulundinskij cavity. The total length of the territory in up-to-date coordinates is more than 1000 km at an average width of 300 km.

Minerals of the Greater Altai are the natural stocks of huge deposits containing non-ferrous and precious metals and other resources. By processing these minerals copper, zinc, lead, gold, silver, platinum, titanium, rare and dissipated elements can be obtained. On the basis of these metals and elements the mining industry infrastructure with plants, towns and villages has been established in this region. For the purpose of mining industry enterprises sustainable work the metals stock should be continuously renewed.

Thus, high rates and volumes of exploration for the last ten years have led to the exhaustion of easily-discovered deposits. The scientific ideas in the field of geology governing for many years, particularly geosynclinal concept of fixism came to their self-exhaustion. That is why the methodology of metallogenic research, forecasting and new deposits exploration needs to be improved. New approaches to the geology, ore deposits depth-geological forecasting criteria development are required, especially in the exploration of loose and concealed deposits (up to 1000-1500 m).

GEOTECTONICS AND METALLOGENY

The overall progress of Earth sciences, new data on paleomagnetism, paleoclimatology, new mobilist hypothesis emerging in the 60th of the last century (plate tectonics, mantle plumes, terranes, new global tectonics) have determined the necessity for reviewing the traditional paradigms on tectonic and metallogenic evelopment of the given region. In this connection on the basis of global mobilism hypothesis the practical material generalization in the field of geology, geophysics and the Greater Altai and adjoining territory metallogeny has been carried out. This provided the identification of new principles for metalliferous geological structures development and the deposits formation under different geodynamic circumstances (Geodynamics…, 2007; Greater Altai…, 1998, 2000; Shcerba et al., 2000). As a result of research activity productive mineragenical levels and patterns have been determined with the perspective for non-ferrous, precious and rare metals exploration and the directions of exploration works have been defined.

Geodynamic Development

The Greater Altai is a holistic geostructure of a vast territory, including the geological structures of the Rudny Altai, Kalba-Narym, West-Kalba, Zharma-Saur and nearby territories of Russia and China. In the tectonic outline it is a linear fault-fold structured pattern in the system of Central-Asiam mobile belt. The space is located between the stable continental arrays – Gornoaltaiskij (on the north-east) and Kazakhstanskij (on the south-west).

From the point of global mobilism, huge geological structures nucleation and formation in Kazakhstan, Siberia, Ural and other regions are connected with the break of the Eurasian continent into separate plates, geoblocks, arrays and xenoliths at the period of late Proterozoic. They were drifting in the Paleoasian ocean and made significant horizontal relocations. According to the paleomagnetic and geological features, the lithospheric plates and xenoliths relocations are supposed to originate from the east to the west (with clockwise rotation). The Greater Altai structures are possibly the xenoliths of the ancient paleo-continent (Eastern Gondvana). In the process of the Paleoasian ocean evolution, some xenoliths joined, perhaps, the Siberian craton, others formed the Kazakhstan microcontinent which then at the Devonian and Early Carboniferous periods were separated from the Gornyi Altai by Irtysh-Zaisan paleobasin (Greater Altai…, 1998).

The Greater Altai as a unified structure has transformed into the stage of the Hercynian collision (in Early Carboniferous and later) as a result of subsidence and interfacing of Kazakhstan continent borders and Siberian sub-continents which were divided by Zaisan sutural zones (Dyachkov et al., 2009). In later tectonic cycles the formed structure was complicated in the process of late-Hercynian intraplate rifting and then stabilized at the Mesozoic and Cenozoic periods. The modern Greater Altai structure is considered as a system of earlier separated blocks that are parallel to the structural-formation zones or collage terranes. The emerged tectonic structures (Altai Mountains, Rudny Altai, Kalba and others) are limited by deep faults, sutural zones and are differentiated according to the development geodynamics, geological structure and metallogeny specifity.

In the process of paleo-geodynamic reconstruction the geodynamic model of the Greater Altai formation has been developed. It is a modern type of structural-formation zone sub-parallel systems distinguished by deep faults. This geodynamic model of lithosphere rhythmic tension and compression having significant horizontal relocations of lithospheric plates by interlayer surfaces of the upper mantle, characterizes the Altai type of the Siberian and Kazakhstan subcontinents global interaction. The Greater Altai unique structures and its metallogeny are due to the geological processes grandiosity and duration.

This article considers the peculiarities of geological development, depth structure and leading types of the Greater Altai deposits having been formed under different geodynamic conditions. The overall geological formations development direction at the Precambrian was at the oceanic rifting mode, then in the early (rifting and insular), medium (collision) and late (post-collision) stages of the Caledonian and Hercynian cycles. This process terminated by the continental rifting and stabilized in the Mesozoic and Cenozoic. The indicators of paleo-geodynamic and landscape-geological conditions are the certain geological formations reflecting their emergence pre-conditions.

In the Precambrian cycle, at the oceanic rifting mode the pre-Riphean crystalline basement of small arrays destruction took place. They were split into separate fragments and blocks and then moved Within the Greater Altai the fragments of the Precambrian basement are fixed in the Charsk-Zimunai zones and Irtysh zone of collapse. They are complicated tectonic-metamorphic structures with intensive dynamo-metaphoric reformations of rocks that are suppressed by folds, thrusts and polycyclic metallogeny. They contain schists, gneisses, amphibolites, granite-gneisses with hyperbasite protrusions and serpentinite melange blocks. In hyperbasites there are magmatic deposits of Cr, Ni, Co (Pt) which form Charsk ore-bearing level with later form gold mineralization.

In the Caledonian volcanic arcs of basalt-andesite-dacite series with iron-manganese and gold-chalcopyrite deposits (Akbastau, Kosmurun, Mizek) were formed at the side structure of Zhingiz- Tarbagatai in rift-insular-setting of an early stage ($\mathcal{E}^1$- O^3) (Table 1).

On the Greater Altai territory and in the Altai Mountains there was a marine mode with the formation of calcareous-siliceous-terrigenous elements. At the medium stage (O_3-S) there emerged the tendency for Irtysh-Zaisan small oceanic-basin degradation due to the growth of accretion zones in Gorno-Altaiskij and Kazakhstan continental arrays. The collision magmatic front was located in the doming spreading zones with the formation of gabbro-diorite-granodiorite intrusions (O-S, S_2) at the focal parts of Zhingiz Tarbagatai and Rudnoaltaysko-Ashalinskij zones. The mineralization is represented by Fe, Cu, Zn, Mo, Au. The most productive level of pyritic copper-zinc ores was located in the active Kazakhstan suburbs and is connected with O_{2-3} insular-arc basalt-andesite volcanism (Mines Akbastau, Kosmurun, Mizek in Zhingiz Tarbagatai).

Table 1. Geodynamic settings and main epochs of ore formation in East Kazakhstan

Cycle	Geodynamic setting	Zhingiz - Tarbagataj	Zharma-Saur	Shar Zone	West Kalba
Precambrian	Oceanic rifting			Ultramafic rocks, Shar group (Cr, Ni. Co	
Hercynian and Caledonian	Rifting and insular arc formation	Middle-Late Ordovician basalt-andesite-dacite, Akbastau (Cu, Zn, Au)			
	Collision		Middle-Late Carboniferous gabbro-granoid-diorite-granite, Sekisovskij (Au, Ag, Te)	Metamorphic and hydrothermally altered rocks, Kyzyl-Shar (Au, Hg)	
			Late Carboniferous rhyodacite and grano-diorite, Vasilevskij (Au)		Late Carboniferous Grano-diorite-plagiogranite, Bakyrchik (Au, As, Ag)

Cycle	Geodynamic setting	Zhingiz - Tarbagataj	Zharma-Saur	Shar Zone	West Kalba
	Post-collision within plate reactivation		Late Permian-Early Triassic alkali granite, Upper Espe, (REE, Nb, Zr)		Late Permian Granite and leucogranite, Delbegetei (Sn, Be)
Kimmerian	Continental rifting and stabilization		Late Cretaceous-Eocene kaolinite-hydromica zone of weathering, Zhanan (Au)	Late Cretaceous-Eocene nontronite zone of weathering, Belogorskiy, (No, Co)	Late Cretaceous-Eocene kaolinite zone of weathering, Karaotkel (Ti, Zr)

At the final stage (S_1-D) the unified caledonian structure of the Greater Altai under general rotation of the Siberian continent and its adjacent folded structures in a northern direction was formed. In the process of the Greater

Altai borders intraplate activation in the Altai Mountains, granodiorite-granite intrusions D_2 (Mo, W, Bi) emerged, and in Chenghis Tarbagatai - arrays of granit- granosienite -leykogranitovoy series (D_2-D_3) with a poor rare-metal mineralization appeared.

The Hercynian cycle was marked by a repeated division and spreading of the Caledonian continental borders and the formation of a secondary Irtysh-Zaisan oceanic basin with a large magmatism and mineralization. At the early stage the main ore-bearing structures of the Rudny Altai formed under the rifting and insular-arc geodynamic conditions in the process of echeloned deep faults system activation. This contributed to the inflow of mantle basaltic magmas and mineralizing fluid flows into the upper parts of the granule cells. Major industrial chalcopyrite and pyrite-polymetallic deposits are genetically connected with an intensively displayed Devonian basalt-andesite-rhyolite volcanism. The most productive deposits are volcanic arcs of round structure flanking the Caledonian paleo-elevations and having long volcanic process and mineralization (Ridder Sokolnoe, Maleevskoe, Nikolaev, etc.)

At the medium stage (C_1-C_3) as a result of geodynamic mode change and prevailing compression (in the Early Carboniferous and later) the unified geological structure of the Greater Altai formed after the Gorno-Altaiskij and Kazakhstan continental borders collision, their connection took place in Zaisan sutural zone. This stage was implanted by syncollision gabbro-diorite-granodiorite intrusions, small hypabyssal intrusions and dikes of gabbro, diorite, granodiorite containing Cu, Ni, Au, Ag. In Zaisan sutural zone in the mode of rhythmic-pulsating tectonic movements of tension and compression the belts of gold-bearing small intrusions and dikes of Kunushskij complex (C_3) are located. The diagonal systems of cross-cutting faults of late-collision stage with gold objects in the terrigenous-carbonate and black shale sequences (Bakyrchik, Suzdal, Kuludzhun, etc.)were the ore-controlling elements. At the late stage (C_3-T_1) under the conditions of intraplate activation of deep faults the granitoid belts with rare metal and rare-metal-rare earth mineralization - Ta, Nb, Be, Cs, Sn, W, Mo, Zr, Tr et al (deposits Bakennoe, Belya Gora, Verkhnee Espe, etc.) have formed.

The Cimmerian and Alpine cycles of the Greater Altai epihercynian structure development had a noticeably relative mobility in the Triassic and Jurassic, were significantly stable in the Cretaceous-Paleogene and during activation in the Neogene-Quaternary. This period is characterized by the formation of deposits in weathering crusts (Au, Ni-Co, Zr-Tr), placer gold, ilmenite, tantalite, cassiterite and other minerals. So, the geodynamic development of the Greater Altai mobile belt reflects a long and complicated history of geological structures formation in the process of collisional displacement of the Siberian and Kazakhstan subcontinents and the Irtysh-Zaisan paleo-basin degradation. It also emphasizes the intensity of ore-magmatic processes and metallogeny.

As a result of tectonic-magmatic processes polycyclic development the main periods of mineralization have been determined. They reflect vertical and lateral mineralization zoning in the frames of ore belts and the region of the Greater Altai.

Deep Structure

The deep structure of the given region is considered according to the complex of geological and geophysical data (gravimetric, magnetic prospecting, seismic exploration and other activities) and is characterized by multilayered crust (MC) and upper mantle (M). According to the modern interpretation of geophysical material MC has the power equal to 50-55 km and includes heterogenic linear –mosaic blocks complicated with deep faults (Greater Altai..., 1998; Shcerba et al., 1984).

The upper mantle is characterized with the inhomogeneous structure and stratified to a depth of more than 250 km. In its structure there is supposed to be an undepleted mantle (pyrolytic) and drained including spinel containing dunites and harzburgites, garnet eclogites and alpine hyperbasites. The surface of the M is under the depth of 40-55km and is immersed in different directions. According to the works of V.N.Lyubetskoj, G.P. Nakhtigal and other researchers there is a scheme of asthenosphere layer topography where the ore-belts are marked and the asthenosphere topography location. The highest elevation of the asthenosphere layer (Semipalatinsk, Zaisan) of the north-west direction is in the Zaisan sutural zone, i.e. in the Kazakhstan and Gornoaltaiskij borders collision zone (Fig.2). The elevation of the Mokhorovich surface is fixed on the north-western flank of the sutural zone (Gornostaevskij) and on the south-eastern part (Zaisanskij) (Fig.3). Stratified asthenospheric zones in the upper mantle, deep faults system dissection are obviously the consequence of magmatic chambers generation, their metallogenic specialization and activation of geodynamic processes in the multilayered crust. The consolidated multilayered crust of the Greater Altai comprises four layers: meta basalt, meta diorite, meta granite and sedimentary.

The power of the meta basalt layer is 20-24 km and is up to 28 km at the crest thickenings of the north-western direction. Here there are also linearly elongated modern thermal anomalies. Deep faults (Charsk-Zimunaysky, Terektinsky and Baiguzin-Bulak, Siretasky), falling in different directions penetrate into the upper mantle and form marquee-type structures. The meta diorite layer is represented by deeply metamorphosed rocks of the Precambrian (the average density of 2,8 g/cm^3, the wave velocity 6,4-6,6 km/s). Its greatest power (12-16 km) is noticed in Kalba-Narym, Beloubinskij- Sarymsaktinskij and Sirektas-Sarzanskij zones). In Zaisan suture the power is minor and is 4-12 km. According to G.P. Nakhtigal in the meta diorite layer there are the lower edges of the most magnetically active objects and the root parts of the large granitoid plutons.

The meta granite layer embraces the metamorphosed Caledonian formations and major arrays of granitoids. It is exposed on the surface of Zhingiz Tarbagatai, Aleisk and the Rudny Altai and Sinyuhinskij blocks, fragmentary it can be viewed in the Irtysh, Sirektas and Sirektas- Sarsazan zones, and also in the Charsk block. The maximum power of the layer (up to 12-14 km) is in Kalba-Narym and Beloubinskij-Sarymsaktinskij zones.

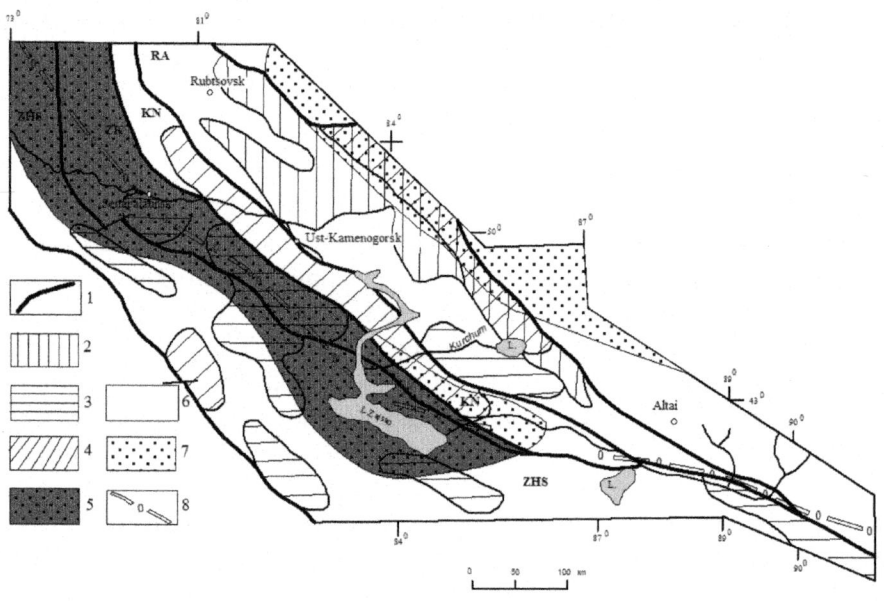

Figure 2. Great Altai ore belts and asthenosphere layer topography 1- borders of the region, 2-3: Hercynian stage ores (1- rifting stage and island-arc borderland stage, 2- collision stage, 3- postcollision stage); 5-8: asthenosphere layer topography (5- elevation of the asthenosphere layer, 6-slope of the elevation, 7-depression, 8- depression axis. Ore belts: RA- Rudny Altai, KN-Kalba-Narym, ZK- West Kalba, ZHS- Zharma-Saur.

Sedimentary layer is formed by non- metamorphosed Paleozoic and Mesozoic and Mesozoic-Cenozoic sediments. According to F.S. Moiseenko (1981) it is divided into two sub-layers: volcano-genic-sedimentary and loose sedimentary cover. The powers of the hercinides volcano-genic-sedimentary sub-layer.is 0-9 km with minimum values (0-3 km) on the elevations (Rudnoaltayskij, Kurcch-Kaldzhirskij, Tersayrykskij) and with the maximum values in troughs. The loose sediments with the power of 50-100 m and up to 400 m and moreare in hollows and troughs (Zaisanskij, Kulundinskij, etc.).

Zoning

The Greater Altai territory as it was mentioned above embraces the geological structures of the Rudny Altai, Kalba-Narym, West Kakba, Zharma-Saur and adjacent regions of Russia and China. The bordelines are north-western deep faults on the north-east (Loktevsko-Karairtyshskij fault). It divides the structure of the GA from the Gornyj and Chinese Altai and borders with Zhingiz – Tarbagatai on the south-west of Zhingiz - Saur. The total length of the geological structures is more than 1000km at the width of 300 km.

According to the geotectonic zoning and considering the adjacent territories of Russia and China, the Greater Altai is subdivided into two major sub-regions:

- South –Western Altai Xinjiang, formed in the active borders of the Sibirean platform (on the north-east) and

- Zharma-Saur-Baganur, located on the board of Kazakhstan microcontinent (on the south-west)

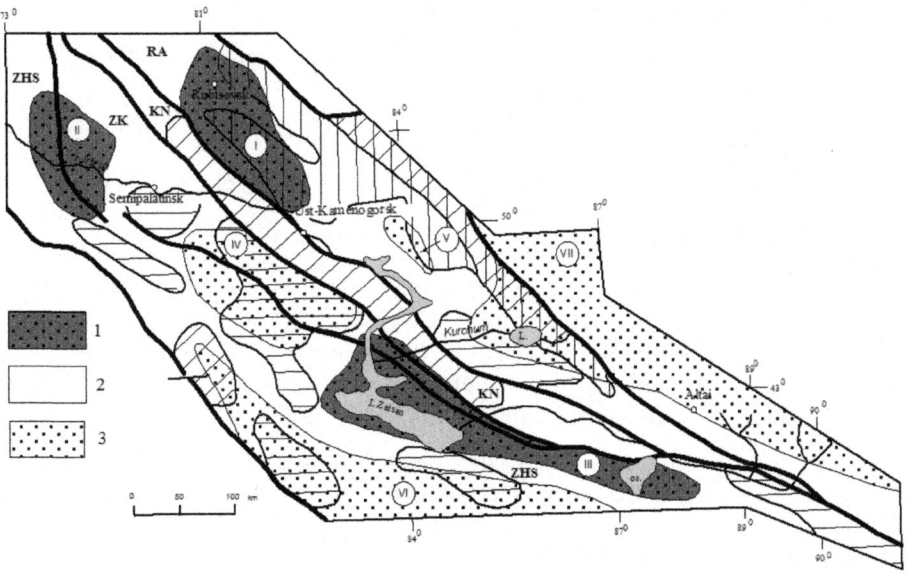

Figure 3. Greater Altai, ore belts, and relief of the Moho surface1-3 the Moho layer topography: 1-elevations, 2- slopes of elevations, 3- depressions; elevations: I-Rubtsovskij, II- Gornostaevskij, III- Zaisaskij; slopes: IV-Kalbinskij, V- Bukhtarminskij, VI-Saur-Manrakskij, VII- Gornoltaiskij.

The borderline between the given sub regions lies on the Charsk-Zimunai deep fault or sutural joint. The junction area coincides with Zaisan sutural zone or Irtysh Zaisan, Ob-Zaisan according to other authors (V.N.Lyubetskoj, B.S. Uzhkenov, A.V.Smirnov).

The South-Western Altai-Xinjiang subregion embraces the geological structures of the Rudny Altai, Kalba-Narym zones and Western Kalba. In the Rudny Altai are three tectonic zones (from the north-east to the south-west): Beloubinsko-Sarymsakty-Kurtinskaya, Rudnoaltaysk-Ashalinskaya and Irtysh- Fuynskaya and from the Kazakhstan side their reflections are Beloubinsko-Sarymsaktinskaya, Rudnoaltayskaya and Irtysh zones. The last one with the terrane tectonic position corresponds with the sutural zone, dividing heterogeneous tectonic blocks (terranes) of the Rudny Altai and Kalba - Narym. The Zharma-Saur-Baganur subregion is divided into three tectonic zones: Sirektas-Sarsazan-Kobukskaya, Zharma-Saur-Haratunguskaya and Charsk-Zimunayskaya zones. On the territory of East Kazakhstan they are Sirektas-Sarsazanskaya, Zharma-Saur and Charskaya zones respectively.

The Zhingiz Tarbagataiskiy belt bordering with the GA on the south-west comprises two tectonic zones: West Zhingiz and East Zhingiz. On the north-east of the Altai Mountains the Charyshskaya, Holzun-Chuysko-Sitsiheskaya and Tsunhu-Chinhenskaya zones flank the Greater Altai.

There are four ore-belts according to the metallogenic zoning within the Greater Altai:

1. Rudny Altai copper-polymetallic (Fe, Mn, Cu, Pb, Zn, Au, Ag, etc.).

2. Kalba-Narym rare-metallic (Ta, Nb, Be, Li, Cs, Sn, W).

3. West-Kalba gold-ore (Au, Ag, As, Sb).

4. Zharma-Saur polymettalic (Cr, Ni, Co, Cu, Au, Hg, Mo, W, Tr).

In the structure of ore belts the ore bearing minor structures are considered: metallogenic zones (sub-zone), ore area, ore zone, ore node and ore field and also, deposits, ore occurrences and points of mineralization.

The main features of the geological structure and minerals of the given ore belts are displayed in the work (Dyachkov et al., 2009).

Metallogeny Peculiarities

The discussed region is characterized by the variety of mineral types differentiated in genesis, age, scale of mineralization and other features. Below the peculiarities of the Greater Altai main ore-bearing structures metallogeny are being considered (Table 1).

In the Precambrian cycle in the oceanic rifting mode the destruction of the pre-Riphean small arrays crystal basement took place, they were divided into separate blocks-fragments and moved on the weakened slip surfaces with further complicated litho-mélange accretion. Within the GA the fragments of the Precambrian basement are fixed in the Charsk-Zimunai zone and the Irtysh zone of collapse. These are complicated tectonic-metamorphic structures with intensive dynamic-metamorphic transformations of the rocks, intense folding, thrust faulting, and polycyclic metallogeny. The crystalline schist, gneiss, amphibolite, granite gneiss with protrusions and hyperbasite blocks of serpentinite melange are typical of these structures.

The Precambrian metallogeny has not been studied well yet. Under the oceanic rifting the hyperbasite bodies with the primary mineralization (Fe, Mn, Cr, Co, Ni, Cu, etc) have been formed in the deep faults. In the process of collision they were squeezed into the upper floors of the structure in the form of hyperbasite and plates serpentinite melange protrusions (Charsk and Irtysh zones).

In the Charsk zone magmatic chrome formation has grown. It is associated with the serpentinized hyperbasites, blocks of ophiolites and metamorphic rocks in the structures of the melange. It is represented by small blocks and lenses of chromite and disseminated ores in serpentinites, which are considered as fractured fragments of larger bodies, the initial bedding has been defined yet (Andreevskoe, Suuk-Bulak, etc.). The estimation of these deposits is still incomplete. Insufficient studies of the Charsk mélange, identification of new hyperbasite bodies by geological and geophysical data, possibility for ore bodies clustering in sutural area allows to further study the Charsk zone.

In the Irtysh zone of collapse there is a golden metamorphic formation represented by gold-manifestations of Polevaevskij-Predgornenskij ore zone (Zolotar, Polevaevskij, Avrorinskij, etc.) Earlier they were considered as small quartz vein sites with limited prospects. Later, they were referred to a more prospective metamorphic-hydrothermal type of gold mineralization connected with dynamo-metamorphic and contact-hydrothermal regeneration of shale (increased content of carbonate and gold) in the process of the Irtysh zone of collapse transformation into the collision stage. According to the conditions of mineralization these objects are corresponded to the metamorphic types of deposits by Ya.N.Belevtsov (1982).

The acquired data testify the fact the greenschist sequences are the primary source of gold that then was mobilized in the process of regeneration. Another source is ore-bearing fluids connected with the intrusive magmatism (small granodiorite intrusions, dikes of plagiogranite porphyry, quartz porphyry and albitophyre). Besides the quartz veins containing pyrite, chalcopyrite and gold, the great importance is given to mineralized zones of increased gold content. They are silicified, tourmalined, pyritized and carbonated host rocks of a significant length with abnormal ore elements contents (Ag, Cu, Pb, Zn, etc). Significant sizes of these zones in length and depth provide a new approach to the evaluation of the Irtysh zone of collapse in gold mineralization. The main prospects of this zone are not quartz vein objects, but more prospective deep-bedded gold-sulfide zones. In this connection the re-evaluation of the known ore objects depth, new deposits survey in the concealed structures is expedient.

In the Caledonian cycle in the border structure of the Zhingiz –Tarbagatai under the rifting and insular-arc geodynamic conditions of an early stage ($\mathcal{C}_1$-O_3) the volcanic arcs of basalt-andesite-dacite series of iron-manganese and gold-chalcopyrite deposits (Akbastau, Kosmurun, Mizek) have formed. On the Greater Altai territory there is a marine mode with the formation of calcareous-siliceous- of basalt and terrigenous structures.

In the middle collision stage (O_3-S) the overall tendency for the Irtysh-Zaisan small oceanic basin degradation from the growth of the accretion zones on the Gornoaltaiskij and Kazakhstan continental arrays. Collision magmatic front is located in arc up-lifts of spreading zones with the formation of gabbro-diorite-granodiorite intrusions (O-S, S_2) in local parts of the Zhingiz –Tarbagatai and Rudnoaltaiskij-Ashalinskij zone. The mineralization is represented by Fe, Cu, Mo, Au.

At the final stage (S_1-D) the unified Caledonian structure of the Greater Altai has formed under the conditions of total rotation of the Siberian continent and adjacent folded structures in the north direction. In the process of intraplate activation in bordering parts of the GA granodiorite-granite intrusions D_2 (Mo, W, Bi) have located in the Altai Mountains and arrays of granit - granosienite -leykogranite series (D_2-D_3) with poor rare-metal-rare earth mineralization in the Zhingiz Tarbagatai.

In the early Hercynian stage the most powerful tectonic and magmatic processes of the Devonian times took place in the Rudny Altai being active continental suburbs of Altai-Sayan folded area. Here under the influence of tectonic stress (compression and tension) the system of contiguous sub parallel deep faults of north-west direction, penetrating in the activated upper mantle, has formed. The activation of deep faults (pushings) in the rifting and insular-arc geodynamic conditions was accompanied by a powerful basaltic volcanism and unique copper-pyrite and pyrite-polymetallic deposits with rich complex ores formation (Cu, Pb, Zn, Au, Ag, Pt, etc.). That is why the Rudny Altai is considered as a unique gold-copper-polymetallic belt including original rudnoaltayskii pyrite deposits together with the famous world types (Kuroko, Ural, Cyprus, filizchaysky and others).

As a result of the Altai geologists long-term research the certain tendencies of ore-bearing geological structures and pyrite-polymetallic deposits formation and location have emerged.

• In the Devonian the suburbs of the ancient Altai continent underwent the destruction and were fragmented into a grid of deep faults, separate longitudinal plates and blocks were spread, displaced and drawn apart giving way to the deep fluid flows, mantle and intracortical magmas. Subsequent oncoming movment of the Kazakhstan array was accompanied by convergence of micro-continents and blocks, insular arc, their accretions and bonding into a single continental formation (Zoneinshein, 1990). Such "altai" type of inter-continental structures formation was due to the formation of longitudinal astenoval in the upper mantle as a result of moving lithospheric blocks trans-pressing and astenomass injection and fluid-flow into the deep faults (Greater Altai..., 1998). This process was accompanied by formation of prevailing complicated homodrom magmatites and local antidrom bimodal volcanic series connected with intracortical basaltic centers. The emerged geological structures in the final geodynamic modes (collision compressions, sub-vertical tensions, horizontal movements in the tectonic zones, etc.) were significantly transformed and complicated with pyrite-polymetallic ores (Shcerba et al., 1984).

The location giant halos of magmatites and deep zones of earth crust according to the geophysical data (Lyubetskij et al., 1994) allows us to presuppose the existence of local fluid-saturated systems at the depth of 60-80 and 160-180 km with complicated real differential in the past, possibly, these are north-western removals of the Northern-Asian superplumes (Greater Altai..., 1998; Yarmoluk et al., 2000). Accumulation of fluids and concentrated migration of ore saturated flows is supposed to be the main feature for the potential productivity index of the ore-forming system of the Rudny Altai. The existence of favorable structural elements contributed to the deposition of ore material and the deposits formation.

• The main pyrite deposits are concentrated in the rod Rudnoaltaiskij-Ashalinskij zone, limited by the Irtysh and North-Eastern zones of collapse. The given zone is characterized by increased femic index of the EC section, has high magma saturation and mineralization density. The location of the pyrite mineralization is clearly correlated with the elevation of the upper mantle, meta-basalt layer, and the Proterozoic and Caledonian basement blocks. The main pyrite-polymetallic and polymetallic zones in the EC deep section are related with the thickened parts of the meta-basalt layer (power is 20-24 km) and concentrated over the crest show of the mantle surface relief (depth is 40-43 km).

The ore-formation model reflects the connection of the pyrite deposits with the Devonian volcanism, multi-staged ore process and multi-layered mineralization distribution (with vertical ore range up to 1000-1500 m). The deposits are bound with the group of basalt-andesite-rhyolite formations differentiated and forming several ore-bearing geo-chronic levels from D_1e to D_3fm_1. The most productive formations are – emskij (Fe, Mn, Pb, Zn (deposits Kholzunskij, Ridder-Sokolnyj)), ems-eifelskij – Zn, Pb, Cu, Au, Ag (Tishinskij, Zyryanovkij), eifel-zhivetskij – Cu, Zn, Pb (Orlovskoe, Maleevskoe), zhivetskij – Cu, Zn, Pb, Au (Artemyevskoe, Nikolaevskoe) (Shcerba et al.,1984: Bespaev et al., 2000).

The ore formation took place under the sub-marine conditions, obviously at ascending vadose-hydrothermal system of solutions with povenil source of metals (Fe, Cu, Pb, Zn, S, Au, Ag, etc) and dissolved gases (CO_2, N_2, H_2S, S, Cl, etc.). There are two ore types according to the formation way: 1) stratiform (volcanic-sedimentary) and 2) hydrothermal-metasomatic, represented in a majority of industrial pyrite-polymetallic deposits (Zyryanovskoe, Maleevskoe, Tishinskoe, etc.). The volcanogenic-ore centers, volcanic domes and volcano-tectonic depression, the nodes of faults intersection, horizons of carbonaceous clay and calcareous shale and others are referred to the ore-bearing structures. The most productive structures among them are the Devonian volcanic arcs of a ring structure surrounding the Caledonian paleo–elevations (Sinyushinskoe, Revnyushinskoe, Alejskoe) and characterized by the length of the volcanic processes and ore-formation (deposits Ridder-Sokolskoe, Tishinskoe, Zyryanovskoe, Nikolaevskoe, etc.).

The main role in the deposits location is given to sub latitudinal ore-controlling faults (Leninogorskij, etc.), especially in their intersection nodes with the breaks of other directions, where there were volcanogenic-ore centers. The characterizing feature of this phenomenon is the linear-nodal distribution of Devonian volcano-tectonic structures with pyrite-polymetallic deposits in the longitudinal ore zones (Leninogorskaya, Zyryanovskaya, Orlovsko-Belousovsky, etc.) (Fig.4). The spacing of ore nodes (at the intersection of the north-western faults with meridional and sublatitudinal faults) is 20-40km. Reconstructed ore zones of considerable sizes (longer than 100 km, at width of 10-20 km), with ore nodes contain major reserves of Cu + Pb + Zn of all known deposits of the Rudny Altai.

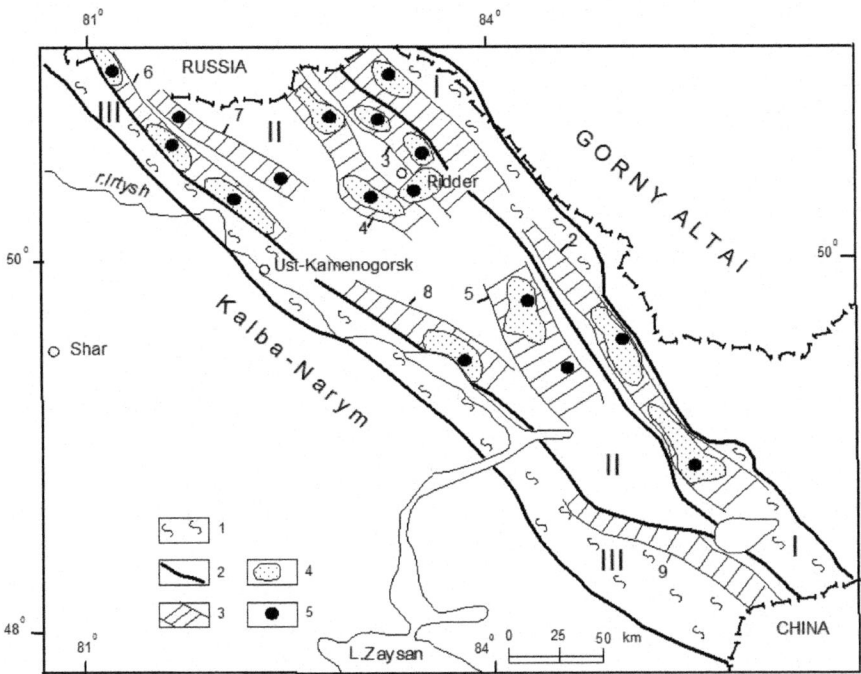

Figure 4. Metallogenic demarcation of Rudny Altai1- Shear zone, 2- boundaries of metallogenic zones, 3- ore zones, 4- ore districts, 5- copper-base-metall polymetallic depositsMetallogenic zones: I-Belay Uba-Sarymsaqtinskij, II-Rudnoaltay-Ashalinskij, III-Irtysh-Fuyungskij.Ore zones: 1-Gyslaykovskij, 2-Yuzhno -Altaiskij, 3-Leninogor-skij, 4-Tishinskij, 5- Zyryanovskij, 6- Orlovsko-Belousovskij, 7- Shemonaikhinskij, 8-Bukhtarminskij, 9- Dzhaltyrsko-Alexandrovskij. Major ore area are: Leninogorskij, Zyryanovskij, Priirtyshsk,ij Rubtsovskij and Zmeinogorskij contain major reserves of copper, lead and zinc (Ridder-Sokolnoe, Tishinskoe, Orlovskoe, Nicholaevskoe, Arte-myevskoe, etc.). In the south-eastern extension in China there are well-known deposits such as Ashaly, Timurty (Fe, Pb, Zn), Koktal (Fe, Pb, Zn) and others. This is consid-ered to be an important verification for the unity of ore-bearing structures in the border region of Kazakhstan and China (Bespaev et al., 1997,Shcerba et al., 1984).

Forecasting-metallogenic studies show that in the Rudny Altai the industrial pyrite deposits formed under the rifting and insular-arc conditions in the process of a powerful rhythmic-pulsating volcanogenic mineralization of the Devonian period. The pyrite-polymetallic and copper-polymetallic have belt distribution in the longitudinal ore zones of considerable sizes. The former preserves significant prospects for mineral resources increase at certain ore-bearing geo-chronic levels, within a flank and deep horizons of known ore nodes, ore fields and deposits. The medium stage of the Hercynian cycle (C_1-C_3) differed by the sharp change of the geodynamic mode (prevailing compression), closing of the Irtysh-Zaisan paleo-basin, exertion of folding and thrusting major phases, accretion of structures and collision of Kazakhstan and Gornoaltayskij arrays. The result of it was the collision of two continental suburbs having formed a single coherent structure of the Greater Altai. In large volumes the sin-collision gabbro-diorite-granodiorite intrusions have infiltrated, volcanic-plutonic belt and molasse formation have localized. In the region geological structure under the collision geodynamic conditions the magmatic copper-nickel, copper-porphyric and hydrothermal gold deposits formed.

In the Rudny Altai the most powerful intrusive magmatism emerged in the focal part of the deep mobile zone (above the elevations of the meta-basalt layer and anomalous upper mantle). The large arrays of multi-phase gabbro-diorite-granodiorite of plagio-granite series have formed (Zmeinogorsk complex C_2-$_3$). The tectonic broken apical zones of separate gabbro-granitoid massifs have undergone hydrothermal-metasomatic changes and are prospective for the detection of copper-porphyric and gold deposits.

The Sekisovskoe deposit refers to the gold-telluride formation of propylites Berezite-type. It is represented by brecciated gold-sulphide mineralized zones and stockworks in the magmatites of the Zmeinogorsk complex. The faults of the north-western direction associated with the Irtysh zone of collapse had the ore-controllling importance. According to Yu.A. Kostin, G.G. Freiman, G.I. Dudukalov and other researchers, increased tectonic dislocations of rocks fixed by brecciated and crushed textures of rocks are characteristic of the ore fields. Ore-explosive breccias in the form of gabbro, diorite and plagiogranites debris cemented by the bulk of propilite and berezite content with the gold-sulphide mineralization are also one of their features. The types of the ore-bearing meta-somatites depended on the original rocks content: 1) propylites-sulphide (by gabbroids, diorites) and 2) beresites-sulfides (by plagiogranites and porphyries), on which the ore-bearing quartz and carbonate-sulfide veins have laid.

The ore-formation was accompanied by Fe, Cu, Pb, Zn, Au, Ag, Bi, SO_3, CO_2 entry and Si, Na exit. Gold and silver in beresites and propylites content are positively correlated with Pb, Zn, Bi, Cu, and in pyrite-quartz veins content and quartz meta-somatites have another correlation of Au Mo (+0,76). Ore bodies in the mineralized zones are characterized by lenticular or column forms due to the conjugation breaks nodes; mineralization is disseminated and veinlet-disseminated. The main ore minerals are pyrite, sphalerite, chalcopyrite, gray ores and gold, minor are arsenopyrite, magnetite, ilmenite, rutile, tetradimit, bismuthinite. Non-metallic minerals are quartz, calcite, sericite. All types of ores are enriched with pyrite (over 17 kg / m). The gold is represented by two types:: 1) free, having irregular, vein-like and elongated forms, performing fissures in pyrite and quartz, and 2) finely dispersed in pyrites (Au - 10 g / t, Ag - 100 g / t, Bi - 290 g / t). The deposit has an industrial value.

On the south-eastern flank of the Rudny Altai drastic narrowing of the structural-formational and metallogenic zones, associated with tectonic compression in the process of Jungar and Siberian plates collision is notices.

The emerged structures of conjugation, converging into a single virgational beam in the south-eastern-part of the Lake Markakol differ by complicated geodynamic development, intense exertion of magmatism and metamorphism, and diversified mineralization (Fe, Cu, Pb, Zn, Au, Ni, W, etc.). In the Irtysh sutural zones quartz-vein gold deposit of Mank in genetic linkage with small intrusions and dikes of diorite-granodiorite composition (C_3) emerged, the analogue of it is gold-ore deposit Dolonosai on the Chinese territory.

This fact allows to mark the single ore field of Mank-Dolonosai on the bordering territory and this area should be thoroughly studied.

In the Zaisan sutural zone in the process of complicated geodynamic development the Charsk-Gornostaevskij ophiolite belt of a planetary rank has formed. It fixes the zone of the mantle deep fault. Here, there are also zones of melanging, fold-thrust structures, olistostrome complexes, auriferous small intrusions and dikes infiltrated. The collision geodynamic conditions turned out to be favorable for different gold-ore deposits formations. The leading deposit is gold-arsenic-carbon type in the black shales, covering the largest reserves of gold deposits in Western Kalba (Bakyrchik, Bolshevik, Glubokij Log, etc.) (Bakyrchic..., 2001). The ore-containing deposits are sediments of molasse limnic coal-bearing formations (bukonskaya formation of C_{2-3}), exposed to intense dynamometamorphic and hydrothermal-metasomatic changes in the zone of deep faults (Fig.5). The ore bodies are represented by broken, fragmented and silicified black shales with abundant inclusions of gold-bearing pyrite and arsenopyrite. The main ore-minerals are pyrite, arsenopyrite and gold. The gold content is 8-9 g/t at average. The deposits have an industrial importance.

Gold-*vein-disseminated type* also has a practical value. It is characterized by the concentration of gold in the hydrothermally altered carbonate rocks of the island-arc type volcanic-carbonate-terrigenous formations (deposits Suzdalskoe, Mirazh, etc.) (Greater Altai..., 1998; Dyachkov et. al., 2009; Kovalev et. al., 2009). Ore bodies have formed as a result of dzhasperoid silicification of limestones, and are represented by mineralized zones, nests, veins and stockworks. The main minerals are pyrite, arsenopyrite, gold, and rarely –stibnite. The gold content in indigenous ores and weathering crusts is 8-10 g / t. According to the range of features the deposit is compared with the famous "karlinskij" type of ore-mineralization in the carbonate formations (Rafailovich, 2009).

In Zharma-Saur intense tectonic-magmatic activity of the collisional stage, was accompanied by intensive development of the gabbro-granitoid intrusionswhich were located in ring structures in the focal part of the deep mobile zone (Saur complexes C_1, maksutskij C_2-$_3$, saldyrminsky C_3). Magmatic formations are characterized by metallogenic specialization on Cu, Ni, Co, Au, Ag, Mo. The wide development of ore-bearing intrusive formations indicates at the possibility to detect hidden and buried deposits (Bazarskij deflection, Severnoe Prizaysanye, etc.).

There are certain prospects for exploring gold in Zhanan-Boko-Zaisan ore zone where the gold-sulfide deposits exploration with primary ores as well as in weathering crusts is forecasted (sites Kempir, Zhanan, Akzhal-Boko and others).

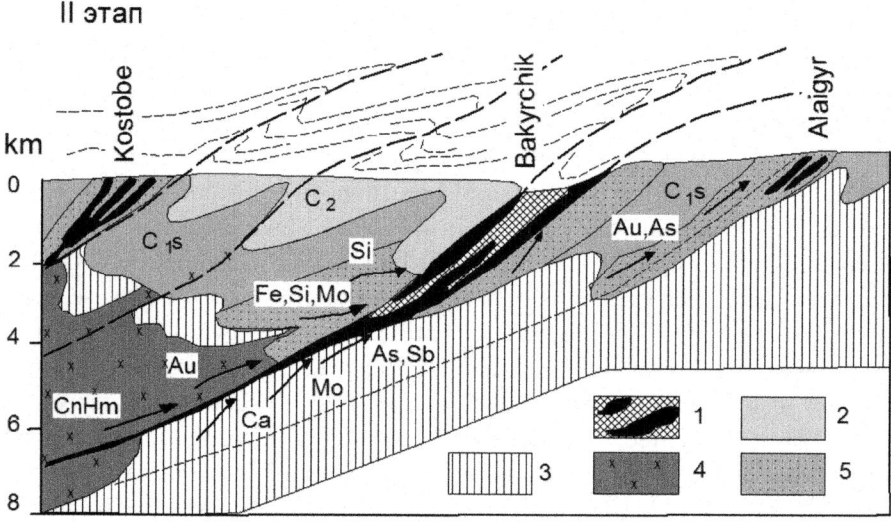

Figure 5. The model of Baqyrchik deposit forming1-surface of gold bearing carbonaseous molasses (C_2-$_3$), 2- submurin molasses (C_2)., siltstone-sandstone formation (C_1), 3- volcano-silica, volcano-terrigenous formation (D_3-C_1), 4- plagiodranit-granodiorite formation (C_3), 5- zones of mineralization (Greater Altai …2000).

In Kazakhstan, the copper-porphyry deposits are an important source of copper and molybdenum (Kounrad, Bozshakol, etc.). In Eastern Kazakhstan the largest deposits are porphyry- copper fields of Aktogai, Aydarly, as well as other known objects - Shar, Kyzyl Kain, Kensal, etc. These deposits are characterized by significant resources of copper, molybdenum, perhaps gold, and can make a significant contribution to the economy of the region. Low metal contents did not give the opportunity to assign them to categories of industrial deposits in the past. The emergence of new technologies opened new possibilities of exploiting deposits with poor content of ores.

Copper-nickel mineralization is genetically related to small intrusions gabbronorite-diorite-diabase (C_{2-3}). On the Maksut deposit the ore-bearing fields are melanocratic gabbro and gabbro-diabases, which are associated with solid and disseminated pyrrhotite-chalcopyrite-pentlandite ores in the form of tabular and lenticular deposits, connected with the bottom of the cup-shaped array. The main ore-field has the length of 1000 meters and the power of 21.5 m at an average content of Cu 0,47%, Ni 0,35%. The exploration works are carried out at the moment. There are real prospects of detecting new sulfide Cu-Ni deposits within Maksut-Petropavlovskij-Kharatungskij belt of disseminated gabbroid intrusions (C_{2-3}), bordering with a large deposit of Kharatung or Kolotong on the south-east flank (Bespaev et al., 1997, Shenghao et. al., 2003). This increases the possibilities for the similar deposits exploration on the Kazakhstan territory as well.

The facts mentioned above indicate that in the middle Hercynian formation range in the Greater Altai region the sin-collision intrusive formations were prevailing (up to 60-80% from the total volume) over volcanogenic and sedimentary elements. The magmatism evolution in each tectonic rhythm was carried out in homodrome sequence (from basalts, gabbroids to rhyodacite, granodiorites and plagiogranites) and the mineralization change in time range respectively: Cu-Mo-Au (C_1) Cu-Ni-Au (C_{2-3}) Au-Ag-Sb-Bi (C_3).

The later stage of the Hercynian cycle was marked by the shift of collision geodynamic mode to post-collision or orogenic mode (P_1-T_1). At this stage ascending crest-block movements, intensive processes of intraplate tectonic-magmatic activation, accompanied by the outbreak of granitoid magmatism with rare metal and rare-earth profile of mineralization were dominating (Ta, Nb, Be, Li, Cs, Sn, W, Mo, etc.). The largest granitoid belts are Kalba-Narymskij, Tigereksko-Chernevinskij and Akbiik-Akzhaiyauskij. They formed in zones with increased siality of the continental Earth crust (Kalba-Narymskaya, Beloubinsko-Sarymsaktinskaya and Sirektas-Sarzasanskaya). The granitoids of the Permian age (Kalbinskij, Monastyrskij, Zharminsky and other complexes) are prevailing in their composition. The magmatism evolution was in the overall tendency for the leucocratic alkalinity of granitoid series increase. Every petrochemical type of granitoids formed under certain geological conditions and was characterized by its ore potentia (Dyachkov et. al., 2009).

The Kalba-Narymskij granitoid belt is the main ore-bearing structure of the region having regional development (length is more than 500 km). It is characterized by sialitic profile of the Earth crust section with an increased power of meta-granite layer and of the Earth crust in general. The large scale of granitoids distribution emphasizes the fact that there are rather a lot of energy and material resources for rare-metal mineralization processes. The structural metallogenic model of the given belt reflects the connection of ore-magmatic systems with deep zones in the Earth crust and the upper mantle and, consequently, the granitoid belt has formed as a result of long-term deep evolution of the lithospheric element.

In the Kalba-Narym there are following ore-formation types of deposits: 1) pegmatitic rare metal (Ta, Nb, Be, Li, Cs, Sn), represented by major industrial deposits (Bakennoe, Belaya Gora, Yubileinoye, etc.); 2) pegmatitic beryl, microcline, in which minerals are blocking microcline and quartz, muscovite, beryl and columbite (deposits Asubulak, Lobaksaj, Nizhny Laybulak etc.); 3) albite-greisen tin-tantalum (apogranite) in hidden granite dome, potentially prospective for the identification of Ta, Be, Li, Sn (Karasu deposit); 4) greisen-quartz-vein tin-tungsten represented by ores of wolframite, scheelite and cassiterite (deposits Cherdoyak, Palattsy, Kaindy, etc.); 5) clastogene Tantalum - Tin - Tungsten forming placers of tantalite, cassiterite, wolframite, scheelite and monazite.

The carried out studies show that ore-generating ability of granitoids together with petrologic factors depend on geodynamic conditions of arrays formation and scale ore-bearing melts degassing. From these positions there are various ore-magmatic parts determined. They are close to ore-bearing fluid systems (Letnikov, 2000). There are the following favorable conditions and criteria for rare-metals mineralization forecasting:

mobile geodynamic conditions of granite arrays formation and this contributed to the intensification of mineralization process under non-equilibrium PT conditions and the formation of industrial deposits of rare-metal pegmatites (Central Kalba ore region);

determination of ore-controlling role in latitude deep faults of prolonged activation, particularly favorable nodes are their intersections with the faults of other areas, where the most important ore fields have formed (Asubulakskoe, Belogorsk, Ognevsko- Bakennoe and others); apical parts and over-intrusive zones of granite arrays, their apophasis, hidden domes and tectonically weakened zones, ore nodes in the thickened parts of the granite intrusions, over the magma-feeding roots or at their peripheries are the most perspective structures for the mineralization concentration; establishing the genetic relationship of rare metals from each intrusive phase of Kalba complex at the spatial confinement of the main rare-metal - pegmatite mineralization (Ta, Nb, Be, Li, Cs, Sn) to the granites of I phase with increased base (deposits Bakennoe, Yubileinoye, Belaya Gora, Verkhnyaya Baymurza etc.) identified petrographic, petrochemical and mineralogical and geochemical criteria for ore content evaluation of different age granitoid complexes and their intrusive phases.

On the continental border, in zones of sub-continental earth crusts (West-Kalba, Charsko-Gornostaevskaya) and in the faults of sutural zones on the surroundings of the Greater Altai (North Eastern zone of collapse, Zhingiz- Saur suture) the subalkaline granite granosienite (Zr, Ti, Sn, Be) and alkali granite (TR, Zr, Nb, Ta) intrusions have localized. They are connected with deeper crust and mantle centers. The known ore objects refer to the epi magmatic Niobium - Zirconium - rare-earth formations (Verhnee Espe, Azutau).

At present the most important objective of the region is exploring new deposits of rare and rare-earth metals considering modern tendencies of world geological science and market economy development.

The Cimmerian cycle (T-Pg2). The Cimmerian formations emerged under the continental rifting mode and had an autonomous development. They are represented by Semeitau volcanic-plutonium association (T_1) of a contrast content and high alkalinity (trachybasalt-trahiriolites and subvolcanic analogues), trap formations J_2 (lugovskij complex), northeastern belts of colored dikes from gabbro-diabase to quartz porphyries (mirolyubovsky, Bugaz complexes) and molasse coal-bearing formations (T_3-J_1,J_2). The latter unify sedimentary-coal-bearing strata imposed animations (Kendyrlykskaya, Zhemeneiskaya, Abaevskaya, etc.). At the ending stage of the tectonic stage under the epigerian plate stabilization mode the silt-clay variegated hematite - kaolin deposits K_2-Pg_2 have accumulated on the vast territory and have the power of 200 m (north-zaisan formations). In a humid subtropical climate the Paleozoic and older rocks formed with crust of weathering. In the Charsk-Gornostaevskij ore zone the Ni-Co and mercury bearing weathering crusts of a nontronite type have formed on the serpentinized hyperbasites as well as ore-bearing crusts in ore-containing rocks and ores (Semipalatinsk Priirtyshie). In the West- Kalba belt the Zr-Ti weathering crusts of kaolin type developed with sub-alkaline granites (Karaotkel deposit). The Cimmerian cycle had a destructive character. Huge masses of loose material together with the disseminated ore element were taken into the Kulundinskaya and Zaisanskya depression, West Siberian Plain and smaller intermountain depressions.

Alpian cycle (Pg2-3-Q). During the Alpian cycle, the given region was an area with intense denudation with removal of products of destruction to the West Siberian Plain and large depressions forming continental sedimentary formations (Paleogene-Neogene-Quaternary period). They are widely developed in the Zaisan basin, Semipalatinskoe Priirtyshye, intermountain depressions and lake basins differing by some lithologic ore peculiarities and deposits power.

In the sedimentary cover there are also sustainable ore minerals placers (gold, ilmenite, tantalite, cassiterite, etc.) which are subdivided into alluvial, deluvial and proluvial according to the genesis. Some of them were explored (Kurchumskaya – Au, Satpaevskaya – Ti, Asubulakskoe – Ta, etc.).

CONCLUSION

So, on the basis of the global mobilism hypothesis, the overall tendencies of the Greater Altai geological and metallogenic structures formation located in the system of the Central Asian mobile belt have been identified. Modern structures – the Rudny Altai, Kalba-Narym, West- Kalba zone and Zharme-Saur are the fragments, xenoliths of the ancient paleocontinent (terranes collage) that were drifting in the paleoasian ocean and joined into one single formation during the Hercynian collision stage in the process of a complicated interaction and the Kazakhstan microcontinent and the gornoaltaiskij border of the Syberian platform interfacing. Now it is a system of sub-parallel structural formational and metallogenic zones divided by deep faults and differing in the geodynamic development and the peculiarities of the geological structure and metallogeny. From the point of mobilism the common trend of the Greater Altai geological structures development and minerageny, under different geodynamic modes and conditions for the long period (from the Precambrian up to Quaternary) has been defined. The identified trends of the deposits connection with certain geodynamic modes based on the principles of ore-formation analysis speak for the fact that there is a genetic connection between geological and ore formations considering the geological formations as the indicators for certain paleogeodynamic and landscape geological conditions.

The research shows that the energy potential of the ore-forming processes in each ore belt theoretically tends to be one certain value. In the Rudny Altai belt in the Early Hercynian stage under the rifting insular arc conditions (on the continental crust) the powerful processes of basaltic vulcanization were accompanied by the major copper-pyrite and pyrite-polymetallic deposits formation in East Kazakhstan(Fe, Cu, Pb, Zn, Au, Ag, Pt, etc.). The scope of ore-content of the following medium (collision) and late (post-collision) periods was weakened. The West- Kalba zone had the largest tectonic-magmatic activity during the medium collision stage, and it was rich with ore minerals (Au, Ag, As, Sb), under appropriate depressed development of other stages metallogeny. The example of this is the Kalba-Narym rare-metal belt where the energy potential was accumulated in the early and medium stages and then it passed into the late stage (post-collision) in the form of granite magmatism flash and the deposits of rare-metals related to it (Ta, Nb, Be, Li и др.). In this regard, the study of mineragenetic specialization of geodynamic conditions along with detailed structural - elemental studies of geological formations and ore sites, is one of the main methods of forecasting and search for new deposits, particularly in poorly studied, and closed areas.

REFERENCES

1. Kh. A. Bespaev, V. N. Lyubetskaya, et.al, 2008 Cold ore-bearing belts of Kazakhstan, Ed. by S. Zh. Daukejev (Satpaev's Institute of Geological Sceinces) 5-62802-058-3 [in Russian]

2. Kh. A. Bespaev, N. V. , G. D. Polyansky, O. D. Ganzhenko, et.al, 1997 Geology and Metallogeny of the China (Fylym, 5-62802-058-3 [in Russian]

3. B. A. Dyachkov, D. V. Titov, E. M. Sapargaliev, 2009. Ore belts of the Greater Altai and their ore Resource Potential Geology of Deposits,, Vol. 51, p p 197-211 ISSN 1075-7015 Pleiades Publishing, Ltd.,2009 [in English]

4. B. A. Dyachkov, N. P. Mayorova, G. N. Shcherba, K. A. Abdrahmanov, 1994 Granitoids and Ore Deposits of the Kalba-Narym Belt, the Rudny Altai. Gylym, Almaty, 5-62800-725-0 Russian]

5. B. A. Dyachkov, N. P. Mayorova, Z. I. . Chernenko, O. N. Kuzmina, 2009 To the methods of search and evaluation of ore-bearing deposits of nontraditional types in the carbonate formation of East Kazakhstan. Ores and metals. № 3.1121 0086-5997 [in Russian].

6. Great Altai: Geology and Metallogeny, Book 1: Geology ISBN 5-628-02439-2 (Fylym, Almaty, 1998) [in Russian]

7. Great Altai: Geology and Metallogeny, Book 2: Metallogeny ISBN 9965-520-44-5 (Almaty, 2000) [in Russian]

8. K. L. Kovalev, Y. A. Kalinin, E. A. Naumov, et al. 2009 A mineralogical study of the Suzdal sediment-hosted gold deposit, Eastern Kazakhstan: Implications for ore genesis. Ore geology Reviews, № 35,, 186205 01691368 English]

9. F. A. Letnikov, 2000 Fluid mode of the endogenic processes in the continental lithosphere and problems of metallogeny Problems of global geodynamics. Moscow, GEOS. 204224 . [in Russian]

10. V. A. Narseev, 2001 Baqyrchik: Geology, Geochemistry, and Ore

Mineralization, (TsNLGRI, 5-85657-097-9in Russian].

11. M. S. Rafailovich, 2009 Gold of bowels of Kazakhstan: geology, metallogeny, forecast-search models. 9-96503-493-4 [in Russian].

12. G. N. Shcherba, Kh. A. Bespaev, B. A. Dyachkov, et al. 2000 Geological Evolution of the Creater Altai on the Basis of Geodinamic Reconstructions, in Geodinamics; and Minerageny of Kazakhstan, Almaty, Part 1, 7381 . 9-96552-036-4 Russian]

13. G. N. Shcherba, B. A. Dyachkov, G. P. Nachtigal, 1984 Metallogeny of the Rudny Altai and Kalba, Nauka, Alma-Ata, 4070584 Russian]

14. U. Shenghao, Zh. Zhaochong, et al. 2003 Kalatongke Magmatic Copper-Nickel Sulfide Deposit, CERCAMS, Natural History Museum, London, 131151 [in English].

15. N. P. Yshkin, V. N. Sazonov, 2007 Geodynamics, magmatism, metamorphism, and ore formation, Ed. (institute of geology and geochemistry, Ekaterinburg, 2007) [in Russian].

Chapter 3

GEOPHYSICAL MODELING OF THE SURROUNDINGS OF LA POPA BASIN, NE MEXICO, WITH GRAVITY AND MAGNETIC DATA

Vsevolod Yutsis[1], Antonio Tamez Ponce[2], and Konstantin Krivosheya[1]

[1]Universidad Autónoma de Nuevo León, Facultad de Ciencias de la Tierra, Linares, N.L., Mexico

[2]Posgrado en Ciencias Geológicas, Facultad de Ciencias de la Tierra, Universidad Autónoma de Nuevo León, Linares, N.L., Mexico

INTRODUCTION

Northeast Mexico is in essence the juncture of two distinctly different tectono-stratigraphic provinces, the eastern Gulf of Mexico (Coastal Plane, Sierra Madre Oriental) province and the western Pacific Mexico (Rivera plate, Meso-American trench, Sierra Madre Occidental) province (Goldhammer & Johnson, 2001). Tectonic evolution of northeast Mexico was dominated by divergent-margin regime associated with the opening of the Gulf of Mexico and overprinted by non-igneous Laramide orogenic effects (Pindell et al., 1988). The structural grain of northeast Mexico consists of Triassic to Liassic fault-controlled basement blocks, the development of which reflects in part late Paleozoic orogenic patterns of metamorphism and igneous intrusion (Wilson, 1990). Several tectonic provinces may be recognized by interpretation of the basement and sediment cover: Coahuila Block, La Popa Basin, Sabinas Basin, Burgos Basin, Sierra Madre Oriental (Monterrey Trough), and Parras Basin (Yutsis et al., 2009, Fig. 1).

Mojave-Sonora Megashear and San Marcos Fault (Chavez-Cabello et al., 2007) are two principal fault zones crossing the northeast Mexico in NW-SE direction.

La Popa Basin is located on the foreland of the Sierra Madre Oriental, 85 km northwest of the city of Monterrey, N.L., northeast of Parras Basin and south of the Sabinas Basin (Fig. 2). The basin's development is greatly defined by salt tectonics, and is of great interest to the petroleum industry (Hudec and Jackson, 2007; Jenchen, 2007), due to the fact that many of oil reservoirs in the world are located in salt basins (e.g. Gulf of Mexico, the Persian Gulf, the North Sea, Lower Congo Basin, Campos Basin, etc.) (Warren et al., 1999). La Popa Basin is the best analogue of the Gulf of Mexico (Willis, 2001). Although no oil or gas deposits were found in it so far, understanding of its structure and geological evolution can be applied to similar areas whether they are located in the continent or offshore.

The Bouguer gravity and the total field aeromagnetic data were supplied by Geological Survey of Mexico (SGM), Pemex Exploration and Prospecting, published data (Mickus et al., 1999), and authors field works (Yutsis et al., 2009, Tamez et al., 2010, in press).

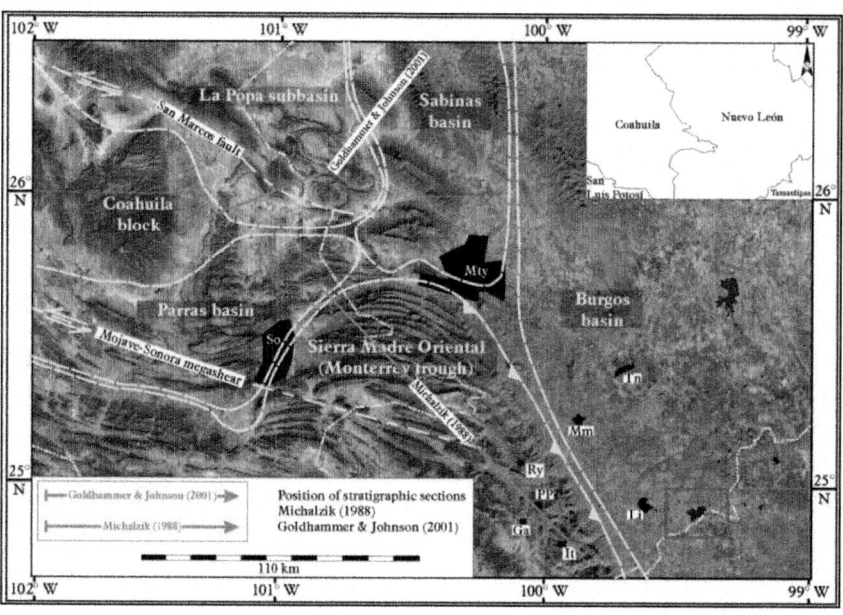

Figure 1. Location of study area in context with major and geologic elements (basins and blocks; orange lines; abbreviations: Mty: Monterrey, So: Saltillo, Tn: Terán, Mm: Montemorelos, Li: Linares, It: Iturbide, Ga: Galeana, PP: Potrero Prieto y Ry: Rayones; satellite image: worldwind.arc.nasa.gov; political limits in the upper right corner)

GEOLOGICAL SETTING AND PALEOGEOGRAPHY EVOLUTION

Geological formations of the NE Mexico range from the Proterozoic crystalline basement to Tertiary sedimentary rocks and Quaternary deposits. Middle Jurassic through Upper Cretaceous tectonic and sedimentary evolution in this area is mainly defined by divergent-margin process/regime associated with the opening of the Gulf of Mexico, and overprinted by nonigneous Laramide orogenic effects. Generally the stratigraphic evolution is interpreted to be dominated mainly by eustacy as thick regional accommodation cycles can be correlated throughout the Gulf of Mexico (Goldhammer & Johnson 2001).

The stratigraphic and structural configuration of Northeastern Mexico reflects it's a complex tectonic evolution. It initiated in the Permian-Triassic with the Ouachita-Marathon orogenic event, followed closely by Late Triassic to Middle Jurassic rifting of Pangea, subsequent opening of the Gulf of Mexico, and passive-margin development through the Late Cretaceous. It culminated with Laramide foreland deformation through the early Tertiary with local associated evaporite tectonism. The structural core of Northeastern Mexico consists of Triassic to Liassic basement fault blocks whose development reflects in part late Paleozoic orogenic events of metamorphism and intrusive magmatism. These early Mesozoic fault blocks in turn controlled Late Jurassic and Cretaceous sedimentation, and also strongly influenced Laramide structural patterns and foreland basin deposition.

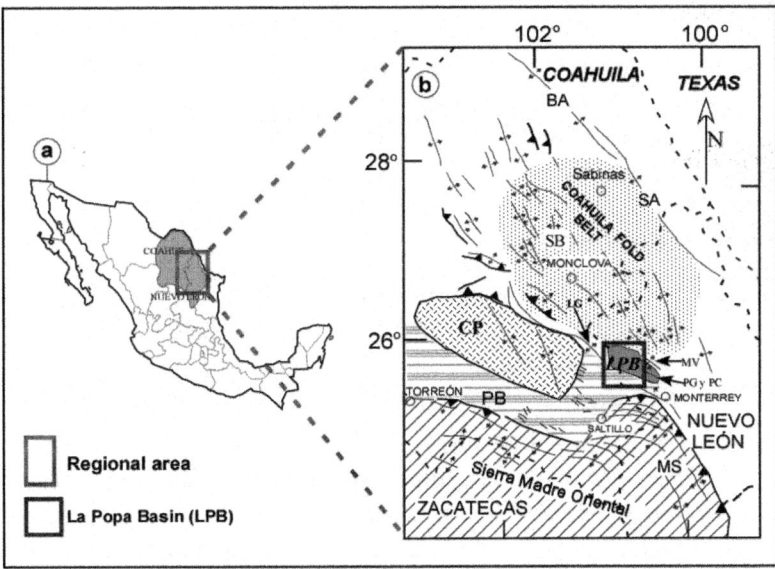

Figure 2. a) Location of foreland of Sierra Madre Oriental in northeastern Mexico and (b) La Popa Basin in the foreland. Explanation: CP, Coahuila Platform; LG, La Gavia anticline; LPB, La Popa Basin; MS, Monterrey Salient of Sierra Madre fold belt; PC, detachment fold at Potrero Chico; PG, detachment fold at Potrero de García; MV, detachment fold at Sierra Minas Viejas; PB,

Parras Basin; SB, Sabinas Basin; SA, Salado arch; BA, Burro arch. Inset blue rectangle in La Popa (Lawton et al., 2001).

Rifting and initial segmentation of Pangea are evidenced by attenuated basement in Northeastern Mexico expressed as basement highs (Coahuila Block, Burro-Salado Arch, Tamaulipas Arch) and lows (Sabinas Basin, the "Monterrey trough"; Fig. 1). These shallow, rigid basement blocks are held up primarily by granite to granodiorite intrusions of Permo-Triassic age, created south of the Ouachita-Marathon orogenic belt by closure of Gondwana and North America and stranded by subsequent rifting (Goldhammer & Johnson 2001).

The Sierra Madre Oriental folds and thrust belt is of Laramide age (Late Cretaceous to Eocene). The belt is characterized by elongated anticlines that trend east to west, curving to the south (further east). The anticlines have very steep, vertical limbs, and some are overturned to the north. Folds are arranged in a series of nappes and may be bounded by thrusts. The deformed section consists of essentially the entire Upper Triassic to Cretaceous rift to passive-margin sequence (e.g., Prost et al. 1994, Gray & Johnson 1995, Marrett & Arranda-García 1999).

Late Cretaceous to Early Tertiary foreland basins include the Parras, La Popa, and Burgos Basins (Fig. 1). The Parras Basin is confined between the Coahuila block and the Sierra Madre front. The sediments in these basins consist of nearly 5000 m of Campanian to Maastrichtian shallow marine and deltaic terrigenous siliciclastics of the Difunta group and up to 10000 m in the Burgos basin (Lawton et al. 2001, Ortiz-Ubilla & Tolson 2004).

Mesozoic Sedimentation

The Gulf of Mexico depicts a Mesozoic divergent margin basin formed through rifting and extension of Pangea, followed by breakup, seafloor spreading, and migration of various cooling and thermally subsiding tectonic plates (further reading Buffler & Sawyer 1985; Pindell 1985; Pindell & Barrett 1990).

During Triassic to Middle Jurassic (pre-Callovian) times, red beds associated with volcanism accumulated in fault-bounded graben systems, to which Mixon et al. (1959) assigned a Late Triassic age. These deposits unconformably overlie Late Paleozoic metasedimentary or Permo-Triassic granite basement. Thicknesses are in the order of 300 to 2000 m, restricted to rift basins (Michalzik 1988; Goldhammer & Johnson 2001).

Widespread Jurassic evaporite deposition occurring in the whole Gulf of Mexico area continues into the more restricted portions of Sabinas Basin and "Monterrey trough" (Fig. 1). In the Monterrey-Saltillo area, the Minas Viejas evaporite outcrops as deformed masses of gypsum unconformably overlying the Huizachal red beds and/or Paleozoic basement.

In Northeastern Mexico, the La Caja Formation is of Kimmeridgian to mid-Berriasian age and consists of rhythmically bedded, thin, calcareous shales, siltstones, and fine sandstones, with thin limestones toward the base. Locally, the La Casita Formation (of late Kimmeridgian to Hauterivian age), shows a period of major clastic influx. Age and thickness of these sedimentary packages vary geographically, in part as a function of proximity to the exposed Coahuila block and Tamaulipas arch.

The Taraises Formation is the Lower Cretaceous (mid Berriasian through Hauterivian) deeper-water, offshore facies equivalent to the middle and upper units of the La Casita and consists of rhythmic-bedded, black, cherty, pelagic lime mudstones and intercalated shales.

Goldhammer & Johnson (2001) do not report a clastic influence during the Lower Cretaceus (Taraises Formation) derived from Tamaulipas arch, however, Michalzik (1988), Ocampo-Díaz (2007) and Ocampo-Díaz et al. (in review) can prove a clastic influence to the Galeana, Potrero Prieto and Rayones areas coming from a contemporaneous paleo-high. These clastic intercalations may also have influenced the subsurface in the study area.

From Hauterivian to early Aptian, the enormous carbonate platform system of Northeastern Mexico maintains a low-relief, reef-rimmed shelf margin with a platform interior shallow-marine facies. The carbonate strata of this Cupido Formation (700 to 1200 m) predominates today's Sierra Madre Oriental.

The Lower Tamaulipas Formation (late Hauterivian to early Aptian) is the basinal equivalent to the Cupido Formation and crops out primarily to the south and east of the Monterrey-Saltillo area. It consists of up to 600 m of dark gray to black, thin- to medium-bedded, cherty lime mudstones to wackestones. The La Peña Formation separates the Cupido/Lower Tamaulipas Formations from the overlying upper Tamaulipas Formation The formation varies in thickness from a few meters to 200 m (Goldhammer & Johnson 2001), and consists of thin-bedded, dark, argillaceous, cherty limestones and black shales.

The upper Tamaulipas Formation (Albian; 100 to 200 m thick) is the basinal equivalent of the Aurora Formation (Fig. 2). Outcropping basinal facies are assigned to the Cuesta del Cura Formation which is latest Albian to Cenomanian and consists of 60 m deep-water carbonates and shales. The Agua Nueva and the San Felipe Formations result from the upfolding diachronous Laramide phase of deformation (from the end of Cenomanian) and the displacement of the depocenter towards the Gulf of Mexico (Smith 1981; Winker & Buffler 1988; Ice 1981; Goldhammer & Johnson 2001). This package of deep-water deposits averages 300 to 400 m in thickness and consists of pelagic lime mudstones to wackestones. At the time of their deposition, the Sierra Madre Oriental fold belt developed and migrated from west to east. In Northeastern Mexico, Maastrichtian foreland basins developed in front of the advancing Sierra Madre (e. g. Sabinas area, La Popa, Parras and Burgos Basins).

Digital Elevation Model

As the geographical situation of the studied area in the north of the Sierra Madre fold belt entails a lack of outcrops, thicknesses and facies of the different stratigraphic units (mainly igneous and sedimentary Mesozoic and Tertiary rocks and Quaternary soils; Lawton et al., 2001) had to be estimated and substantiated by geophysical and satellite data.

Digital Elevation Model (DEM) taken from The Consortium for Spatial Information (www.csi.cgiar.org) was used to compare the surface structures and basement blocks (Figs. 2, 3).

This model consists of a database of 72,202,000 points and covers the major structures such as La Popa Basin and its surrounding structures: The Monterrey Silent, anticlines Minas Viejas, Potrero Garcia, Potrero Chico, Las Gomas - Bustamante, La Gavia, Enmedio, Lapazos - Sabinas, Coahuila Island, Parras Basin, Sabinas Basin, and Monclova Island (Eguiluz, 2001), and Candela - Monclova Intrusive Belt (Fig. 2). Figure 3 shows main geological structures distinct in the Digital Elevation Model.

Surface Geology

The surface of study area is formed by blocks of marine sediments deposited between the Middle-Late Jurassic and Cretaceous (Fig. 3) (Padilla y Sanchez, 1986, Salvador, 1987, Pindell et al., 1988; Winkler and Buffler, 1988; Wilson, 1990, Pindell and Barrett, 1990 and Dickinson and Lawton, 2001), and the deformation observed in these blocks and in many parts of northeastern Mexico is complicated by evaporite sequences (Minas Viejas Formation), local incorporation of the basement in the deformation, and by reactivation of some basement faults such as San Marcos Fault (Padilla y Sanchez, 1986 and McKee et al., 1990). Part of this fault is covered by the gravity and magnetic data (Yutsis et al., 2009).

Foreland La Popa Basin is located in front of the Monterrey Silent as part of the Sierra Madre Oriental, a province that represents the structural high. The Parras Basin is located west of the Monterrey Silent and South Coahuila Block, which in turn is adjacent to the Sabine Basin in the northern part of the area. La Popa and the Parras Basin's sedimentary rocks contain fine-grained siliciclastic deep-water carbonates of Late Cretaceous - Early Tertiary age. These units overlie carbonates of Early Cretaceous platform. Late Cretaceous rocks underlie the Difunta Group and the Parras shale (McBride et al., 1974, Vega-Vera and Perrilliat, 1989 and Ye, 1997). This sedimentary sequence was deposited in front of the Sierra Madre Oriental during Laramide Orogeny (Vega-Vera and Perrilliat, 1989, Vega-Vera et al., 1989 and Ye, 1997). There are also some tectonic blocks or Permo-Triassic paleo-highs such as the Coahuila block (Tardy, 1980 and Charleston, 1981), the Monclova Island, representing basement high (Golhammer and Johnson, 2001).

The most important fault in this area is the San Marcos Fault, which was defined by Charleston (1981), who suggested the lateral movement along it in the Late Jurassic and normal fault behavior in the Early Cretaceous. In addition, the San Marcos Fault structurally separates the Coahuila Block and Coahuila Folded Belt (McKee and Jones, 1990); within the latter there are a number of intrusive bodies of nearly EW orientation which in general are called Intrusive Belt Candela-Monclova (Fig. 3).

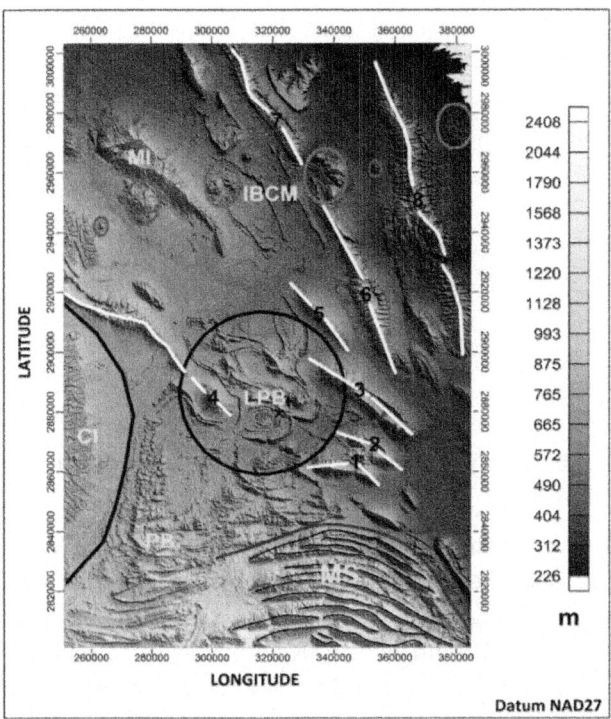

Figure 3. Geological structures recognized in the Digital Elevation Model: CM, Monterrey Silent, CP, Parras Basin, CI, Coahuila Island, CLP, La Popa Basin, IM, Monclova Island, CICM, Candela-Monclova Intrusive Belt (circled in red). Anticlines: 1-

Potrero Garcia, Potrero Chico 2 - 3 - Minas Viejas, 4 - En Medio, 5 – La Gavia, 6 - Las Gomas, 7 — Bustamante, 8 - Lapazos – Sabinas.

GRAVITY AND AEROMAGNETIC DATA

Gravity data used in this study consists of a database of 9857 measured points on the surface with a 500 m interval (Fig. 4). This data analysis prompted to recognize the Mesozoic sedimentary features. It was also confirmed that the basement of La Popa Basin is composed by evaporates of the Minas Viejas Formation.

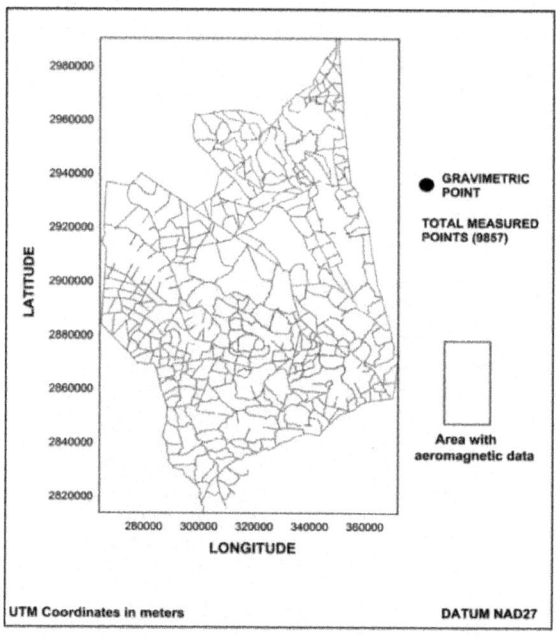

Figure 4. Area covered by gravity net (total 9857 measurements). The same region and all blank parts are covered by aeromagnetic data grid.

The total-field aeromagnetic data were supplied by the Mexican Geological Service (EMS), with the International Geomagnetic Reference Field removed. This method was used because its principle is based on changes in the magnetic properties of rocks in the subsurface (Reynolds et al., 1990 andBlakely, 1995). To correlate the aeromagnetic map with the crystalline basement, we examined the contrast of the physical property of sedimentary and igneous rocks.

The area of gravity and aeromagnetic data is located between coordinates 260000 E and 375000 E, 2810000 N and 29900000 N (UTM coordinates in meters, DATUM: NAD 27) but information is lacking for some parts.

Bouguer Gravity

The complete Bouguer anomaly map was obtained using a density of 2.4 g/cm3 which was obtained using the method of Nettleton, Terrain correction was applied according to the method of Hammer (Burger, 1992) as there is variation in elevations from 300 to 1500 m. This gravity map shows that the values from west to east are more negative in a range of -43 to -165 milligals (Fig. 5b), which is interpreted as a change in cortical thickness between 33 to 38 km by Bartolini et al. (2001). Also, the regional-residual separation of the Bouguer anomaly was applied (Reynolds, 2007). In order to analyze the structures, the shallow residual Bouguer anomalies were used. They were obtained by applying a high pass filter with a cutoff wavelength of 50.000 m to the map of the Bouguer anomalies, which resulted in a map with values ranging from 5 to -17 mGals (Fig. 5d).

This showed that negative residual gravity anomalies are concentrated at the peripheries of La Popa Basin marking their border, and bounded by some blocks of Permo-Triassic basement such as Coahuila Block, Monclova Island, the Arch de Tamaulipas, Tamaulipas Archipelago (Goldhammer and Johnson, 2001). This phenomena is possible to explain taking into account that salt being initially deposited in the Upper Jurassic, later began to spreading to the periphery of the basin.

It is also noted that these minimums are surrounded by gravimetric maximum reflecting the density contrast between salt and denser rocks such as limestone, shale and sandstone.

Residual Bouguer Anomaly

To study shallow bodies in the area we analyzed the map of residual Bouguer anomaly. It was observed that gravimetric minima ranging from -2 to -17 mGals almost perfectly match with some anticlines such as the anticlines Potrero Garcia, Potrero Chico, Minas Viejas, Las Gomas - Bustamante and the anticline La Gavia (Fig. 5d), Since the anticlines are considered to be nucleated by evaporites, the gravity minima may correlate with some of the most important accumulations of evaporites in the area under study.

Magnetic Anomaly

The analysis of the crystalline basement of the study area was done using magnetic anomaly map. The total-field magnetic data were supplied by the SGM, with the International Geomagnetic Reference Field removed. The sedimentary cover in the NE of Mexico is generally considered to be almost non-magnetic, and the anomalies are sourced overwhelmingly in the crystalline basement. Local intra-sedimentary anomaly sources may be related to depositional concentrations of magnetic minerals in some clastic rocks, or to secondary magnetization of sedimentary rocks by circulating brines. Analysis of the total magnetic anomaly map, with variations in the range of 252 to -89 nT, shows that the La Popa Basin is characterized by magnetic minimum with values ranging from -8 to -14 nT; these minimums are found throughout the central part of the map in the E-W direction. The observed high magnetic values, ranging from 10 to 200 nT, increase to the northern part of the basin, and the southern part of the map are characterized by the low magnetic field In addition, an isolated high can be clearly seen in the southern part of the La Popa Basin.

This analysis shows that La Popa Basin can be limited to the north and south by high magnetic fields (Fig. 6a).

The aeromagnetic map shows a quiet magnetic field in the area of the Basin. The general trend of the magnetic field reduced to pole is NW–SE in which background anomalies of northeast trend are obviously traced. However, local magnetic anomalies have mosaic character and, being morphologically extended in a NE direction, they are grouped in chains of northwest trend. The analysis of magnetic data allows assuming a series of linear elements focused in a NW direction. The NE part of the area is occupied by a series of positive magnetic anomalies intensively up to 120-160 nT. The maximum of this anomalies is the same as of gravity high (Fig. 5,6).

REGIONAL GEOLOGICAL-GEOPHYSICAL MODELS

Geophysical data interpretation includes two-dimensional gravity and magnetic modeling. The observed anomalies contain the effects of both shallow and deep bodies, reflected in the gravity response as the sum of short and long wavelengths. We performed the separation of regional and residual components of the Bouguer anomaly and magnetic anomaly, to conduct a qualitative analysis, and selected two profiles were selected for 2.5D modeling which represent the study area to a maximum depth of 15 km.

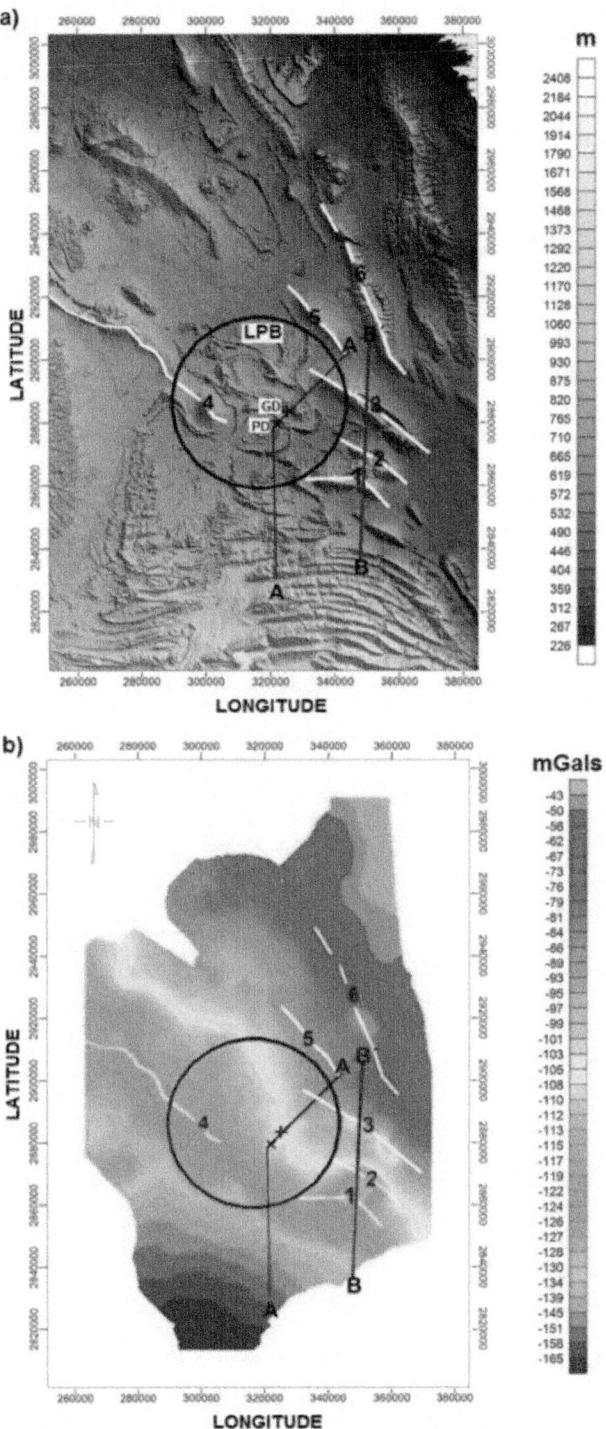

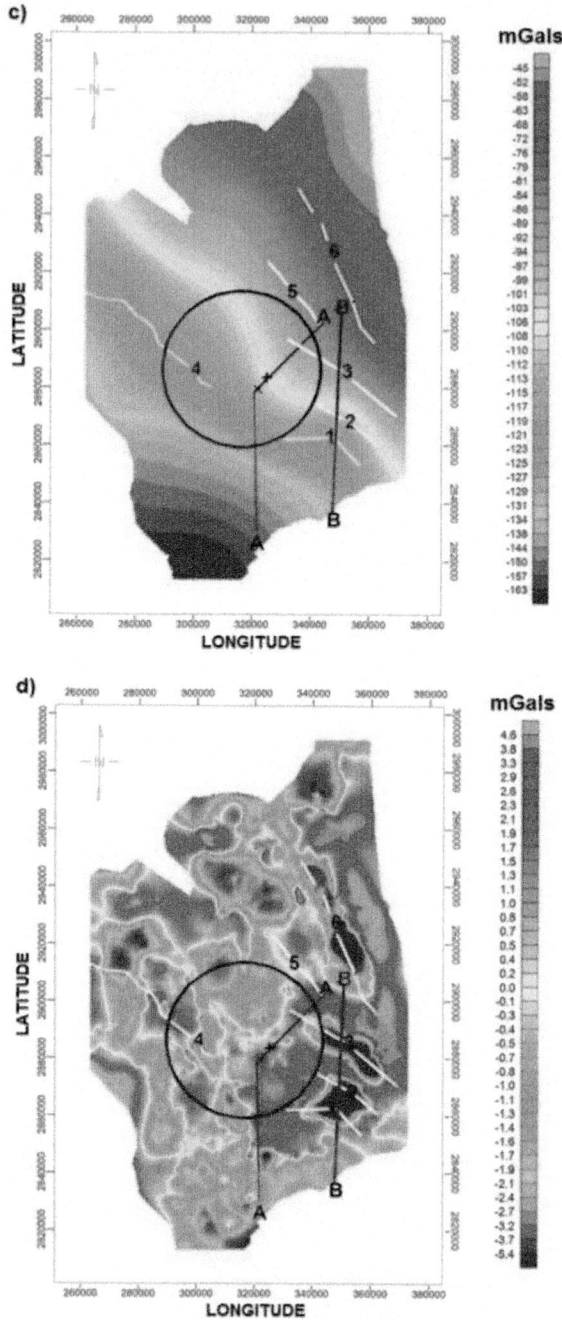

Figure 5. a) Digital elevation model where: La Popa Basin (LPB), the Diapirs El Gordo (DG) and El Papalote (DP), profiles (A - A ') and (B - B ') as well as some anticlines (see Figure 2 for the names of the numbered structures). b)

Map of the Bouguer anomaly 2.4 g/cm3, c) Map of regional anomaly obtained from the Bouguer anomaly, d) Map of residual anomaly obtained from the Bouguer anomaly

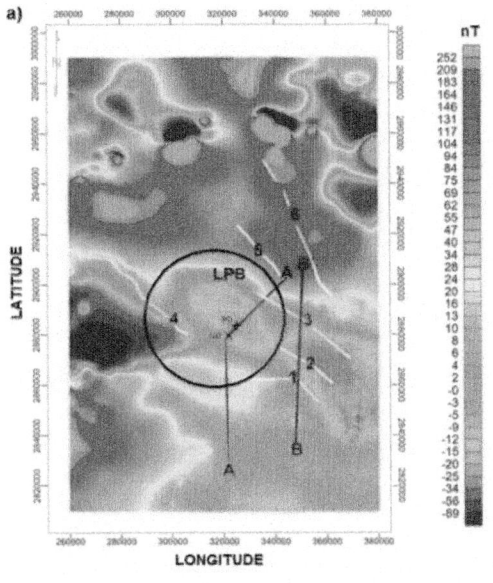

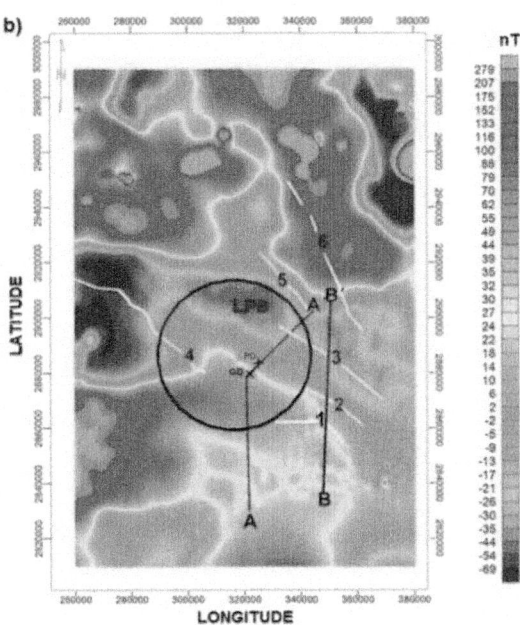

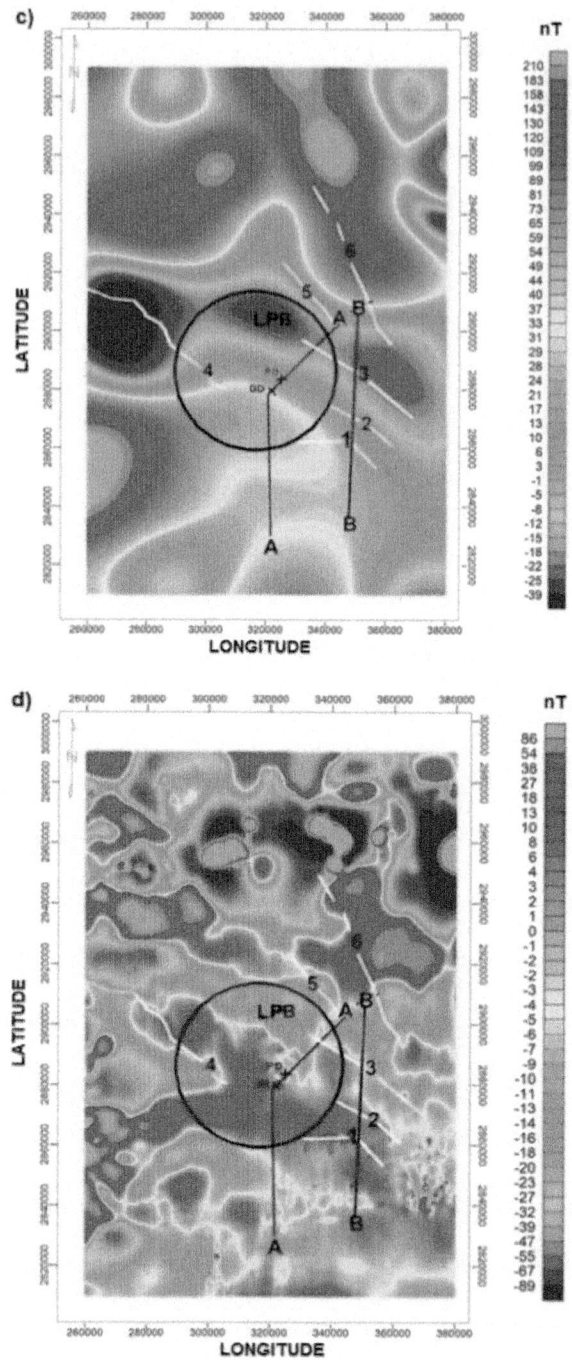

Figure 6. a) Map of the total magnetic anomaly which is located La Popa Ba-

sin (LPB) within which are the diapirs El Gordo (GD) and El Papalote (PD), profiles (A - A ') and (B - B ') as well as some anticlines (see Figure 2 for the names of the numbered structures). b) Total magnetic anomaly map reduced to pole, c) Map of the regional component obtained from total magnetic anomaly reduced to pole, d) Map of the residual component obtained from total magnetic anomaly reduced to pole.

The models presented here were developed based on the total magnetic anomaly, Bouguer anomaly and residual Bouguer anomaly, and were supported by the work of Aranda et al. (2008), who provided a compilation of published structural sections by Echanove (1962), Padilla and Sanchez (1982), Gray et al. (2001), Millán-Garrido (2004), Higuera et al. (2005) and Latta and Anastasio, (2007).

To highlight local anomalies, the regional component of the gravity or magnetic anomaly field is commonly subtracted from the data, generating a residual map. The definition of regional vs. local anomalies is subjective. Regional-local separation can be achieved by band pass wavelength filtering, but as previously mentioned, this procedure requires assuming the cut-off wavelengths, which can smear the separation due to non-vertical filter roll-off, and can contaminate the data by Gibbs ringing. It is more intuitive to compute from the gridded data the best-fit smooth surface, of an optimal low order, and then remove that smooth surface as the regional component. Good results in northern Mexico, including La Popa Basin, are obtained by subtracting from the data a third-order best-fit surface. Gravity data benefited from this procedure the most, whereas no significant improvement was obtained for the magnetic data.

The procedure and technology of data interpretation was generally described in publications of one of the co-authors (Yutsis et al. 2004; Yutsis et al. 2009). In this case the data were extracted from the complete Bouguer anomaly chart and magnetic anomaly map. The profiles were located across and along the structural elements of the La Popa Basin. The resulting models obtained are shown in Figs. 7-8.

The gravity line demonstrates a significant elevation of anomaly from -33 mGal in the SW up to -24 mGal in the NE part of the profile. It is possible to recognize a zone of relatively high gravity gradient in the south part of the model. This zone is also characterized by topographic low.

Magnetic anomaly showing amplitudes between -250 and -150 nT. The central part of the model (corresponding to the central part of the Basin area) is characterized by relatively high intensive anomalies. The highest magnetic anomaly is located in the NE part of the basin. Two high-gradient zones are located in the same areas as topography and gravity irregularities.

Integrated geological-geophysical interpretation was based on gravity and magnetic models constructed using Geosoft and WingLink software.

Selected Profiles for Modeling

The methodology to select two modeled profiles consisted on the selection of representative structures of the area of study and to that they could also verify if the structural sections compiled by Aranda et al. (2008) agree with the geophysical response observed in these structures.

Later, they were located on the MDE and gravity and magnetic maps that represent the effect of the structures that cross the profiles; with this it was possible to select the maps that isolate the effect of deep structures by gravity and to see the effect of Mesozoic sedimentary block and the magnetic response of the crystalline basement.

The profile A - A ' (Fig. 7) extends from the far North of the Curvature of Monterrey, the northeast part of the Parras Basin, the El Gordo and El Papalote diapirs in the La Popa Basin, ending in the marginal part of the Minas Viejas anticline. The profile B - B ' (Fig.8) crosses the anticlines of Potrero Garcia, Potrero Chico and Minas Viejas (Fig. 6a).

Model (A-A'), "Detached Diapirs"

The profile A - A ' (Fig. 7) is representative of the area as it allows us to observe 6 styles of folding in northeastern Mexico coexisting. Anticline Los Muertos corresponds to a fold-off of Jurassic evaporites, the anticline Venado of the Parras Basin suggests a hybrid fold, where you firstly had a fold-off, and then its northern flank elevation is modified by salt developing a drape fold. The Delgado syncline developed by evacuation of evaporites, the anticlines El Lobo and El Gordo are a fold hybrid, where structure possibly was formed by a wall halo-kinetic salt and subsequently amended during contraction to a fold of two detached diapirs, one on each side of it (El Gordo y El Papalote). The halo-kinetic structure La Soldadura (Giles and Lawton, 1999) has been considered as a wall of salt (La Popa Fold), and finally the model proposes a basement inversion which should help the La Soldadura development (Aranda et al., 2008).

The model shows that the Mesozoic sedimentary sequence lies on the salt. This sequence is altered by the plastic properties of salt and its incompressibility prompting the inversion of densities in the area causing this migration to the periphery of the La Popa Basin,

Model (B-B'), "Cut Off Anticlines"

The profile B - B ' (Fig. 8) is located east of the La Popa Basin, crossing perpendicularly the anticlines of Potrero Garcia, Potrero Chico and Minas Viejas, which represent folds in the Jurassic evaporites of the Minas Viejas Formation. Similar structures have been reported in the Sabinas Basin by Peterson-Rodriguez et al. (2008).

The Bouguer anomaly map was used for this model, because the anticlines of interest correspond to high frequency negative anomalies related to shallow bodies, specifically the accumulation of evaporites in the cores of the structures.

The total magnetic anomaly reduced to pole for the crystalline basement mapping was used. This information was supported by structural sections constructed by Aranda et al. (2008) based on the work of Gray et al (2001) and Latta and Anastasio (2007), as well as lithological map of the basement (Albarran et al., 2008).

The cores of anticlines covered by this profile are eroded, so we combined information of the evaporite outcrops, data of the well Minas Viejas 1, which drilled a section of evaporites of 4.500 m (Lawton et al., 2001) and interpretation of the low values of the residual Bouguer anomaly. Finally, we obtained the thickness of evaporite at the core of the anticlines more than 4 km (Potrero Chico, Potrero Garcia anticlines). The anticline Minas Viejas also shows in its core a thickness of more than 5 km and a length of 1 km. The morphology of the basement in the model corresponds to a graben-like depression, being composed of schist and granite (Fig. 8).

CONCLUSIONS AND DISCUSSION

Gravity data in the northeast Mexico basins are sensitive to local vertical offsets across high-angle faults, where rocks with different densities are juxtaposed. Yet, high densities in some Mesozoic sedimentary rocks just above the basement may smear out the subtle gravity traces of basement faults. Notably, in the Coahuila block in northwestern part of the area, where vertical basement-fault offsets reach tens and hundreds of meters, the associated gravity anomalies are not strong. The total-field magnetic data were used, with the International Geomagnetic Reference Field removed. The sedimentary cover in northeast Mexico is generally considered to be almost non-magnetic, and the anomalies are sourced overwhelmingly in the crystalline basement.

Local intra-sedimentary anomaly sources may be related to depositional concentrations of magnetic minerals in some clastic rocks, or to secondary magnetization of sedimentary rocks by circulating brines.

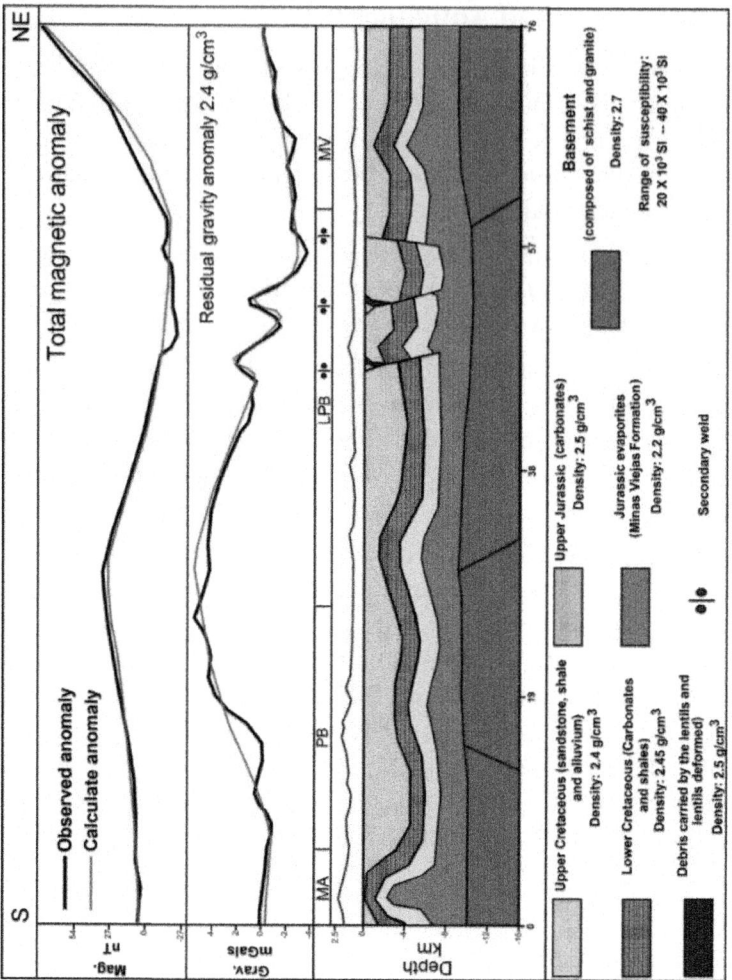

Figure 7. Geological-geophysical model 2.5D (detached diapirs), by inversion of the total magnetic anomaly and the residual Bouguer anomaly 2.4 g/cm³, Abbreviations: AM, anticline of Los Muertos, PB, Parras Basin, LPB, La Popa Basin, MV, anticline Minas Viejas.

Steep, straight faults are commonly expressed as subtle potential-field lineaments, which can be gradient zones, alignments of separate local anomalies of various types and shapes, aligned breaks or discontinuities in the anomaly pattern, and so on. Many large magnetic and gravity anomalies represent the ductile, ancient, healed basement structures, obscuring the desirable subtle features. Subtlety of the desirable lineaments need detailed data processing, using a wide range of anomaly-enhancement techniques and display parameters. So data processing includes Fourier transformation, wavelength filters, upward and down ward continuation, vertical and horizontal derivates, analytic signal analysis, etc.

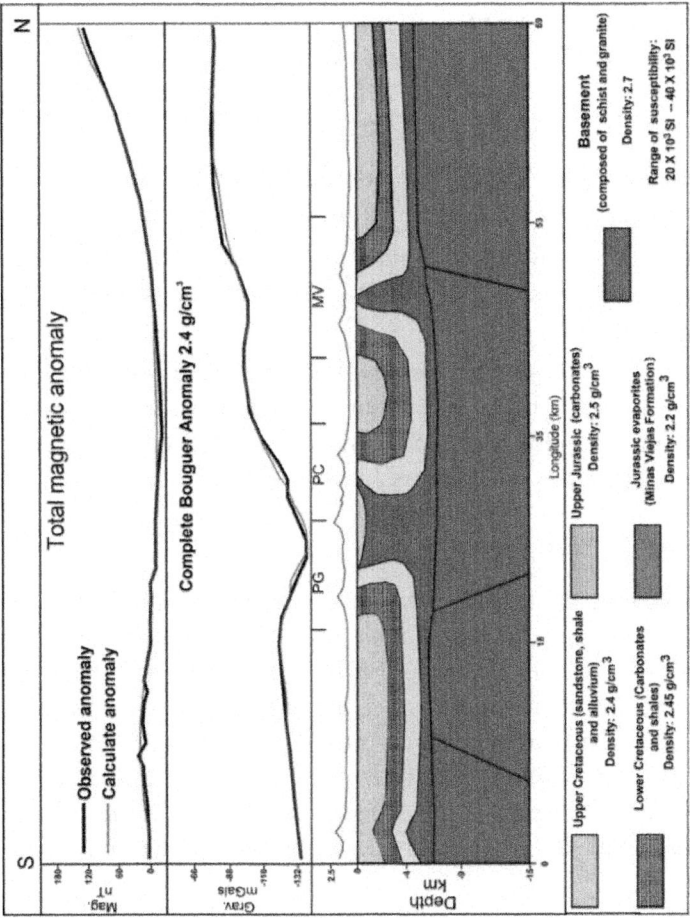

Figure 8. Geological-geophysical 2.5D model (detachment anticline), by inversion of the complete Bouguer anomaly 2.4g/cm³ and total magnetic anomaly. Abbreviations: PG, Potrero Garcia anticline; PC, Potrero Chico anticline; MV, Minas Viejas anticline.

The qualitative analysis of geophysical potential field data and its relation to the geological structures in the area of study shows that some of the residual minima of the Bouguer gravity are related to the topographic highs and generally positive tectonic features, specifically to anticlines such as Potrero de García, Minas Viejas and Potrero Chico. It is well known that the nucleus of anticlines here are characterized by accumulation of evaporates, so we propose that the observed gravity minima correspond to the most important accumulations of salt in the area. This phenomenon is shown in model (B - B

'). Also, it was discovered that these minima are located on the peripheries of La Popa Basin, so it is proposed that the La Popa Basin acted as the depocenter of salt sedimentation during Late Jurassic and finally "salt basement" was formed. Then more recent Mesozoic sedimentary rocks were deposited on the salt basement and caused a significant lithostatic pressure. Then, due to such physical properties of evaporites as plasticity and investment, salt began to migrate to the periphery of La Popa Basin.

Folds in the Minas Viejas Formation delineate the structures of some sub-domes and indicate the possibility that it is still rising.

Evidence of significant accumulation of salt is shown by the diapirs El Gordo, El Papalote (the model A - A ') that are detached diapirs with debris and lentil at the edges.

Magnetic analysis showed that the La Popa Basin is bounded on the north and south by magnetic highs and it may be modeled as graben-like depression. This model corresponds to the hypothesis of Aranda et al. (2008) and Hernandez (2008) who named this structure as Jurassic Pit.

ACKNOWLEDGEMENTS

The authors wish to thank PEMEX PEP for supplying the gravity data presented in this study. Special thanks to the Poza Rica, Veracruz geophysical team, especially to Mr. Gerardo Basurto Borbolla and Dr. Pedro Tomas Cabrera Gómez for their support since the beginning of this project. The many invaluable and significant geological contributions toward this work provided by Dr. Uwe Jenchen. We would like to express our gratitude to Prof. Vadim A.Galkine (Canada) for his constructive review of this manuscript. Finally, we would like to thank Dr. Evgenii Sharkov for his important comments.

REFERENCES

1.   J. J. Albarrán, Pineda. J. F. González, Cisneros. R. Muñoz, y. Rosales, J. Rodríguez, 2008 Actualización petrológica del basamento cristalino de la Cuenca de Sabinas. En Aranda-García, M., y Peterson-Rodríguez, R. H., (2008): Estructuras del Arco de Monterrey y Cuenca de la Popa, en la Sierra Madre Oriental y sus analogías para la exploración de Hidrocarburos en el Golfo de México. Monterrey, Nuevo León, México. Guía de campo y artículos relacionados. Congreso Mexicano del Petróleo 2008, Asociación Mexicana de Geólogos Petroleros.

2.   M. Aranda-García, Antuñano. S. Eguiluz De, R. H. Peterson-Rodríguez, G. Chávez-Cabello, 2008 Estratigrafía Jurásico-Cretácica y Estructuras del Arco de Monterrey de la Cuenca de La Popa en el Golfo de Sabinas.

En Aranda-García, M., y Peterson-Rodríguez, R. H., (2008): Estructuras del Arco de Monterrey y Cuenca de la Popa, en la Sierra Madre Oriental y sus analogías para la exploración de Hidrocarburos en el Golfo de México. Monterrey, Nuevo León, México. Guía de campo y artículos relacionados. Congreso Mexicano del Petróleo 2008, Asociación Mexicana de Geólogos Petroleros.

3.   C. Bartolini, K. Mickus, 2001 Tectonic Blocks, Magmatic Arcs, and Oceanic Terrains: A Preliminary Interpretation Based on Gravity, Outcrop, and Subsurface Data, Northeast-central Mexico, Tectonics, sedimentary basins, and petroleum systems: AAPG Memoir 75, 2943 .

4.   R. J. Blakely, 1995 Potential theory in gravity and magnetic applications: Cambridge, Cambridge University Press, 441 p.10.1017/ CBO9780511549816

5.   H. R. Burger, 1992 Exploration geophysics of the shallow subsurface. Prentice Hall, 323331 .

6.   G. Chávez-Cabello, J. J. Aranda-Gómez, R. S. Molina-Garza, T. Cossío-Torres, I. R. Arvizu-Gutiérrez, y. González-Naranjo, G. , 2005 La Falla San Marcos: Una estructura jurásica de basamento multi-reactivada del noreste de México. Boletín de la Sociedad Geológica Mexicana. Eds. Alaniz-Álvarez, S.A., y Nieto-Samaniego, A.F. Número Especial del Primer Centenario de la Sociedad Geológica Mexicana, Boletín de la Sociedad Geológica Mexicana, 2752 .

7.   S. Charleston, 1981 A summary of the structural geology and tectonics of the State of Coahuila, Mexico, in: Schmidt, C.I. & Katz, S.B., eds., Lower cretaceous stratigraphy and structure, northern Mexico. West Texas Geological Society Field Trip Guidebook, Publication, 81-74, 2836

8.   W. R. Dickinson, T. F. Lawton, 2001 Carboniferous to Cretaceous assembly and fragmentation of Mexico. Geological Society of American Bulletin, 113 n. 9, 11421160 .

9.   E. O. Echanove, 1965 Informe Fotogeológico del área de Monterrey, Nuevo, León, Hoja G-6, NEM-960, Petróleos Mexicanos, Informe Inédito.

10.   Antuñano. S. Eguiluz De, 2001 Geologic Evolution and Gas Resources of the Sabinas Basin in Northeastern México, in: Bartolini, C., Buffler, R. T., and Cantú-Chapa, A., eds., The western Gulf of México Basin: Tectonics, sedimentary basins, and petroleum systems. American Association of Petroleum Geologists Memoir 75, 241270 .

11.   K. A. Giles, T. F. Lawton, 1999 Attributes and evolution of an exhumed salt weld, La Popa basin, northeastern Mexico: Geology, 27 323326 .

12. K. A. Giles, T. F. Lawton, 2002 Halokinetic sequence stratigraphy adjacent to the El Papalote Diapir northeastern Mexico, American Association of Petroleum Geologist Bulletin, 86 (5), 823840 .

13. R. K. Goldhammer, C. A. Johnson, 2001 Middle Jurassic-Upper Cretaceous Paleogeographic Evolution and Sequence-stratigraphic Framework of the Northwest Gulf of Mexico Rim: American Association of Petroleum Geologists, Memoir, 75 4581 .

14. F. González, R. Puente, Eduardo. González, A. Camprubi, 2007 Estratigrafía del Noreste de México y su relación con los yacimientos estratoligados de fluorita, Barita, Celestina y Zn Pb. Boletín de la Sociedad Geológica Mexicana, LIX. 4362 .

15. G. Gray, Gray, I. Mahon, Pottorf. J. Keith, Pevear. R. Robert, Yurewiccz. A. David, Chuchla. J. Donald, Richard, 2001 Thermal and chronological record of syn- to post- Laramide burial and exhumation, Sierra Madre Oriental, Mexico, in C. Bartolini, R. T. Buffler, and A. Cantú-Chapa, eds., The western Gulf of Mexico Basin: Tectonics, sedimentary basins, and petroleum systems: AAPG Memoir 75, 159181 .

16. W. J. Hinze, 1990 Geotechnical and environmental Geophysics. Society of exploration Geophysicts.

17. M. R. Hudec, M. P. A. Jackson, 2007 Terra infirma: Understanding salt tectonics. Earth- Science Reviews 82. 128 .

18. U. Jenchen, 2007 La Popa Basin, NE Mexico, an analog for near salt deformation and hydrocarbon trapping.- Guide Book edited for Force Norway- Field Trip to the La Popa Basin (October 24- 28, 2007), 60 pp.

19. E. F. Mc Bride, A. E. Weidie, Jr , J. A. Wolleben, R. C. Laudon, 1974 Stratigraphy and structure of the Parras and La Popa basins, northeastern Mexico. Geological Society of America Bulletin, 85 16031622 .

20. J. W. Mc Kee, N. W. Jones, 1979 A large Mesozoic Fault in Coahuila, Mexico. Geological Society of America, Abstracts With Programs, 11 476

21. J. W. Mc Kee, N. W. Jones, L. E. Long, 1990 Stratigraphy and provenance of strata along the San Marcos fault, central Coahuila, Mexico, Geological Society of America Bulletin, 102 593614 .

22. E. F. Mc Bride, A. E. Weidie, Jr , J. A. Wolleben, R. C. Laudon, 1974 Stratigraphy and structure of the Parras and La Popa basins, northeastern Mexico. Geological Society of America Bulletin, 85 16031622 .

23. H. Millán-Garrido, 2004 Geometry and kinematics of compressional growth structures and diapirs in the La Popa basin of northeast Mexico:

Insights from sequential restoration of a regional cross section and three-dimensional analysis. Tectonics, 23

24. K. D. Latta, D. J. Anastasio, 2007 Multiple scales of mechanical stratification and decollement fold kinematics, Sierra Madre Oriental foreland, northeast Mexico, Journal of Structural Geology, 29 14211255 .

25. T. F. Lawton, F. J. Vega, K. A. Giles, C. Rosales-Dominguez, 2001 Stratigraphy and origin of the La Popa basin, Nuevo Leon and Coahuila, Mexico, in Bartolini, C., Buffler, R.T., and Cantú-Chapa, A., eds., Mesozoic and Cenozoic evolution of the western Gulf of Mexico basin: Tectonics, sedimentary basins and petroleum systems: American Association of Petroleum Geologists Memoir 75, 219240 .

26. y. Padilla, R. J. Sánchez, 1986 Post Paleozoic tectonics of northeast México and its role in the evolution of the Gulf of México. Geofísica Internacional, 25 157206 .

27. R. H. Peterson-Rodríguez, M. Aranda-García, A. J. Alvarado-Céspedes, 2008 Etapas y estilos de deformación que desarrollaron trampas estructurales en el sector centro-oriental de la Cuenca de Sabinas Coahuila, México. En Aranda-García, M., y Peterson-Rodríguez, R. H., (2008): Estructuras del Arco de Monterrey y Cuenca de la Popa, en la Sierra Madre Oriental y sus analogías para la exploración de Hidrocarburos en el Golfo de México. Monterrey, Nuevo León, México. Guía de campo y artículos relacionados. Congreso Mexicano del Petróleo 2008, Asociación Mexicana de Geólogos Petroleros.

28. J. L. Pindell, S. C. Cande, W. C. Pitman, D. B. Rowley, J. F. Dewey, J. Labrecque, W. Haxby, 1988 A plate-kinematic framework for models of Caribbean evolution. Tectonophysics, 155 121138 .

29. J. L. Pindell, S. F. Barrett, 1990 Geological evolution of the Caribbean region; A plate tectonic perspective, The Geology of North America, v. H, The Caribbean Region. The Geological Society of America, 405432 .

30. R. Rasmussen, L. B. Pedersen, 1979 End corrections in potential field modelling. Geophysical Prospecting 27 749760 .

31. J. M. Reynolds, 1997 An Introduction to applied and environmental geophysics. Editorial Wiley.

32. R. L. Reynolds, J. G. Rosenbaum, M. R. Hudson, N. S. Fishman, 1990 Rock magnetism, the distribution of magnetic minerals in the earth's crust, and magnetic anomalies in geological applications of modern aeromagnetic surveys. United States Geological Survey Bulletin, 1924 2446 .

33. A. Salvador, 1987 Late Triassic-Jurassic paleogeography and origin of

Gulf of Mexico basin. American Association of Petroleum Geologists Bulletin, 71 419451 . Schmidt, C. I., y Katz, S. B., Eds., Lower cretaceous stratigraphy and structure, northern Mexico. West Texas Geological Society Field Trip Guidebook, Publication, 81-74, p. 28-36.

34. M. Talwani, J. R. Heirtzler, 1964 Computations of magnetic anomalies caused by two dimensional bodies of arbitary shape, in Parks, G. A., Ed., Computers in the mineral industry, Part I, Stanford Univ. Publ., Geological Sciences, 9 464480 .

35. M. Talwani, J. L. Worzel, M. Landisman, 1959 Rapid Gravity Computations for Two-Dimensional Bodies with Application to the Mendocino Submarine Fracture Zone, Journal of Geophysical Research, 64 4961 .

36. Ponce. A. Tamez, V. Yutsis, Flores. E. R. Hernández, A. A. Bulychev, K. Krivosheya, 2010 Rasgos tectónicos de la Cuenca de la Popa y de las estructuras que la rodean en el NE de México derivados de campos geofísicos potenciales. Boletín de la Sociedad Geológica Mexicana, in press

37. M. Tardy, 1980 Contribution a l'étude geologique de la Sierra Madre Oriental du Mexique: Tesis doctoral, Université Pierre et Marie Curie de Paris, 445 p. Reynolds, J.M. 1997. An Introduction to applied and environmental geophysics. Editorial Wiley.

38. W. M. Telford, L. P. Geldart, R. E. Sheriff, D. A. Keys, 1990 Applied Geophysics 2n edition. Cambridge University Press.

39. F. J. Vega-Vera, M. C. Perrilliat, 1989 La presencia del Eoceno marino en la cuenca de la Popa (Grupo Difunta), Nuevo León: orogenia post-Ypresiana: Universidad Nacional Autónoma de México, Instituto de Geología, Revista, 8 6770 .

40. F. J. Vega-Vera, L. M. Mitre-Salazar, y. Martínez-Hernández, E. , 1989 Contribución al conocimiento de la estratigrafía del Grupo Difunta (Cretácico superior-Terciario) en el noreste de México: Universidad Nacional Autónoma de México, Instituto de Geología, Revista, 8 179187

41. J. Warren, 1999 Evaporites: Their Evolution and Economics. Blackwell Science, Oxford. 438 pp.

42. J. J. Willis, B. E. Lock, D. A. Ruberg, K. C. Cornell, 2001 Field Examination of Exposed Evaporite-Related Structures, United States and Mexico: Relations to Subsurface Gulf of Mexico Examples. Gulf Coast Association of Geological Societies Transactions, LI.

43. J. L. Wilson, 1990 Basement structural controls on Mesozoic carbonate facies in Northeastern México.- a review in: Contribuciones al Cretácico

de México y América Central. Actas de la Facultad de Ciencias de la Tierra / Universidad Autónoma de Nuevo León, Linares, Nuevo León, México. 4 545 .

44. C. D. Winkler, R. T. Buffler, 1988 Paleogeographic evolution of early deep water Gulf of Mexico and margins, Jurassic to Middle Cretaceous (Comanchean): American Association of Petroleum Geologists Bulletin, 72 318346 .

45. I. J. Won, M. G. Bevis, 1987 Computing the gravitational and magnetic anomalies due to a polygon: Algorithms and Fortran subroutines. Geophysics, 52 232238 .

46. H. Ye, 1997 The arcuate Sierra Madre Oriental orogenic belt, NE Mexico: Tectonic infilling of a recess along the Southwestern North America continental margin: in: Structure, stratigraphy and paleontology of Late Cretaceous-Early Tertiary Parras-La Popa foreland basin near Monterrey, northeast Mexico. American Association of Petroleum Geologists Field Trip # 10, 85115 .

47. V. V. Yutsis, Oesterreich. D. Masuch, A. Martinez, 2004 Application of GIS and GPS technologies in gravimetrical and hydrogeological studies of the Santa Catarina Basin, NE Mexico. Proceedings of 8th World Multiconference on Systemic, Cybernetic and Informaticas.Orlando, Florida, 373378 .

48. V. V. Yutsis, U. Jenchen, H. de León-Gómez, Valdez. F. Izaguirre, K. Krivosheya, 2009 Paleogeographic development of the surroundings of Cerro Prieto water reservoir, Pablillo basin, NE Mexico, and geophysical modeling of the reservoirs subsurface. Neues Jahrbuch für Geologie und Paläontologie, 253 1 July 2009, 4159

Chapter 4

OPTIMIZING HYDRAULIC FRACTURING TREATMENT INTEGRATING GEOMECHANICAL ANALYSIS AND RESERVOIR SIMULATION FOR A FRACTURED TIGHT GAS RESERVOIR, TARIM BASIN, CHINA

Feng Gui[1], Khalil Rahman[1], Daniel Moos[2], George Vassilellis[3], Chao Li[3], Qing Liu[4], Fuxiang Zhang[5], Jianxin Peng[5], Xuefang Yuan[5], and Guoqing Zou[5]

[1]Baker Hughes, Perth, Australia
[2]Baker Hughes, Menlo Park, USA
[3]Gaffney, Cline &Associates, Houston, USA
[4]Baker Hughes, Beijing, China
[5]PetroChina Tarim Oil Company, Korla, China

A comprehensive geomechanical study was carried out to optimize stimulation for a fractured tight gas reservoir in the northwest Tarim Basin. Conventional gel fracturing and acidizing operations carried out in the field previously failed to yield the expected productivity. The objective of this study was to assess the effectiveness of slickwater or low-viscosity stimulation of natural fractures by shear slippage, creating a conductive, complex fracture network. This type of stimulation is proven to successfully exploit shale gas resources in many fields in the United States.

A field-scale geomechanical model was built using core, well log, drilling data and experiences characterizing the in-situ stress, pore pressure and rock mechanical properties in both overburden and reservoir sections. Borehole image data collected in three offset wells were used to characterize the in-situ natural fracture system in the reservoir. The pressure required to stimulate the natural fracture systems by shear slippage in the current stress field was predicted. The injection of low-viscosity slickwater was simulated and the resulting shape of the stimulated reservoir volume was predicted using a dual-porosity, dual-permeability finite-difference flow simulator with anisotropic, pressure-sensitive reservoir properties. A hydraulic fracturing design and evaluation simulator was used to model the geometry and conductivity of the principal hydraulic fracture filled with proppant. Fracture growth in the presence of the lithology-based stress contrast and rock properties was computed, taking into account leakage of the injected fluid into the stimulated reservoir volume predicted previously by reservoir simulation. It was found that four-stage fracturing was necessary to cover the entire reservoir thickness. Post-stimulation gas production was then predicted using the geometry and conductivity of the four propped fractures and the enhanced permeability in the simulated volume due to shear slippage of natural fractures, using a dual-porosity, dual-permeability reservoir simulator.

For the purpose of comparison, a conventional gel fracturing treatment was also designed for the same well. It was found that two-stage gel fracturing was sufficient to cover the whole reservoir thickness. The gas production profile including these two propped fractures was also estimated using the reservoir simulator.

The modeling comparison shows that the average gas flow rate after slickwater or low-viscosity treatment could be as much as three times greater than the rate after gel fracturing. It was therefore decided to conduct the slickwater treatment in the well. Due to some operational complexities, the full stage 1 slickwater treatment could not be executed in the bottom zone and treatments in the other three zones have not been completed. However, the post-treatment production test results are very promising. The lessons learned in the planning, design, execution and production stages are expected to be a valuable guide for future treatments in the same field and elsewhere.

INTRODUCTION

Following the success in exploiting shale gas resources by multi-stage hydraulic fracturing with slickwaters or low-viscosity fluid (i.e., linear gel) in horizontal wells in North America, there has been a lot of interest in applying this technique to other regions and other types of tight reservoirs. This is due in part to the fact that conventional gel fracturing treatments have been less successful in some naturally fractured reservoirs due to excessive unexpected fluid loss and proppant bridging in natural fractures, leading often to premature screen-outs. Additionally, the high-viscosity gel left inside the natural fractures causes the loss of virgin permeability of the reservoir in the case of inefficient gel breaking. However, the challenge for doing this is that the physical mechanism responsible for this kind of stimulation is yet to be fully understood and a standard work flow for design and evaluation is yet to be developed. Furthermore, industry so far mainly relies on performance analogs to improve understanding of each shale play, and thus it usually takes years to advance up the learning curve for determining which factors best affect well production [1].

Currently, the general opinion on the mechanism leading to the success of waterfrac in shale gas reservoirs is that a complex fracture network is created by stimulation of pre-existing natural fractures. Although it is difficult to observe the processes acting during stimulation, microseismic imaging has enabled us to understand that both simple, planar fractures and complex fracture networks can be created in hydraulic fracture stimulations under different settings [2]. Fracture complexity is thought to be enhanced when pre-existing fractures are oriented at an angle to the maximum stress direction, or when both horizontal stresses and horizontal stress anisotropy are low, because these combinations of stress and natural fractures allow fractures in multiple orientations to be stimulated [3]. The result of stimulation therefore depends both on the geometry of the pre-existing fracture systems and on the in-situ stress state. It is now generally accepted that stimulation in shale gas reservoirs occurs through a combination of shear slip and opening of pre-existing (closed) fractures and the creation of new hydraulic (tensile) fractures [4-6]. In wells that are drilled along the minimum horizontal stress (S_{hmin}) direction, stimulation generally creates a primary radial hydraulic fracture that is perpendicular to S_{hmin}. Then, pressure changes caused by fluid diffusion into the surrounding rock and the modified near-fracture stress field induced by fracture opening cause shear slip on pre-existing natural fractures. If the horizontal stress difference is small enough, new hydraulic fractures perpendicular to the main fracture can open. Each slip or oblique opening event radiates seismic energy, which, if the event is large enough, can be detected using downhole or surface geophones.

Founded on the idea that productivity enhancement due to stimulation results not just from creation of new hydraulic fractures but also from the effect of the stimulation on pre-existing fractures (joints and small faults), a new workflow dubbed "shale engineering", was established by combining surface and downhole seismic, petro-physical, microseismic, stimulation, and production data [7, 8]. In this new workflow (Figure 1), the change in flow properties of natural fractures is predicted using a comprehensive geomechanical model based on the concept of critically stressed fractures [9-11]. Existing reservoir simulation tools can then be used to model the hysteresis of fracture flow properties that result from the microseismically detectable shear slip, which is critical to the permanent enhancement in flow properties and increased access to the reservoir that results from stimulation. The primary hydraulic fracture created and propped during the stimulation can be modeled using conventional commercial hydraulic fracture models by taking into account fluid leaked into natural fractures in the surrounding region. The propped conductivity is estimated using laboratory-based proppant conductivity data adjusted for the proppant concentration in the fracture. The propped main fracture model and the reservoir model with stimulated natural fracture properties can then be integrated into production simulators to predict production after the slickwater hydraulic fracturing treatment. When available, microseismic data can be used to help define the network of stimulated natural fractures that comprises the stimulated reservoir volume (SRV).

Although this new workflow was developed based on experiences in shale gas reservoirs, we believe it can also be applied to any unconventional reservoir requiring stimulation that has pre-existing natural fractures. Both Coal Bed Methane (CBM) and fractured tight gas reservoirs are examples of where this approach could be applied. In this paper, we will illustrate the workflow using the results of a study conducted in a fractured tight gas reservoir in the Kuqa Depression, Tarim Basin.

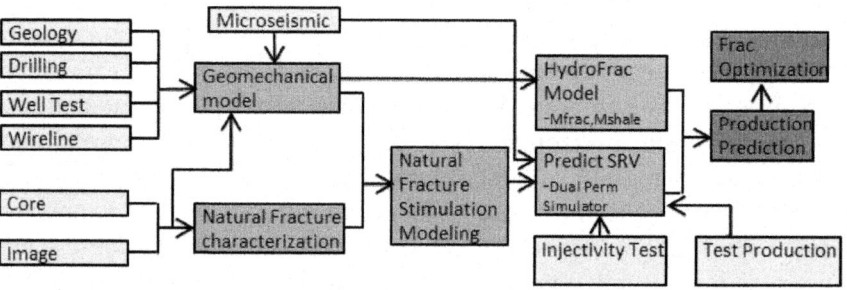

Figure 1. Workflow for predicting the complex fracture network developed by stimulating fractured reservoirs using low-viscosity fluid.

PROJECT BACKGROUND

The project discussed in this paper was initiated to investigate various methods and practices to improve the economics of the field. Conventional gel fracturing had been tested in a few wells with disappointing results. One fault block (see Figure 2) was chosen as the target of a pilot study that included building a geomechanical model, optimizing hydraulic fracturing design, assessing the stability of faults near the target well, and (although it is not discussed here) analyzing wellbore stability for drilling horizontal wells. Three vertical wells were drilled. D2 and D3 are near the crest of the structure and D1 is ~ 2.5-3 km to the west. The main target is Cretaceous tight sandstone occurring at ~5300m to ~6000m depth. Reservoir rock is composed of fine sandstone and siltstones interlayered with thin shales. Average reservoir porosity is ~7% and average permeability is ~0.07 mD. The gross reservoir thickness is ~180-220m in this fault block. Wells D1 and D2 were completed by acidizing and gel fracturing; test production was ~15-27 ×10^{4}m^3/d. The objective of this project was to optimize hydraulic fracturing design for Well D3 based on the geomechanical analysis and investigate whether it is better to conduct slickwater treatment in the D3 well to stimulate and create a complex fracture network or utilize conventional two-wing gel fracturing.

Comprehensive datasets were available for all three wells including drilling experiences, wireline logs, image data, mini-fracs and well tests. Laboratory tests were also conducted on cores from well D2 to estimate the rock mechanical properties of reservoir rocks.

GEOMECHANICAL MODEL

A geomechanical model includes a description of in-situ stresses and of rock mechanical and structural properties. The key components include three principal stresses (vertical stress (S_v), maximum horizontal stress (S_{Hmax}) and minimum horizontal stress (S_{hmin})), pore pressure (Pp) and rock mechanical properties, such as elastic properties, uniaxial compressive strength (UCS) and internal friction. The relative magnitude of the three principal stresses and the consequent orientation of the most likely slipping fault or fracture define the stress regime to be normal faulting ($S_v > S_{Hmax} > S_{hmin}$), strike-slip faulting ($S_{Hmax} > S_v > S_{hmin}$) or reverse faulting ($S_{Hmax} > S_{hmin} > S_v$). The horizontal stresses are highest relative to the vertical stress in a reverse faulting regime and lowest relative to the vertical stress in a normal faulting regime. Hydraulic fractures are vertical and propagate in the direction of the greatest horizontal stress in a strike-slip or normal faulting regime. In a reverse faulting stress regime in which S_v is the minimum stress, hydrofractures are horizontal. These different stress regimes also have consequences for the pressure that is required to open a network of orthogonal hydrofractures by stimulation. In places where the horizontal stresses are low and nearly equal, a relatively small excess pressure above the least stress may be required to open orthogonal fractures. Where the horizontal stress difference is larger, a larger excess pressure is required to open orthogonal fractures. Where the least stress is only slightly less than the vertical stress, weak horizontal bed boundaries and mechanical properties contrasts between layers may allow opening during stimulation of horizontal bedding ("T-fractures").

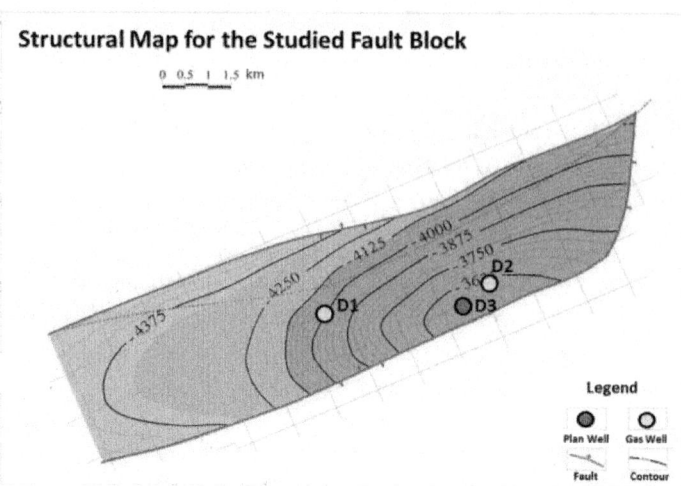

Figure 2. Structural map showing the offset well locations.

Except for the magnitude of S_{Hmax}, other components of the geomechanical model can be determined using borehole data by reviewing a few representative wells in the field. Vertical stress is calculated by integrating formation density, which is obtained from wireline logs. The magnitude of S_v across this fault block is in a similar range. Pore pressure was constrained, mainly by referencing direct measurement data and drilling experiences. This is due to the complex tectonic history. Conventional under-compaction approaches for pore pressure estimation may not apply in the study area. Evidence for this is the over-compacted density profile. In addition, due to the complex lithology changes the log response with depth may reflect lithology changes rather than pressure variation. Well test data from D1 and D2 showed that the reservoir pressure is ~88-90 MPa, an equivalent pressure gradient of ~1.6-1.7 SG, which is abnormally over-pressured.

Rock mechanical laboratory tests were conducted on cores from the sandstone reservoirs and the interlayered shales in the D2 well, and the results were used to constrain a log-calibrated range of UCS and other rock mechanical parameters. Figure 3 shows the match between log-derived rock strength profiles and laboratory test results in D2. Dynamic Young's modulus was calculated from compressional and shear velocities and density and calibrated to static values using laboratory test results. The relationship between dynamic and static Poisson's Ratio was not obvious; the dynamic Poisson's Ratio computed from Vp/Vs matched reasonably well with the laboratory results, so it was used directly in the modeling. Young's Modulus-based empirical relationships were used to estimate the UCS for both sandstone and inter-layered shales.

Minimum horizontal stress (S_{hmin}) at depth can be directly estimated from extended leak-off tests (XLOT), leak-off tests (LOT) or mini-frac tests. No extended leak-off tests were conducted in the field. LOTs and leak-off points from two reliable LOTs were used to constrain the upper limit of S_{hmin} (~2.09 SG EMW at ~4000 m TVD). One mini-frac test was conducted in the sandstone reservoir in D2, with the interpreted fracture closure pressure (closest estimation to S_{hmin}) ~2.064 ppg EMW at ~5400 m TVD. Because LOTs are usually conducted in shaly formations while mini-frac tests are usually carried out in sandstone reservoirs, the LOTs and mini-frac tests are used to construct separate S_{hmin} profiles in shales and sandstones, respectively using the effective stress ratio method (S_{hmin}-Pp/S_v-Pp). The effective stress ratio from LOT is ~0.725 and from mini-frac test is ~0.48, which indicates there is a dramatic stress difference between sandstones and shales (stress contrast). The contrast between different lithology significantly influences hydraulic fracturing design. The relative lower stress in sandstones indicates that a hydraulic fracture should be easily created in the tight sandstone, however, the interlayered shales which have higher stress act as frac barriers and pinch points, thereby complicating fracture propagation and the final fracture geometry and conductivity.

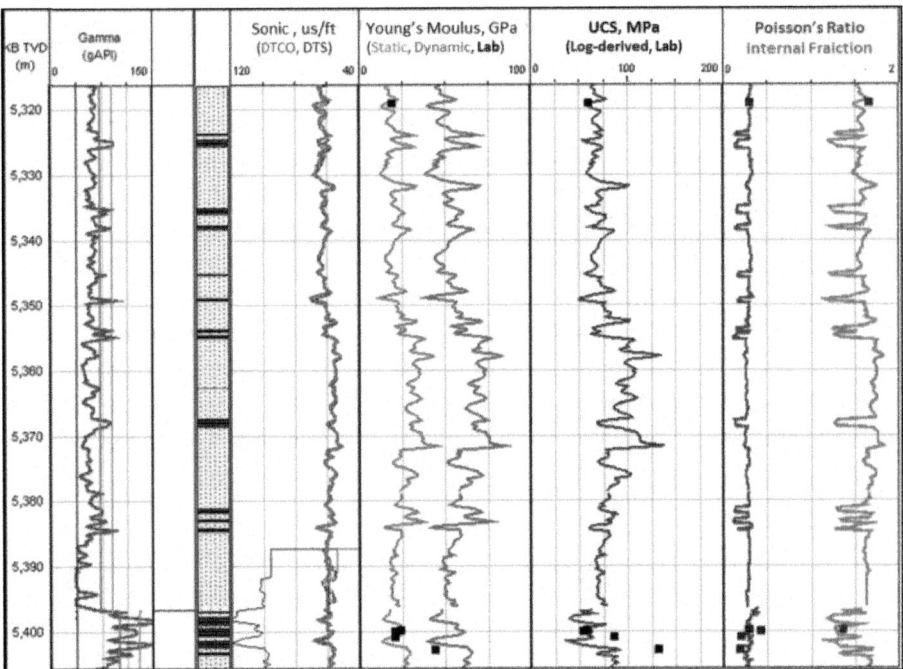

Figure 3. Comparison of laboratory (black squares) and log derived rock mechanical properties in D2 well.

The azimuth and magnitude of maximum horizontal stress (S_{Hmax}) can be constrained through the analysis of wellbore failures such as breakouts and tensile cracks observed on wellbore images or multi-arm caliper data. Wellbore failure analysis allows constraining of the orientation and magnitude of the S_{Hmax} because stress-induced wellbore failures occur due to the stress concentration acting around the wellbore once is drilled. The presence, orientation, and severity of failure are a function of the in-situ stress fields, wellbore orientation, wellbore and formation pressures and rock strength [12]. High-resolution electrical wireline image logs were available in all three study wells. Both breakouts and drilling-induced tensile fractures (DITFs) were observed in the reservoir sections in D2 and D3 wells. Only DITFs were observed in well D1, which could be due to the higher mud weights used during drilling and the poor quality of the image data in lower part of the reservoir.

Figure 4 shows examples of the breakouts seen in the D3 well. The example shows the typical appearance of breakouts observed on images. Here, the average apparent breakout width is ~30-40 degree. The breakouts mostly occur in shales and more breakouts are observed in the lower part of reservoir where the formations become more shaly. The orientation of breakouts is quite consistent with depth and across the block. However, small fluctuations of breakout orientation can be observed locally while intercepting small faults (an example can be seen in the right plot in Figure 4). This may indicate that some of these faults are close to or at the stage of being critically stressed. This has important implications for the stress state in the area and the likelihood of stimulating fractures by injection. Breakouts usually develop at the orientation of S_{hmin} and DITFs in the direction of S_{Hmax} in vertical and near-vertical wells. In the left plot of Figure 4, DITFs can also be observed in the same interval as the breakouts with an orientation that is ~90 degrees from the breakout directions, consistent with this expectation. DITFs are seen more often in sandstone than in shale. Based on wellbore breakouts and DITFs interpreted from the image data in D3, the azimuth of S_{Hmax} is inferred to be ~143° ±10 °. This is similar to the azimuth of S_{Hmax} inferred from wellbore failures observed in the other two wells. It is also consistent with the regional stress orientation from the World Stress Map [13].

The magnitude of S_{Hmax} is constrained by forward-modeling the stress conditions that are consistent with observations of wellbore failures observed on image logs, given the data on rock strength, pore pressure, minimum horizontal stress, vertical stress, and mud weight used to drill the well. Figure 5 is a crossplot of the magnitude of S_{hmin} and the magnitude of S_{Hmax}, which summarizes the results of S_{Hmax} modeling in D3. The magnitude of S_v (~2.49 SG) is indicated by the open circle. The modeling was conducted in both sandstone and shale. The rectangles in different colors are the possible S_{hmin} and S_{Hmax} ranges at every modeling depth.

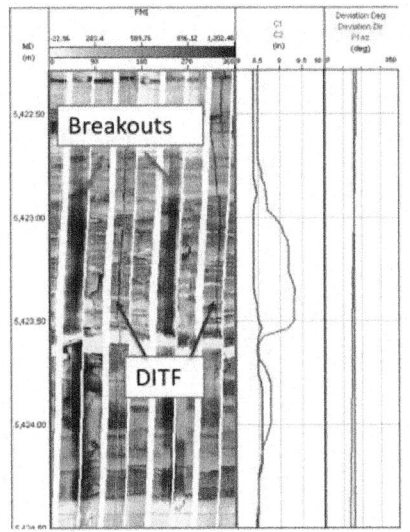 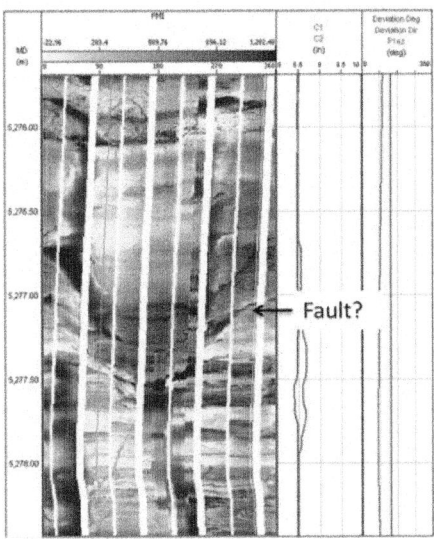

Figure 4. Drilling induced wellbore failures (breakouts & tensile fractures) observed on electrical image in D3 well.

Modeling shows slightly different results for the S_{Hmax} and S_{hmin} magnitudes in the different lithologies. However, both results are consistent with the magnitudes of S_{hmin} inferred from LOTs and mini-fracs. Figure 5 shows that the magnitude of maximum horizontal stress is higher than the vertical stress in both cases, and higher in the shale than in the sand. Thus, the study area is in a strike-slip faulting stress regime ($S_{hmin} < S_v < S_{Hmax}$). The difference between the magnitudes of S_{Hmax} and S_{hmin} is ~0.8 SG in the reservoir section, suggesting high horizontal stress anisotropy. In such a condition, it is unlikely to open the natural fractures by tensile mode. However, the natural fractures might dilate in shear mode depending on their orientations and stress conditions. The final geomechanical model was verified by matching the predicted wellbore failure in these wells with that observed from image data and drilling experiences.

NATURAL FRACTURES CHARACTERIZATION AND STIMULATION MODELING

Natural fractures have been observed on cores and image logs in the study area. The fluid losses during drilling not only suggest the existence of natural fractures but also that some at least of these fractures are permeable in-situ. Based on the core photos shown in Figure 6, open high-angle tectonic fractures can be seen on cores from D2 and D3 wells near the crest of the structure. A fracture network consisting of a group of fractures with different orientations can be seen on the cores from the D1 well, and these fractures appear to have less apertures than high-angle fractures observed in D2 and D3.

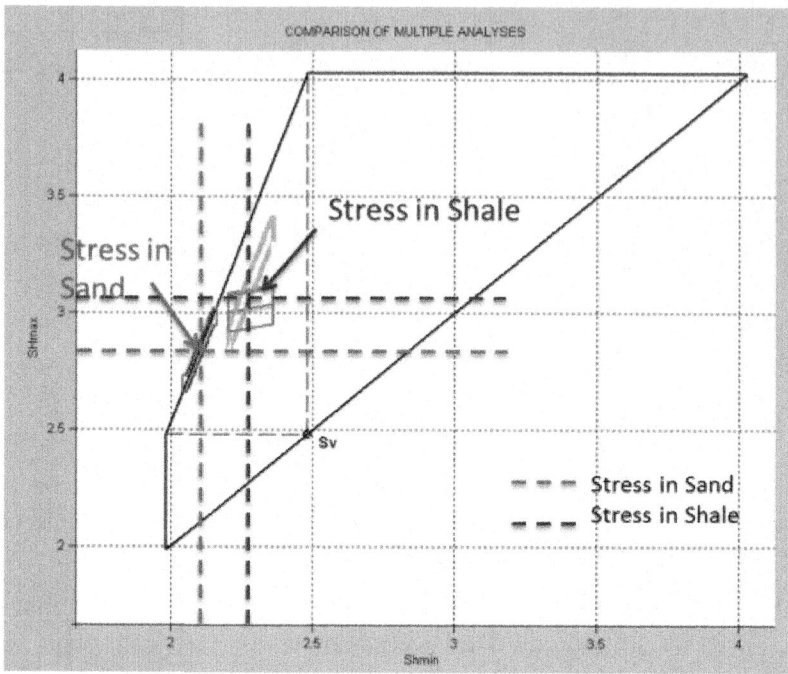

Figure 5. Stress modeling results in well D3. The range of horizontal stress magnitudes are consistent with the occurrence of wellbore failures (breakouts and DITFs) observed on wellbore images.

Natural fractures were interpreted and classified using high-resolution electrical images in all three wells. Based on the appearance on image data, the natural fractures are classified as below:

- Conductive: dark highly dipping planes on image logs

- Resistive: white dipping planes on image log

- Critically Stressed: related to local failure rotation

- Fault: features discontinued across the dipping planes

- Drilling Enhanced: discontinuous and fracture traces are 180 degrees apart and in the direction in tension

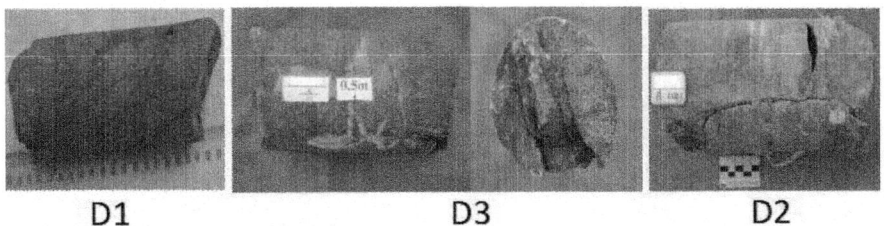

D1   **D3**   **D2**

Figure 6. Core photos showing the natural fractures observed in three offset wells.

Figure 7 shows a few examples of natural fractures observed on the electrical images. The plot on the left shows some examples of high-angle and low-angle conductive fractures that appear to be continuous dark lines on the images. Flexible sinusoids can be fit to the fracture traces and fracture orientation can be determined. The plot on the right shows an example of drilling enhanced natural fractures for which the fracture trace is discontinuous. The fact that parts of these fractures can be detected on the electrical image is due to fluid penetration into the fracture at the orientation around where the rock at the borehole wall is in tension during drilling. The classification of the natural fractures indicates the relative strength of the fractures. For example, the resistive fractures are closed and mineralized. Active faults or critically stressed natural fractures might be open and conductive, even under the original conditions. During stimulation, these fractures are the most easily stimulated. However, it is important to note that the classification of natural fractures is purely based on their appearance on the electrical images, and cannot be used directly to quantify permeability or other flow properties.

Figure 8 shows the fractures orientations on a crossplot of the strike and dip angles of all fractures observed in the three wells. The natural fractures observed can be divided into three groups. The first group is low-angle fractures (dip<20 °), which could be related to beddings. The second group is the major fractures seen in this block that have intermediate dip angles (~25-55°) and strike at an azimuth of ~155°N. The third group consists of fractures with strikes of ~355°N and ~100°N and dip angles ~ 35°-65° and 25°-35°, respectively. Because of their wide range of orientations and cross-cutting relationship, these three groups of fractures could be stimulated to form a well-connected grid with a major fracture azimuth (~155°N) aligned with the direction of maximum horizontal stress (~143°N). This direction is nearly perpendicular to the faults, defining the shape of fault block (see Figure 2). Because the structural trends and the stresses are aligned, it enabled us to create a reservoir model with a grid that is consistent with both.

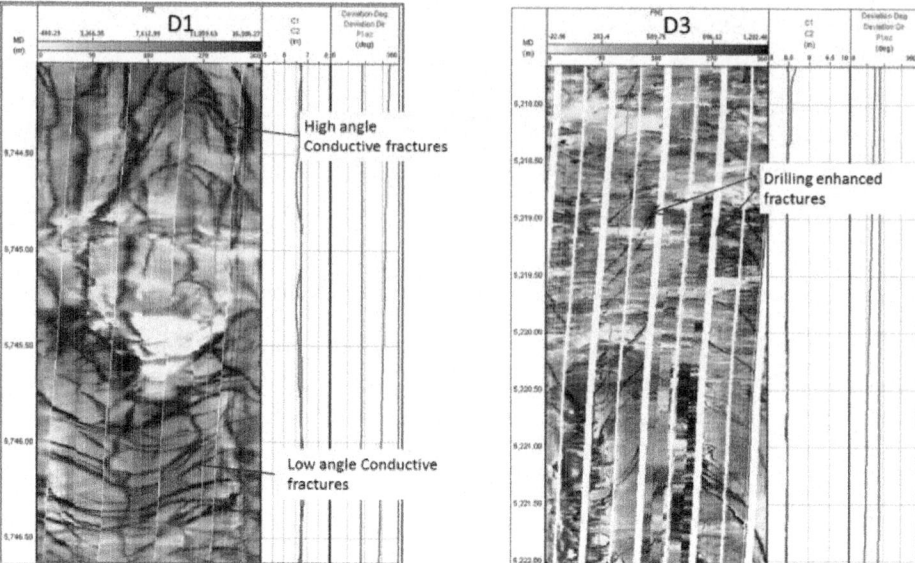

Figure 7. Examples of natural fractures observed on electrical image in D1 and D3 wells.

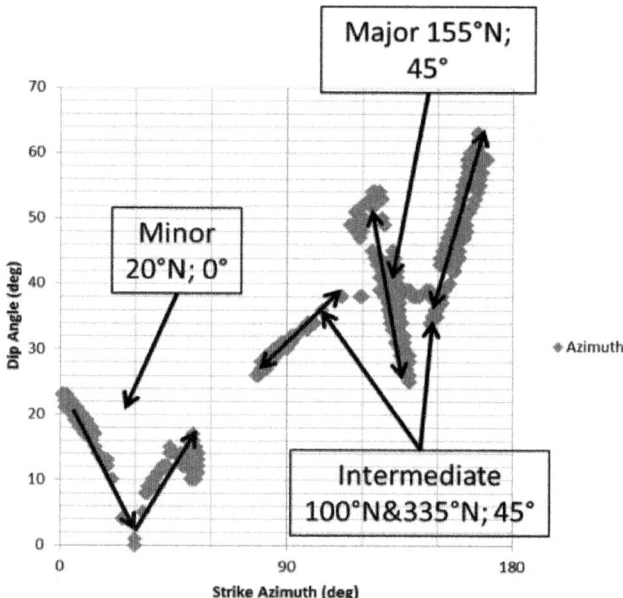

Figure 8. Cross-plot between the strike and dip angles of all the fractures observed in three offset wells.

Effective stresses in the earth are always compressive, and natural processes tend to "heal" fractures through vein filling and other processes. Therefore, the intrinsic fracture aperture of most fractures is likely to be very small or even zero (cases where dissolution creates voids that prevent full closure are a notable exception). Thus, it is increasingly recognized that active processes are necessary to maintain fracture permeability. One such process is periodic slip along fractures that are critically stressed (i.e., those that are at or near the limiting ratio of shear to normal stress to slip). This process, and the influence of effective normal stresses on fracture aperture, can be modeled using a simple equation that describes the variation in aperture as a function of normal stress for a pure Mode I fracture. The same equation with different parameters can also be used to model the same fracture after slip has occurred [9-11].

$$a = \frac{A \bullet a_0}{(1 + 9\sigma'_n / B)}$$
(1)

Equation 1 is one example that describes aperture in terms of an initial aperture ($A \cdot a_0$) and an effective normal stress at which the aperture is only 10% as large (B). A and B both increase due to slip, resulting in a larger "unstressed" aperture and a stiffer fracture caused by "self-propping" due to generation during slip of a mismatch in the fracture faces and/or creation of minor amounts of rubble at the fracture face.

The contribution of fractures to the relative productivity of a well of any orientation can be computed by summing the contributions of all fractures, weighted by the product of their relative transmissivity (which is a function of aperture) and the likelihood of the well intersecting the fracture (which is a function of the difference between the fracture and the well orientation). This relative productivity can be written as [10]

$$P_{well} = \sum_{fracs} \left\{ \max \left(|\hat{w} \bullet \hat{n}_i|, a \right) \times P_i \right\}$$
(2)

where $\hat{w}$ and $\hat{n}_i$ are unit vectors along the axis of the well and normal to the ith fracture, a is a number representing the likelihood of a well intersecting a fracture if it lies in the plane of the fracture, and P_i is the relative permeability of the fracture.

The fractures interpreted from image data are only those that intersect the logged wells that are a function of their orientations, and there is no information about the fracture distribution between the wells. To ensure the most meaningful representation of the fractures in the reservoir, the fractures interpreted from all three wells were combined and the distribution was corrected to account for the likelihood of each fracture intersecting the well at the point where it was observed. This combined fracture data set was then used to model the productivities of wells in their natural condition and the change in productivity due to the shear-slip of natural fractures.

Figure 9 shows relative productivity for wells of all orientations based on the fractures observed in all three wells. Natural fractures are shown as poles to the fracture surfaces (black dots). Different apertures and strengths were assumed for the different types of fractures based on their classifications described above (Table 1). The plot on the left shows the relative productivity under pre-stimulation conditions, while the plot on the right shows the relative productivity calculated using equation 2 after the fractures were stimulated with a pressure 20 MPa above the original reservoir pressure. It can be seen that the maximum productivity increases by a factor of 5 if all fractures see the same 20-MPa pressure increase, which is obviously not the case during real stimulation. Superimposed on Figure 9are the computed optimal orientations of wells based on the fracture and stress analysis (green circles). If none of the fractures is critically stressed, then the best orientation to drill a well is perpendicular to the largest population of natural fractures. If some fractures have enhanced permeability because they are critically stressed, the optimal orientation shifts in the direction of the greatest concentration of critically stressed fractures. Figure 9 shows there are some fractures already near or being critically stressed, even under ambient condition (left plot), and the maximum productivity is achieved by drilling highly deviated wells with ~20 °N hole azimuth. The optimum wellbore orientation after ~ 20-MPa stimulation is nearly horizontal and in the direction of ~228 °N.

Table 1. Model parameters to calculate relative productivities for different types of natural fractures

Fracture classification	Fracture cohesion (MPa)	Sliding Friction	a0a0	A		B (MPa)	
				Un-stimulated	Stimulated	Un-stimulated	Stimulated
Conductive	5	0.6	10	0.18	0.18	10	100
Resistive	5	0.6	10	0.1	0.18	1	100
Faults	0	0.6	30	0.18	0.18	100	100
Drilling enhanced	0	0.6	10	0.1	0.18	10	100
Critically Stressed	1	0.2	10	0.1	0.2	10	100

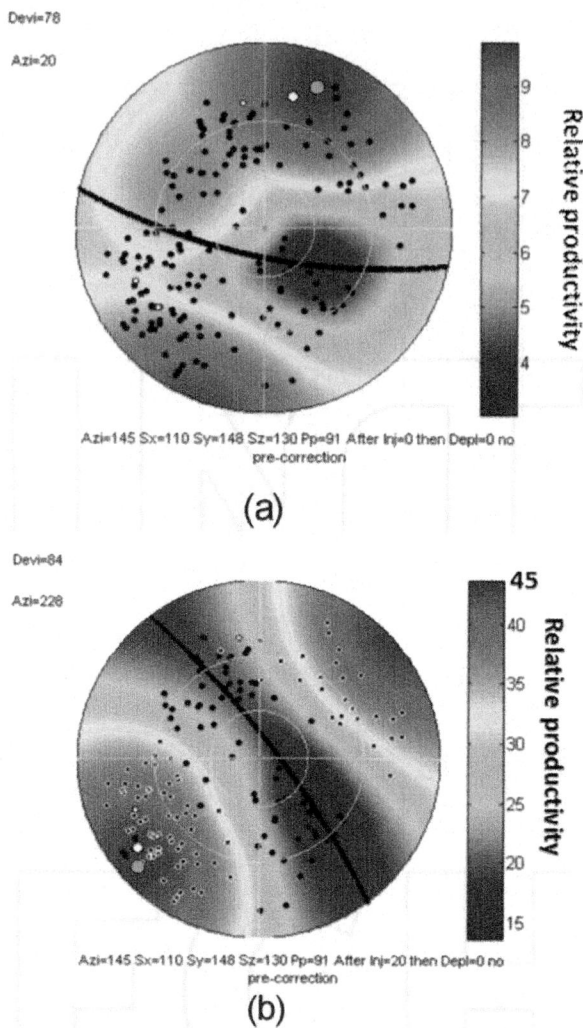

Figure 9. Relative well productivity for wells of all orientations based on the fractures observed in all three wells. (a) Ambient condition. (b) After 20-MPa stimulation. Natural fractures are shown as poles to the fracture surfaces (black dots). Green circles are computed optimal orientations of wells with highest productivity from natural fractures based on the fracture and stress analysis.

Figure 10 shows the general effect of reservoir flow properties changes due to the natural fracture stimulation for studied fault block. Again, all the fractures interpreted from image logs in the three wells are used for modeling. Cross-plots between relative productivity (flow rate/pressure) vs. reservoir pressure are shown for three different cases: under original conditions, after

30-MPa and after 50-MPa stimulation. The blue curves show productivity changes during stimulation when the pressure is increasing, the green curves show the productivity changes during flowback and production. Modeling ends at ~20-MPa depletion. The relative productivity at ~20-MPa depletion increases five-fold after the 30-MPa stimulation (productivity increases from ~4 to ~20). There is no obvious improvement in the relative productivity of natural fractures for 50-MPa stimulation (bottom left) compared to 30-MPa stimulation. The bottom-right plot shows the number of stimulated natural fractures under different pressure conditions. It is clear that nearly all of the natural fractures are stimulated while the pressure increases to ~130 MPa (40-Ma stimulation), which explains why there is little improvement with further stimulation. It is important to note that this result does not take into account the possibility of injecting proppant to maintain the conductivity of fractures which open at pressures above 40 MPa.

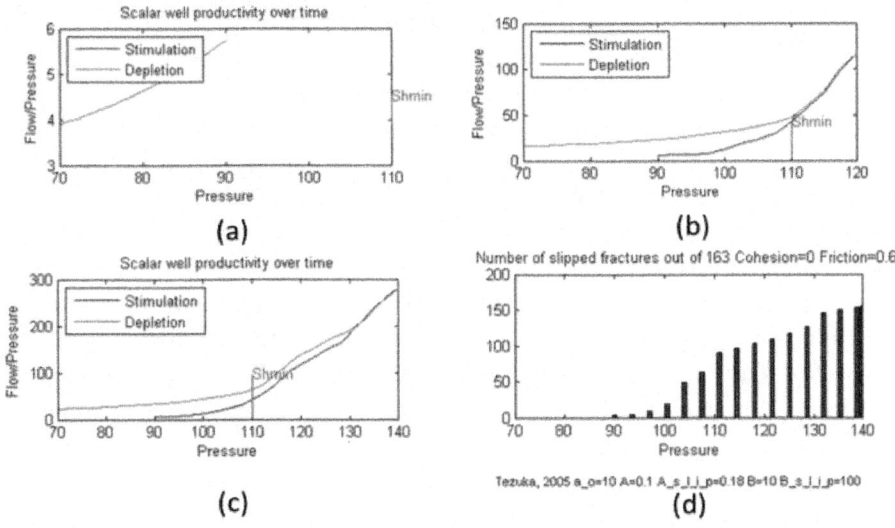

(a) (b) (c) (d)

Figure 10. Reservoir flow properties changes with time due to natural fracture stimulation for studied fault block. The blue curves are showing the productivity changes when pressure increases during stimulation, the green curves are showing productivity changes during flowback and production. (a) no stimulation (b) 30-MPa stimulation (c) 50-MPa stimulation (d) number of stimulated (shear slip) natural fractures. Fracture properties: cohesion=0, sliding friction=0.6.

The above relative productivity modeling of natural fractures shows the conductivity of natural fractures increases significantly if the stimulation pressure is at or above the minimum horizontal stress. This is because many of the natural fractures are non-optimally oriented. Assuming a connected fracture network exists, the conductivity increase could be a factor of five for the stimulated fracture network while stimulation pressure is ~130MPa or higher (assuming the pressure reaches all fractures).

PREDICTING THE SHAPE OF THE STIMULATED RESERVOIR VOLUME

Fracture stimulation modeling showed that the shear slip of natural fractures could be effective in improving reservoir properties. Next, we need to reproduce the affected productive volume in the reservoir using the "shear stimulation" concept to enable more accurate production prediction. At the present no commercial simulator can fully model this process in 3D, although some research simulators have been developed. It was decided to use two different commercial models to simulate both fracture network stimulation created by low-viscosity frac fluid and the growth of the main hydraulic fracture. A commercial dual-porosity, dual-permeability simulator is used to simulate the flow property changes of natural fractures due to the shear slip. A commercial hydraulic fracturing design and evaluation simulator is used to model the geometry and conductivity of the principal hydraulic fracture filled with proppant. The modeling in two separate simulators is coupled by the fluid volume used for stimulation. The fluid volume leaked off in the shear-dilated natural fracture network was estimated in the dual-permeability, dual-porosity flow simulator. By adjusting the pressure-dependent leak-off coefficient, the fluid volume leaked off in the hydraulic fracturing simulator was matched with the fluid volume leaked into natural fractures networks estimated by the flow simulator. The prediction of the stimulated reservoir volume is discussed in the rest of this section and the hydraulic fracturing design will be discussed in next section.

To predict the extent and properties of the stimulated volume by a dual-permeability, dual-porosity simulator, a finely gridded model (Model A) was created based on the original reservoir model. The main function of this model is to simulate the change in flow properties in every single frac stage during and immediately after injection. The model is initialized with average known reservoir characteristics such as matrix porosity and permeability, fracture permeability and initial pressure, characterized from core and log analysis. Although different cases have been tested in the study, only one of the most realistic cases will be discussed here: the average matrix porosity used in the initial model is ~7.4%, matrix permeability is 0.07 mD in all directions, and the initial fracture permeability is ~ 0.2 mD. The initial fracture permeability is set close to the lower bound of fracture permeability based on core and log analysis. The orientations of the principal flow directions were chosen to correspond to the principal directions of the fracture sets and of bedding, which also approximately corresponded to the principal stress directions.

The relative magnitudes of the permeability enhancements in different directions were constrained by the geomechanical analysis. A set of permeability-pressure tables for different directions were then used to describe the hysteretic rock behavior that results from shear fracture activation. Although the fracture properties during stimulation can be estimated as described in the previous section, it is better to calibrate and constrain the permeability-pressure relationship based on real lab or in-situ tests, e.g., using a pre-stimulation injectivity test [4]. The injectivity test should ideally be conducted in the open hole using slow injection to evaluate the potential natural fractures being stimulated, as permeability changes could then be interpreted based on the flow-rate/pressure changes along with the reservoir pressure. Because the D3 well has already been cased it was impossible to conduct such a test in the field before the actual treatment is carried out. Consequently, it was decided to produce a permeability-pressure table based on experience from shale gas reservoirs. Based on this table, on fracture density in different directions and on the stress anisotropy, a composite transmissibility multiplier was produced for the prediction of properties and extent of the stimulated reservoir volume. Transmissibility multipliers were different for each of the I, J and K directions; those directions were aligned as discussed above with the primary structural fabric and stresses. The propagation of the pressure and fluid front in these directions can be controlled by modifying these multipliers.

Figure 11 shows diagrammatically the relationship between the permeability multiplier and the pore pressure (green curve). A slow increase in the permeability multiplier with increasing pressure occurs until fractures begin to slip. Above this pressure, the injectivity increases rapidly as an increased number of fractures are stimulated. During decreasing injection pressure in the injectivity test, the injectivity should decrease more slowly, retaining behind a permanent injectivity increase. The post-stimulation response can also be extrapolated to pressures below the original reservoir pressure. This makes it possible to predict the reservoir's response to depletion, which could lead to improved predictions of production decline. When the pressure during stimulation exceeds the minimum horizontal stress, extensional hydrofracs are created, and the permeability-pressure relationship does not follow the green line. Three different flow paths (A, B, C) were assumed for conditions with pressure above S_{hmin}, and the intermediate path, B was chosen to be used in the simulation.

The result of this modeling work is a 3D induced permeability map that describes the stimulated rock volume as discrete blocks, each with a unique permeability. The stimulated rock volume is therefore described not as a geometrical shape with identical flow properties throughout, but as a rock body with variable induced permeability, as shown in Figure 12.

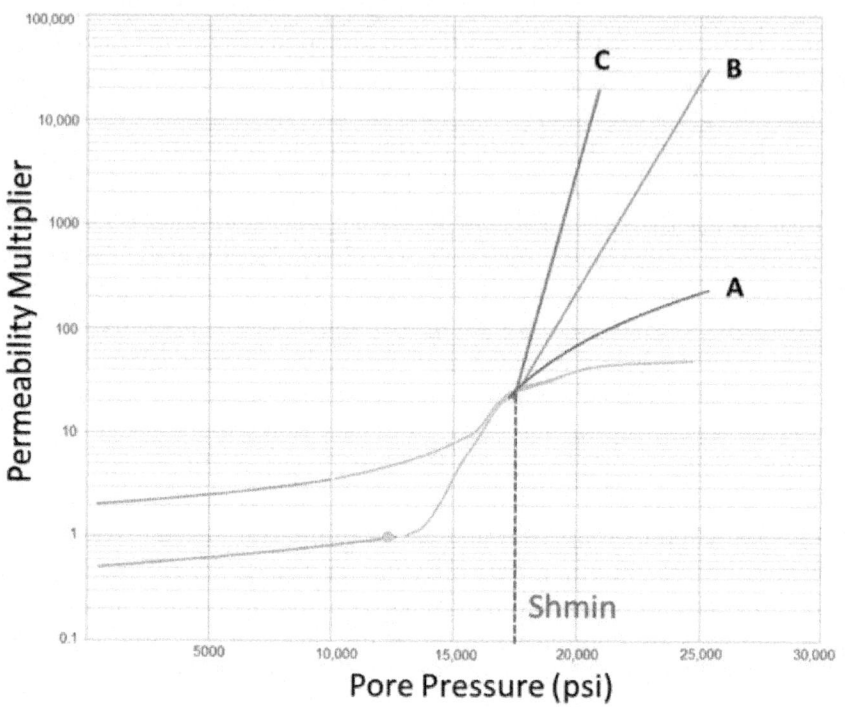

Figure 11. Relationship between the permeability multiplier and the pore pressure (green curve) for natural fractures used in the simulation. Three different flow paths (A, B, C) were assumed for conditions with pressure above Shmin.

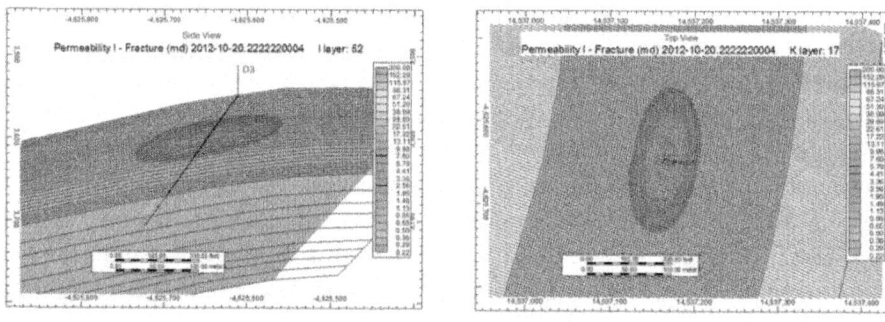

Figure 12. Side view (left) and top view (right) of the predicted 3D permeability map. The property shown in the plots is present fracture permeability.

HYDRAULIC FRACTURING DESIGN AND RESERVOIR SIMULATION

As discussed earlier, a commercial simulator was used to model the hydraulic fracture created during the stimulation along with the stimulated natural fracture network using low-viscosity fluids. Stress profiles and other elastic rock properties estimated in the geomechanical analysis were used as input for the design. To achieve better proppant distribution, a low-viscosity linear gel was combined with slickwater in the treatment. The low-viscosity linear gel was optimized using different concentrations of ingredients for the high reservoir temperature (~126°C) using source water and local ingredients. Due to the high closure pressure and low viscosity of the fluid, high-strength small-mesh proppants were used in the design.

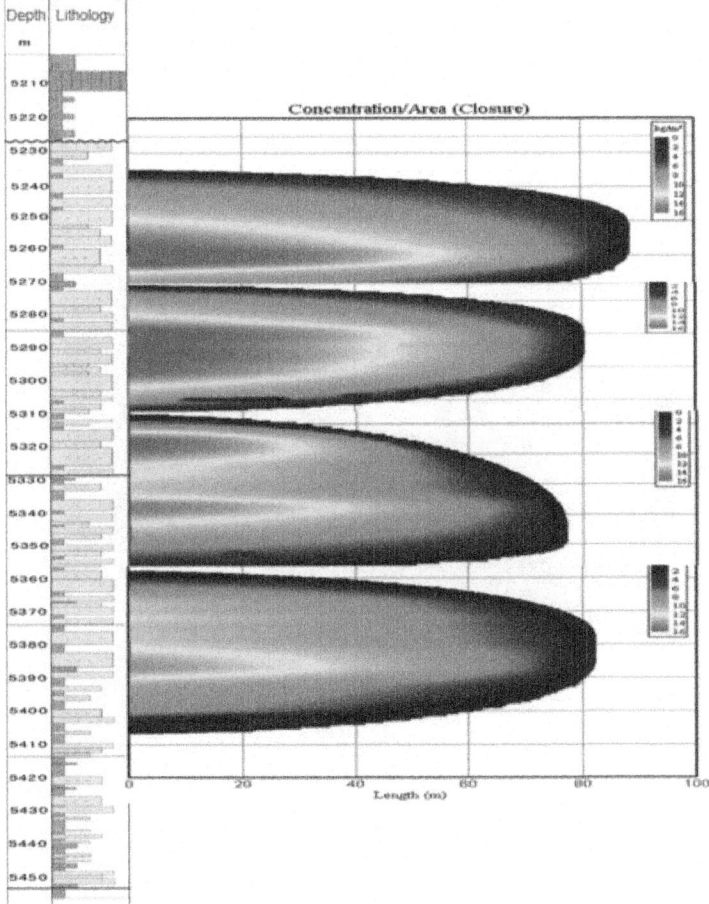

Figure 13. Fracture growth and proppant coverage (colour) for four stage of hydraulic fractures using slickwater/linear gel.

Modeling showed that four stages would be required for slickwater/linear gel treatment to cover the 160 m thick reservoir due to the high leak off of low-viscosity fluids (Figure 13). A reasonable proppant distribution was achieved by using the low-viscosity linear gel.

To predict the production after the stimulation, the propped hydraulic fractures were imported into the reservoir model with flow properties enhanced by stimulated natural fractures (Model A). Because the natural fracture distribution between wells is unknown, the same stimulated Model A was used for all four stages. The left plot of Figure 14 shows a side view of the reservoir model combining four Model A's with stimulated reservoir volumes and four propped hydraulic fractures, which was used for production prediction.

To compare the prediction result from slickwater/liner gel treatment with conventional gel fracturing, a conventional bi-wing hydraulic fracturing design using a high-viscosity gel was also developed. The gel fluid was optimized using different concentrations of ingredients for the high reservoir temperature (~126°C) using source water and local ingredients.

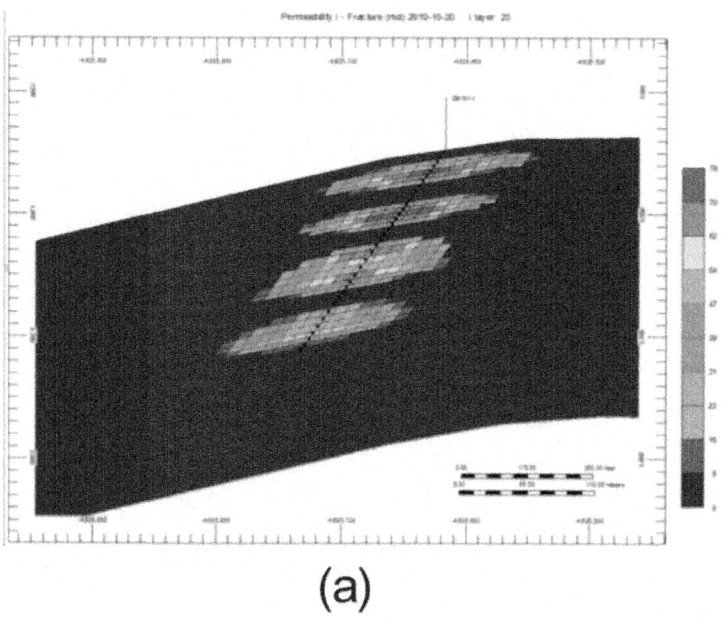

(a)

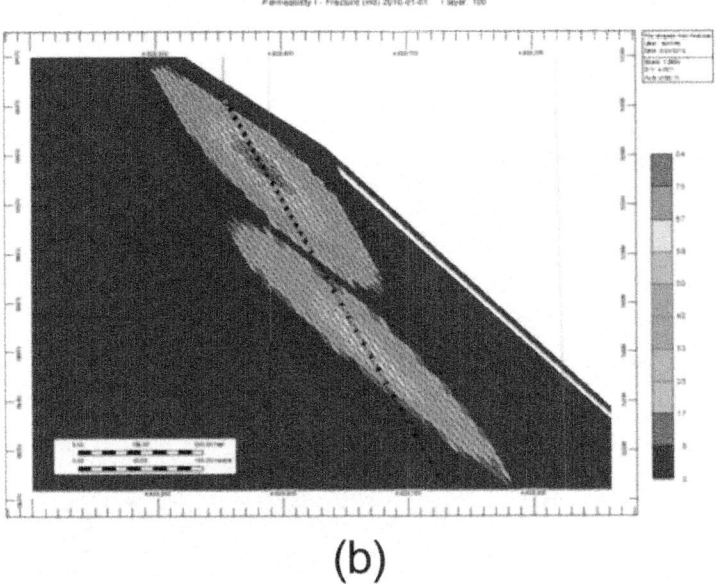

(b)

Figure 14. Side views of reservoir models showing fracture permeability used for production prediction. (a) Reservoir model combining four Model A's with stimulated reservoir volumes and four propped hydraulic fractures using slickwater and linear gel; (b) Original reservoir model and two propped hydraulic fractures using high-viscosity gel fluid.

The same type of proppant used for the slickwater/liner gel treatment was used for the design of gel treatment. The proppant concentrations and amounts will be certainly different in these two types of treatments. It was found that two stages were enough to cover the whole reservoir interval (Figure 15). These two designed hydraulic fractures were then imported into the original reservoir model (right plot in Figure 14) for production prediction and comparison of the production to that predicted after slickwater linear gel stimulation.

Figure 16 shows the production prediction comparison from the two different hydraulic fracturing treatments. The red curve is the production prediction from slickwater/linear gel treatment, which is scaled down to ~2/3 of the initial prediction to account for the heterogeneity of the reservoir model due to a simplified reservoir model used for pre-stimulation condition. The blue curve is the production from conventional two-wing gel fracturing design. It is found that post-frac flow rate from slickwater stimulation is expected to be about three times the flow rate from the gel treatment in the stabilized regime (one year after stimulation). Although actual flow rates from both treatments depends on the applied drawdown, the corresponding flow rates after one year are expected to be ~55 ×× 10^4 m³/d for slickwater treatment and ~ 17 ×× 10^4 m³/d for gel treatment, respectively, with a constant drawdown of 20 MPa.

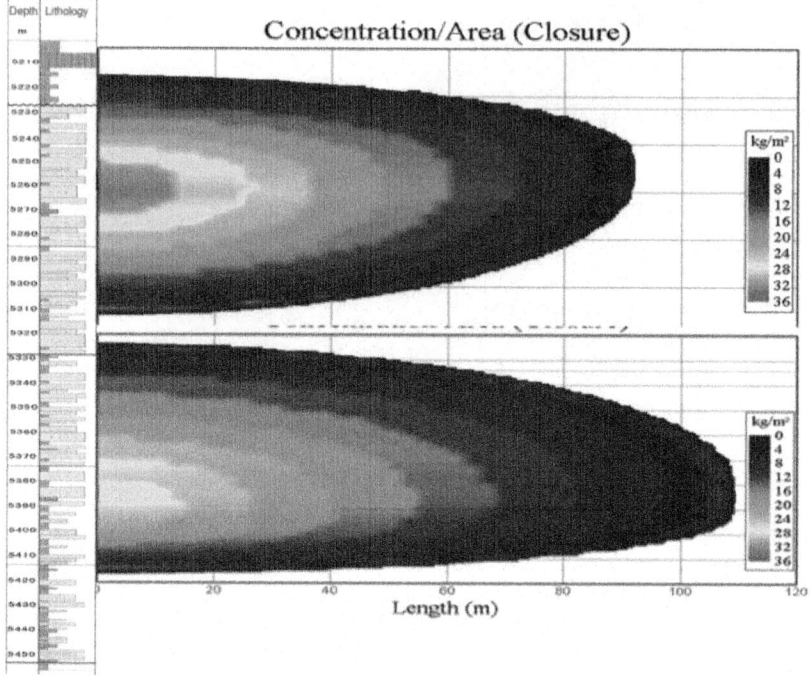

Figure 15. Fracture growth and proppant coverage (colour) for two stage of hydraulic fractures using conventional gel treatment.

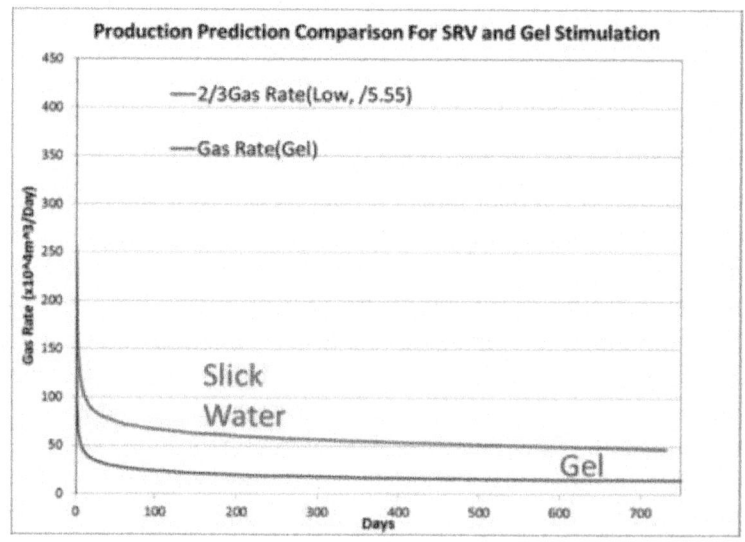

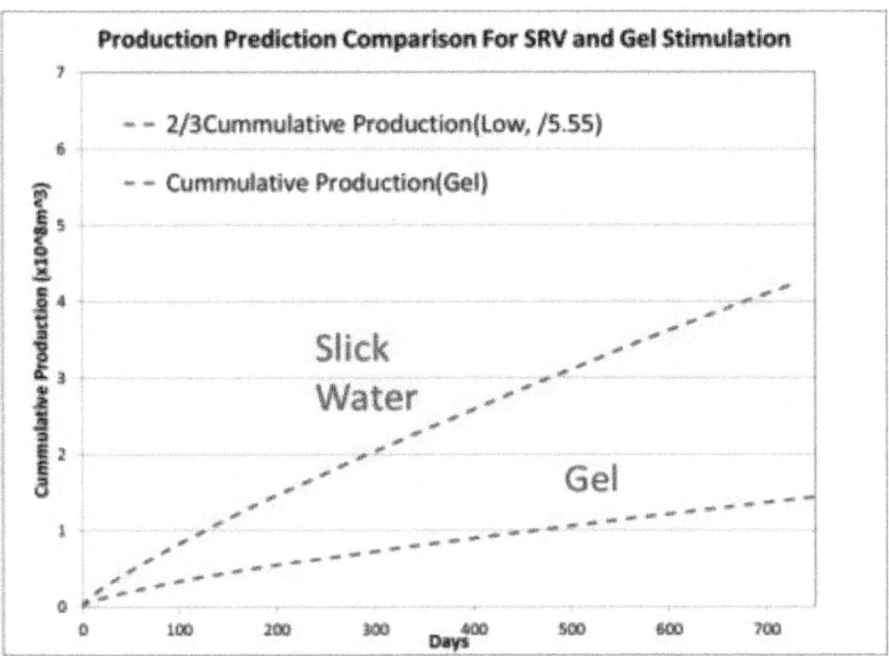

Figure 16. Production prediction comparison of two different hydraulic fracturing treatments. The red curve is the production prediction from slickwater/linear gel treatment; the blue curve is the production from conventional two-wing gel fracturing design.

INJECTIVITY TEST AND STAGE 1 TREATMENT

It was decided to test the slickwater/liner gel treatment in D3 well after the study was completed. A pre-stimulation injectivity test was performed through perforations prior to Stage 1 and after the mini-frac test (Figure 17). Interestingly, the test showed the opposite behavior from what one would expect if the stimulation enhances reservoir permeability. Later-stage injectivity (during step-down) is lower than early stage injectivity (during step-up), rather than higher. Although there might be other reasons affect the test result, i.e., the un-stable injection during the whole test, it is believed the main reason was lack of access to natural fractures in the tested interval and the high closure pressure because the test was conducted in a cased and perforated hole and after a mini-frac.

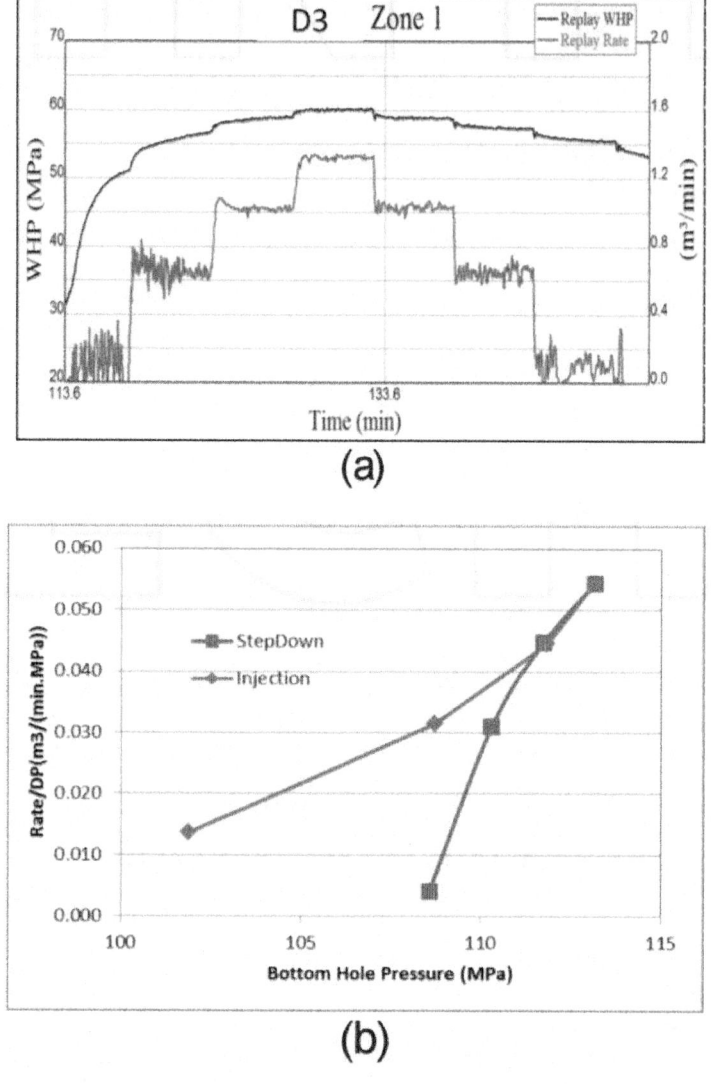

Figure 17. Pre-stimulation injectivity test pressure curve (a) and injectivity interpretation (b).

The Stage 1 treatment was conducted using slickwater and linear gel after the injectivity test. However, a screen out was experienced at the end of the execution and tubing leakage was discovered afterwards. Treatments in the other three zones had not occurred at the date of writing this paper. The stage 1 production test is still very promising, and it has been decided to continue slickwater/linear gel treatment in other three stages after the tubing problem is fixed.

DISCUSSION AND CONCLUSION

In this paper we have outlined a new workflow for simulation of a complex fracture network created by stimulation using low-viscosity fluids in a fractured tight sandstone reservoir. The workflow is based on critically stressed fracture theory. This process of natural fracture stimulation is believed to be the underlying reason for the success in shale gas reservoir stimulation. The results suggested that there would be significantly higher production from this approach compared to conventional two-wing gel fracturing.

There are, however, some uncertainties in the modeling of the natural fracture stimulation for this fractured tight gas reservoir.

- The pressure-permeability relationship used in modeling the permeability enhancement by slickwater stimulation is taken from a shale gas field. It is unclear whether the data from the analogue field drilled through mudstones will be applicable to the modeled fractured tight sandstone reservoir. Post-stimulation production simulation, or a pre-stimulation injectivity test in nearby wells in open hole could help to better constrain this relationship, hence improve the accuracy of the prediction.

- Due to the lack of knowledge of fracture distribution between wells, the fractures interpreted from all three offset wells were used to predict the stimulation behavior of natural fractures, and it was assumed that a similar fracture distribution would be found in all formations. In reality, the fracture distribution is likely to be different, depending among other things on the lithology and structural location. For example, it is already noticed that there are fewer fractures in the lower part of the reservoir than in the upper part in the D3 well. Intervals with dense fracture networks are more likely to benefit from slickwater treatment compared to formations with no or very sparse fractures. A 3D description of the fracture distribution is always preferred.

- Micro-seismic imaging is not available in the study area. No wells are close enough to work as a monitoring well and surface monitoring is also impossible due to the great depth of the reservoir. The lack of microseismic data made it impossible to calibrate the prediction of the shape of SRV.

The main uncertainty in gel frac productivity estimation comes from the propped fracture conductivity estimation. This conductivity is based on proppant testing in the laboratory. The proppant inside fractures involves clogging, crashing and embedment over the production period. There is no analytical method available to model these long-term effects on propped fracture conductivity. An approximate conductivity damage factor has been used in this study to consider these effects. Although there are still some shortcomings with the workflow, it can assist in the assessment of development concepts and the evaluation of stimulation enhancement options. The anisotropy in the slickwater treatment can be reasonably well-predicted and applied into the production simulation, which provides a more robust prediction than a simple isotropy model. The new workflow can be used in naturally fractured shale gas, tight gas/oil and CBM reservoirs.

ACKNOWLEDGEMENTS

The authors wish to thank PetroChina Tarim Oil Company for providing us with the data and for permission to publish this paper, and Baker Hughes internal support to carry out the work.

REFERENCES

1.   N Modeland, D Buller, K. K Chong, Stimulation's influence on production in the Haynesville Shale: a playwide examination of fracture-treatment variables that show effect on production. In: proceedings of Canadian Unconventional Resources Conference, CSUG/SPE 1489401517November 2011Calgary, Alberta, Canada.

2.   S. C Maxwell, T Pope, C Cipolla, et alUnderstanding hydraulic fracture variability through integrating microseismicity and seismic reservoir characterization. In: proceedings of SPE North American Unconventional Gas Conference and Exhibition, SPE 1442071416June 2011Woodlands, Texas, USA.

3.   C Sayers, and Le Calvez, J., 2010Characterization of microseismic data in gas shales using the radius of Gyration tensor, SEG Expanded Abstract.

4.   D Moos, Improving Shale Gas Production Using Geomechanics, Exploration & Production- Oil & Gas Review 2011928488

5.   M. D Zoback, A Kohli, I Das, M Mcclure, The importance of slow slip on faults during hydraulic fracturing stimulation of shale gas reservoirs. In: proceedings of SPE Americas Unconventional Resources Conference, SPE 15547657June 2012Pittsburgh, Pennsylvania, USA.

6.   M Mullen, M Enderlin, Is that frac job really breaking new rock or just pumping down a pre-existing plane of weakness?- the integration of geomechanics and hydraulic-fracture diagnostics. In: proceedings of 44th US Rock Mechanics Symposium and 5th US-Canada Rock Mechanics Symposium, ARMA 10-285, 27-30 June 2010Salt Lake City, UT, USA.

7.   D Moos, G Vassilellis, R Cade, Predicting shale reservoir response to

stimulation in the Upper Devonian of West Virginia. In: proceedings of SPE Annual Technical Conference and Exhibition, SPE-145849, 30 October-2 November 2011Denver, Colorado, USA.

8. G. D Vassilellis, C Li, D Moos, et alShale engineering application: the MAL-145 Project in West Virginia. In: proceedings of Canadian Unconventional Resources Conference, CSUG/SPE-1469121517November 2011Calgary, Alberta, Canada.

9. C Barton, M. D Zoback, D Moos, Fluid Flow Along Potentially Active Faults in Crystalline Rock, Geology1988238683686

10. D Moos, C. A Barton, Modeling uncertainty in the permeability of stress-sensitive fractures. In: proceedings of 42nd US Rock Mechanics Symposium and 2nd U.S.-Canada Rock Mechanics Symposium, ARMA 08312June- 2 July 2008San Francisco, USA.

11. M. M Hossain, M. K Rahman, S. S Rahman, A Shear Dilation Stimulation Model for Production Enhancement From Naturally Fractured Reservoirs, SPE 78355, SPE Journal; June 2002183

12. D Moos, M. D Zoback, Utilization of Observations of Well Bore Failure to Constrain the Orientation and Magnitude of Crustal Stresses: Application to Continental, Deep Sea Drilling Project and Ocean Drilling Program Boreholes, Journal of Geophysical Research 1990959305

13. O Heidbach, M Tingay, A Barth, J Reinecker, D Kurfe, and B Müller, The World Stress Map database release 2008doi:10.1594/GFZ.WSM. Rel2008.http://www.world-stress-map.org

Chapter 5

A NEW APPROACH TO HYDRAULIC STIMULATION OF GEOTHERMAL RESERVOIRS BY ROUGHNESS INDUCED FRACTURE OPENING

Nima Gholizadeh Doonechaly[1], Sheik S. Rahman[1] and Andrei Kotousov[2]

[1]School of Petroleum Engineering, University of New South Wales, Sydney, Australia
[2]School of Mechanical Engineering, the University of Adelaide, South Australia, Australia

ABSTRACT

Hydraulic fracturing by shear slippage mechanism (mode II) has been studied in both laboratory and field scales to enhance permeability of geothermal reservoirs by numerous authors and their success stories have been reported. Shear slippage takes place along the planes of pre-existing fractures which causes opening of the fracture planes by the fracture asperities (roughness induced opening). Simplified empirical relationships, which are derived based on simple fracture experiments or best guess, are used to calculate compressive normal surface traction, residual aperture and shear displacement. This introduces ambiguity into the simulation results and often leads to erroneous predictions of reservoir performance.

In this study an innovative analytical approach based on the distributed dislocation technique is developed to simulate the roughness induced opening of fractures in the presence of compressive and shear stresses as well as fluid pressure inside the fracture. This provides fundamental basis for computation of aperture distribution for all parts of the fracture which can then be used in the next step of modeling fluid flow inside the fracture as a function of time. It also allows formulation of change in aperture due to thermal stresses. The stress distribution and the fluid pressure are calculated using the fluid flow modeling inside the fracture in a numerical framework in which thermo-hydro mechanical effects are also considered using finite element methods (FEM). In this study, fractures with their characteristic properties are considered to simulate rock deformation.

This new approach is applied to the Soultz-Sous-Forets geothermal reservoir to study changes in permeability and its impact on temperature drawdown. It has been shown that the analytical approach provides a more realistic prediction of residual fracture aperture which agrees well with the experience of existing EGS trials around the world. An average increase in aperture due to fluid induced shear dilation has been found to be lower and time required to obtain a sizeable reservoir volume is greater than those previously estimated.

INTRODUCTION

Reservoir Stimulation by Induced Fluid Pressure

Fractured reservoirs in crystalline rocks are usually stimulated by injected fluid pressure. As the injection of fluid continues the pressure inside the fractures increases gradually. The effective stress due to fluid pressure is expressed as:

$$\sigma_{eff} = \sigma_t - p \tag{1}$$

where

σ_{eff} is the effective stress, σ_t is the total stress and p is the pore pressure. With further injection of fluid the effective shear stress, which is a function of effective stress, continuously decreases until it reaches a threshold value at which time it can no longer resist shear displacement of the fracture surfaces. At this stage the shear dilation will occur. During shear displacement rock fails by the shearing (Mode II) instead of opening (Mode I). In Mode II opening, the surface asperities of the rock slide over each other which cause more separation of the fracture surfaces. Such an interlocking of asperities increases the permeability of the rock. Any further increase in pressure can cause the effective closure stress to decrease to zero at which time the separation and interlocking of the fracture surfaces perpendicular to the fracture walls occur. The amount of pressure required to reach zero effective stress is highly dependent on the rock and fracture properties [1]. If the injection continues at some point it will exceed the tensile strength of the rock, which leads to tensile failure of the rock. This means that a certain level of permeability enhancement by shear displacement can be obtained. Mechanical representation of the shear displacement and the normal separation of the fracture surfaces can be described based on a specific failure criterion, such as Mohr-Coulomb (see Fig. 1). As the pressure inside the fracture increases the effective stress decreases: Mohr's circle moves towards the origin. As shown in Fig.1, when the minimum principal stress (closure stress) reaches zero the normal separation of fracture surfaces (Mode I) occurs. However, the shear dilation happens much earlier: when the Mohr's circle encounters the failure envelope (CD) at E. Shear dilation by induced fluid pressure was first detected in the laboratory experiments in 1970s. One of the earliest attempts by [2] showed a significant permeability increase by shear displacement. This observation was confirmed by [3] and [4]. Since then, shear dilation has been comprehensively studied in geotechnical and mining engineering. However, investigation of permeability enhancement by shear dilation in petroleum reservoirs began much later [5]. Since the shear dilation is caused by slippage of the asperities on top of each other, there is maximum dilation that can be reached. The maximum displacement that can be achieved is called characteristic height of the fracture [6]. Based on an experimental study the characteristic height is measured to be of the order of a fraction of a millimeter [7]. Fracture aperture that can be created by conventional hydraulic fracturing is in the order of tens of millimeters [8]. Reservoir rocks with rough surfaces and high shear strength are highly desirable for stimulation by shear displacement to work. One of the most comprehensive attempts to characterize the shear dilation caused by the fracture surface asperities was developed by [9]. In their model, the rock behavior was studied by considering fracture surface and its aperture, normal and shears closure and shear dilation. In another attempt, [10] proposed a methodology to obtain the mechanical aperture of the fractures. The authors used the methodology proposed by [11] to measure the aperture by a tapered feeler gauge using plane sawn surfaces to gain access to the joints. Mechanical aperture can be calculated using an empirical equation as proposed by [10]. Later [10] used the empirical equation proposed by [11] to model the normal closure of fracture surfaces based on the normal stress. [12] proposed an approach to describe the hydraulic and mechanical properties of the fracture including the shear dilation by induced fluid pressure.

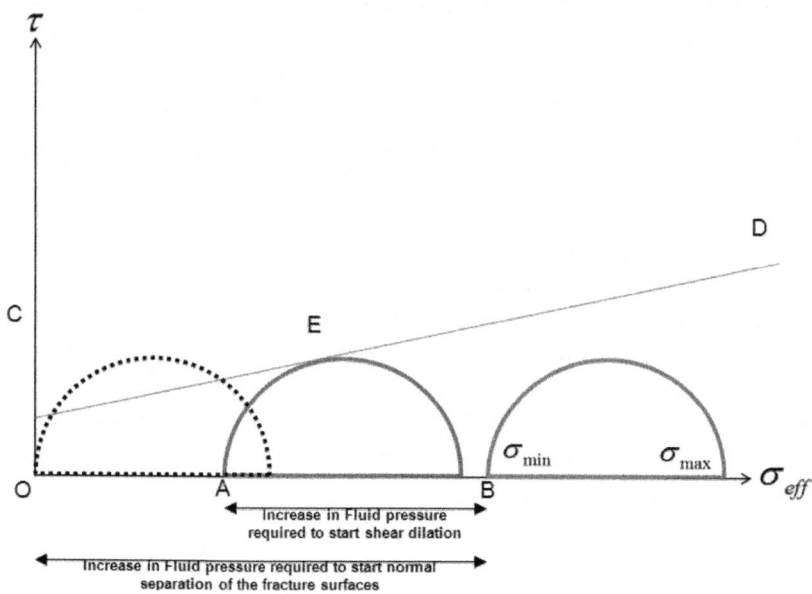

Figure 1. Mohr diagram describing the initiation of the shear dilation and normal fracture surface separation.

Mechanical models for shear displacement include displacement estimation under different stress boundary conditions in which a proper topographical model is used to describe the fracture surface. Also [13] experimentally studied the effect of normal stress and shear dilation on fluid flow properties of a naturally fractured core sample. They have used a servo-controlled axial/torsion load frame to test the fluid flow and mechanical behavior of the fracture surface during normal stress, slip and shear dilation. In another approach [14] proposed a semi-empirical correlation to determine the change in fracture aperture based on the amount of shear displacement between the fracture surfaces and the stress boundary condition. Also [15] extended the previous attempt of [14] by considering the effect of fracture propagation in shear dilation.in another attempt [16] used a linear relationship between shear displacement and the dilation of the fracture surfaces.

In this study, an analytical computational methodology based on distributed dislocation technique proposed by [6] is used to estimate the aperture distribution caused by the shear dilation in a fracture subject to different varying stress boundary conditions [6].

Two major assumptions are used in this approach to characterize the shear displacement of the fracture surfaces. The shear slippage between the fracture surfaces is described by using Coulomb friction law which explains the friction stress during the shear slippage based on the normal stress exerted on the fracture planes with a proportionality contact named friction factor as shown in Eq.

$$\tau = c + f\sigma_n \tag{2}$$

where, τ_0 is the threshold shear stress value to initiate the shear slippage between the fracture surfaces. Also the friction factor, f, is dependent on the material properties, fracture geometry and surface asperities of the fracture [6]. Because a minor change in the fracture aperture causes a significant alteration of the fracture permeability estimation of the shear slippage of the fracture surfaces is of crucial importance in fluid flow simulation. In this study the coupling between the shear displacement and the change in fracture aperture is described by a step function. Fracture displacement normal to the fracture plane is simulated by using virtual springs distributed along the fracture length. Such springs are characterized by a specific spring constant which can be calculated numerically, experimentally or analytically [6]. Also the spring deformations are modeled in an elastic framework which results in the following system of equations describing the stress between the fracture surfaces:

$$\sigma_n = kE(\Delta - \delta_y) \text{ for } \delta_y < \Delta \tag{3}$$

$$\sigma_n = 0 \text{ for } \delta_y > \Delta \tag{4}$$

where, Δ is the characteristic height of the fracture as shown in Eq. (3) and k is the spring constant. Equation (3) is associated with the rock compressibility and gives us the normal stress exerted on the fracture surfaces. After calculating the normal stresses on the fracture surfaces, the normal displacement of the surfaces is calculated by the distribution dislocation concept. Also the methodology proposed by [17] is used to calculate the spring constant based on a bed of nails as [17]:

$$k = E\frac{b\Delta}{L} \tag{5}$$

where, E id the Young modulus of elasticity, L is the fracture length and b is a constant less than unity. Also Eq. (4) implies the complete separation of the fracture surfaces in which no contact exists between the fracture asperities.

The complete set of boundary conditions for a fracture as shown in Fig. 2 are listed below [6]:

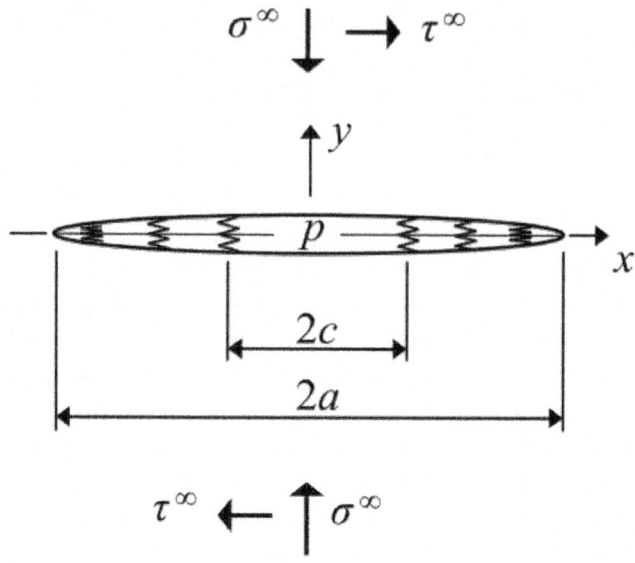

Figure 2. A schematic representation of a fracture subject to in-situ stress boundary conditions.

$$\sigma'_y = \sigma^\infty \text{ for } x^2 + y^2 \to \infty \tag{6}$$

$$\sigma'_y = kE(\Delta - \delta_y) + p \text{ for } c < |x| \le a \tag{7}$$

$$\sigma'_y = p \text{ for } |x| \le c \tag{8}$$

$$\tau_n = \tau_0 + f \cdot kE(\Delta - \delta_y) \text{ for } c < |x| \le a \tag{9}$$

$$\tau_n = 0 \text{ for } |x| \le c \tag{10}$$

$$u_x = 0 \text{ for } |x| \ge a \tag{11}$$

$$u_y = 0 \text{ for } |x| \ge a \tag{12}$$

where

σ'_y is the effective normal stress exerted on the fracture surfaces, k is the spring constant, p is the pore pressure, Δ is the characteristic height of the fracture, δ_y is the displacement of the fracture surfaces, E is the Modulus of elasticity, τ_n is the shear stress exerted on the fracture surfaces, τ_0 is the threshold stress requires to start the shear displacement of the fracture surfaces and u is the displacement. *As* mentioned above, the aperture distribution along the fracture surface is calculated based on an analytical methodology in which fracture geometry, stress distribution and fluid pressure inside the fracture are needed to be known as a priori. For this purpose a thermo-poro-elastic model is developed to simulate the fluid flow in the reservoir scale.

SIMULATION OF FLUID FLOW AND HEAT TRANSFER

Three distinct approaches exist in the literature to simulate the fluid flow in naturally fractured reservoirs namely: single continuum, dual continuum and discrete fracture approach. In single continuum, the fractured medium is represented by an equivalent homogeneous system using a specific permeability tensor. In dual continuum approach the whole domain is divided into two interacting domains: fractures and matrix where by matrix (represented by sugar cubes) provides the storage and fractures (having regular pattern) the permeability. In discrete fracture approach, fractures are explicitly discretized in the domain. These approaches are briefly discussed below followed by the proposed methodology which is used in this study.

Hybrid of Single Continuum and Discrete Fracture

Different approaches have been used in the literature to incorporate the fractures into the flow modeling. Each of these techniques has its own drawbacks and benefits. In this study a hybrid methodology combining the single continuum and discrete fracture networks model is used to increase accuracy and efficiency of the fluid flow simulation. In the proposed methodology a threshold value is defined for the fracture length. Fractures which are smaller than the threshold value are used to generate the grid based permeability tensor using boundary element technique.

Fluid flow simulation is carried out by using the single continuum approach in the nominated blocks. Fractures which are equal to and longer than the threshold value are explicitly discretized in the domain using appropriate elements and the fluid flow is modeled using the discrete fracture approach. Such an approach provides a more accurate and realistic framework to consider the effect of long fractures on the fluid flow in fractured medium.

Domain Discretization Using the Hybrid Methodology

In this study the medium and long fractures ($l \geq 50\text{m}$) are discretized using triangular elements and the contribution of flow by fractures ($l < 50\text{m}$) are taken into account by calculating permeability tensor for each discretized element. A schematic representation of the domain discretization for a fractured reservoir is shown in Fig 3 (a) and (b).

Permeability tensor for each block is expressed as:

$$K = \begin{bmatrix} k_{xx} & k_{xy} \\ k_{yx} & k_{yy} \end{bmatrix}$$

(13)

Permeability tensors are calculated by simulating fluid flow in individual fractures in each element. The concept of permeability tensor was first introduced by [18] by considering a set of parallel fractures in a Representative Elementary Volume (REV) with zero matrix permeability [18]. In another attempt [19] developed a methodology for calculation of permeability tensor for arbitrary oriented fractures using superposition technique [19].

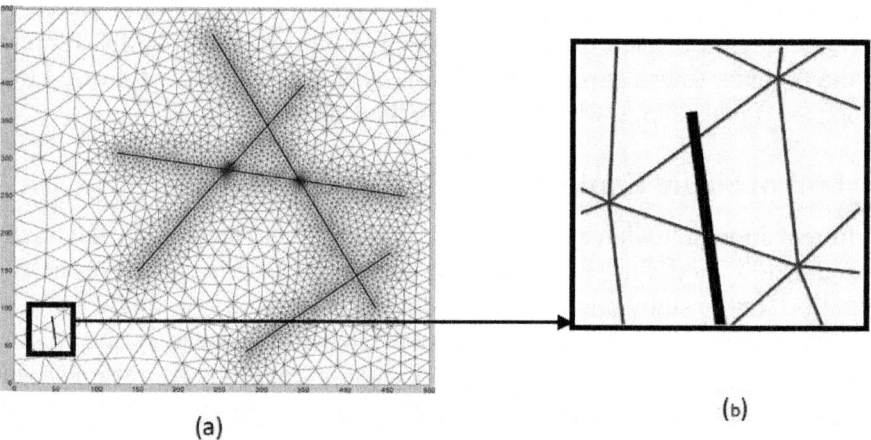

(a)

(b)

Figure 3. Domain discretization by using the hybrid of the single continuum and discrete fracture approach. (a) fractures equal to and longer than 50 m are explicitly discretized in the reservoir domain by using the triangular elements. (b) after the discretization of the long fractures, the effect of short fractures (<50m) are taken into account by calculation of the permeability tensor of the corresponding blocks which are cut by the fractures.

In this study the authors have considered interconnected fractures with fracture surface as infinite plate without roughness. In another approach [20] estimated permeability tensor by assuming fractures as a planar sink/source term [20]. Also [21] extended the approach and studied the effect of vertical fracture/ matrix permeability ratio on the permeability tensor. In a separate study, [22] used a numerical technique (BEM) to calculate the permeability tensor of the REV containing medium sized fractures considering fractures as a sink/source term [22]. Following this work [23] presented an analytical model to calculate the permeability tensor of the blocks containing infinite parallel fracture sets [23]. Also [24] improved the efficiency of their previous approach by considering the effect of short fractures using the analytical method proposed by [24]. In another approach [25] presented the first comprehensive methodology to calculate the permeability tensor for arbitrary oriented fractures in different length scales. In this study permeability tensor was determined by discretizing the solution domain into different subdomains depending on the length of the fractures using BEM [25]. Short fractures are considered as part of matrix porosity to improve the matrix permeability inhomogeneity. However, medium and long fractures are discretized explicitly in the domain and fluid flow is simulated using BEM. Then [26] extended [25] by increasing the efficiency of the BEM so that fluid flow in greater number of fractures can be simulated. The authors also presented for the first time effective permeability tensor calculation for the fractured REV by using the BEM. The effective permeability model was validated using laboratory derived data.

Rev Discretization for Permeability Tensor Calculation

To calculate the effective permeability tensor, the fractured REV is divided into three distinct regions: matrix (region 1), fracture (region 2) and region around the fractures (region 3) as shown in Fig. 4.

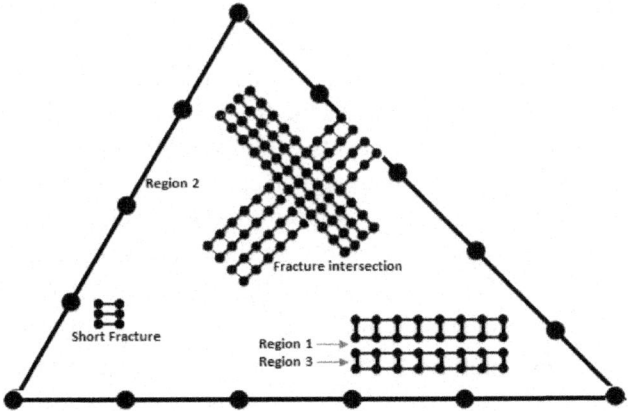

Figure 4. Domain discretization based on different fracture lengths.

Flow inside the fractures (region 2) is modeled using the cubic law. With the assumption of smooth fracture surfaces, cubic law can accurately simulate the flow inside the fractures [19, 27]. In matrix regions close to the fractures (region 3), the Darcy equation Eq. (14) is coupled with the mass conservation equation to consider the effect of the fracture on the flow of fluid in the region close to the fractures. Size of this region depends on the size of the fracture. Also fluid flow simulation in the matrix (region 1) is described as follow:

$$V_f = -K_f \nabla P_f \tag{14}$$

$$\frac{\partial}{\partial L}(k_f \frac{\partial p}{\partial L}) + Q_f + q_{ff} = 0 \tag{15}$$

where p_f is the fluid pressure inside the fractures and p is the pore fluid pressure. For the short fractures which are considered as part of the matrix porosity, the Laplace equation is solved using the following boundary conditions:

$$p_{mi} = p_{fi} \tag{16}$$

$$v_{mi} = v_{fi} \tag{17}$$

Where, p_{mi} is the matrix pressure and p_{fi} is the fracture pressure at the matrix/fracture interface and v_{mi} is the normal fluid velocity at the ith fracture node along the fracture surface. Since the pressure on the matrix fracture interface is unknown, periodic boundary condition is applied in an iterative scheme to calculate the pressure values.

Reservoir Scale Fluid Flow Simulation

Fluid flow in long fractures (l>50m) is coupled with discretized element based permeability tensor in poro-thermo-elastic environment by using local-thermal non-equilibrium.

Different numerical techniques have been used to model thermo-poro-elastic phenomena in fractured porous media. To have a detailed understanding of the complex geomechanical aspects of the fractured rocks and the induced perturbation, such as thermal drawdown caused by the cold injection fluid in geothermal reservoirs an appropriate numerical technique should be used which is capable of (a) adequately applying the boundary and initial conditions and (b) accurately representing the system geometry. In order to take the aforementioned issues into account, FEM is used in the current study.

Weighted residual method and the Green's theorem are applied to discretize the mass, momentum and energy conservations equations [28]. As mentioned before, the finite element method is used in this study for the numerical simulation purpose. Therefore the state variables namely: displacement, pore pressure and temperature are defined using proper shape functions as:

$$u = N_u \bar{u} \tag{18}$$

$$p = N_p \bar{p} \tag{19}$$

$$T = N_T \bar{T} \tag{20}$$

Where N is the corresponding shape function and $\bar{u}$, $\bar{p}$ and $\bar{t}$, are the nodal values of the corresponding state variable. By applying the Galerkin's method and replacing the weighting functions by the corresponding variables' shape functions, the discretized form of the conservation equations can be written as follow [29, 30]:

$$(K + \frac{G}{3})\nabla(\nabla \cdot u) + G \cdot \nabla^2 u - \alpha \nabla p - \gamma_1 \nabla T_m = 0 \tag{21}$$

$$\alpha(\nabla \cdot \dot{u}) + \beta \dot{p} - \frac{k}{\mu}\nabla^2 p - \gamma_2 \dot{T} = 0 \tag{22}$$

$$\dot{T} + v(\nabla T) - c^T \nabla^2 T = 0 \tag{23}$$

where, K is the bulk modulus of elasticity, G is the shear modulus, γ_1 and γ_2 are the thermal expansion coefficient of the fluid and solid respectively; k is the permeability T_m is the matrix temperature, T is the fluid temperature and μ is the fluid viscosity.

FRACTURE NETWORK GENERATION

Simulation of naturally fractured reservoirs offers significant challenges due to the lack of a methodology that can utilize field data. To date several methods have been proposed in the literature to characterize naturally fractured reservoirs. In this study a hybrid tectono-stochastic simulation is proposed to characterize a naturally fractured reservoir [31]. A finite element based model is used to simulate the tectonic event of folding and unfolding of a geological structure. A nested neuro-stochastic technique is used to develop the inter-relationship between different sources of data (seismic attributes, borehole images, core description, well logs etc.) and at the same time the sequential Gaussian approach is utilised to analyze field data along with fracture probability data. This approach has the ability to overcome commonly experienced discontinuity of the data in both horizontal and vertical directions.

RESULTS AND DISCUSSIONS

The proposed methodology is used to generate the discrete fracture map of the Soultz geothermal reservoir at the depth of 3650 m. the statistical parameters used to generate the discrete fracture map is shown in Table 1.

Table 1. Statistical data used for the discrete fracture network generation. After [32]

| Fracture set | Azimuth | | | Dip | | | | Fracture No. | Radius (m) | Transmissivity (m2\s) |
	Distribution Law	Mean	Half-Width	Distribution Law	Mean	Half-Width	Dip Direction			
F1	Normal	2	16	normal	70	7	NW	1.3E-7	187	6E-6
F2	Normal	162	19	normal	70	7	NE	3E-9	150	6E-6
F3	Normal	42	6	normal	74	3	NW	1.76E-8	95	4E-6
F4	Normal	129	6	normal	68	3	SW	3.3E-8	112	2E-6
F5	Uniform	0	180	normal	70	9	-	1E-8	100	5E-7

The discrete fracture map, the corresponding mesh generated for the reservoir domain and the permeability tensors for each triangular element (a sample region which is cut by a fracture of length<50m) are shown in Fig. 5 (a), (b) and (c) respectively.

Also the reservoir properties used for the stimulation purpose are shown in Table 2. The reservoir is pressurized by injecting fluid through the injection well (GPK2). The pressurization was carried out over a period of 52 weeks. During the pressurization, the change in fracture width for each individual natural fracture and the resulting permeability tensor were calculated. Following stimulation of the reservoir, a flow test was carried out over a period of 14 years. During the flow test, changes in fracture apertures due to thermo-poro-elastic stresses and the consequent changes in permeability were determined. Also estimated were the thermal drawdown, produced fluid temperature and production rate of the Soultz EGS.

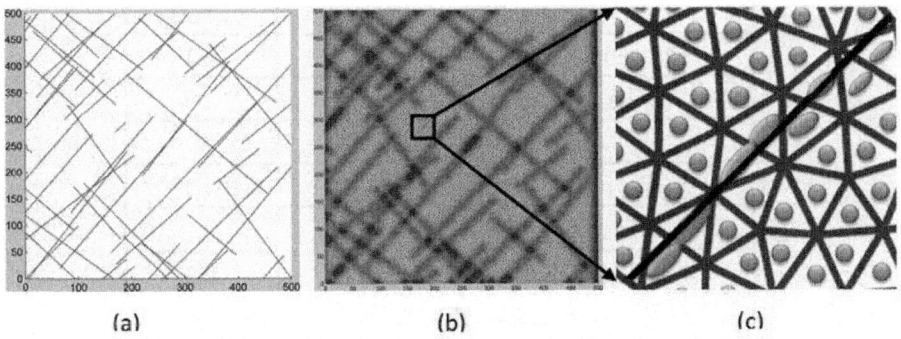

(a)   (b)   (c)

Figure 5. a) discrete fracture network at the depth of 3650 m (b) the corresponding discretization for the fractures longer than 50 m and (c) permeability tensor for a sample fracture (<50m).

Results of shear dilation are presented as average percentage increase in fracture aperture (see Fig. 6). From Fig. 6, it can be seen that there exists three distinct aperture histories: 0-40 weeks, 40-50 weeks and 50 weeks and above. Until about 40 weeks, a slow but linear increase in occurrence of dilation events due to induced fluid pressure of 51.7 MPa (bottom hole) and reaches a value of about 18% (average increase in aperture). Following this time, the rate of occurrence of dilation events increases sharply until about 50 weeks, thus reaching 60% increase in average fracture aperture. After which, no significant dilation events can be observed (a plateau of events is reached). When compared with previous study [29], in which shear dilation events are estimated based on a semi-empirical model (Willis-Richards et al, 1996), it can be seen that the time required to overcome the threshold stress is 40 weeks which is about 12 weeks longer than the previous studies. Also the time requires for an increase in the average fracture aperture of 58 % is about 8 weeks longer than that predicted by the previous study. During the flow test, changes in fracture apertures due to thermo-poro-elastic stresses and the consequent changes in permeability were determined. Also estimated were the thermal drawdown, produced fluid temperature and production rate of the Soultz EGS.

Table 2. Stress and reservoir data for strike-slip stress regime at Soutlz geothermal reservoir.

Rock Properties	
Young's modulus (GPa)	40
Poisson's ratio	0.25
Density (kg/m³)	2700
Fracture basic friction angle (deg)	40
Shear dilation angle (Deg)	2.8
90% closure stress (MPa)	20
In situ mean permeability (m²)	9.0×10^{-17}
Fracture properties	
Fractal Dimension, D	1.2
Fracture density (m²/m³)	0.12
Smallest fracture radius (m)	15
Largest fracture radius (m)	250
Fracture Permeability	0.3×10^{-15}
Stress data	
Maximum horizontal stress (MPa)	78.9
Minimum horizontal stress (MPa)	53.3
Fluid properties	
Density (kg/m³)	1000
Viscosity (Pa s)	3×10^{-4}
Hydrostatic fluid pressure (MPa)	34.5
Injector pressure, stimulation (MPa)	51.7
Injector pressure, production (MPa)	44.8
Producer pressure, stimulation (MPa)	N/A

Producer pressure, production (MPa)	31.0
Other reservoir data	
Well radius (m)	0.1
Number of injection wells	1
Number of production wells	2
Reservoir depth (m)	3650

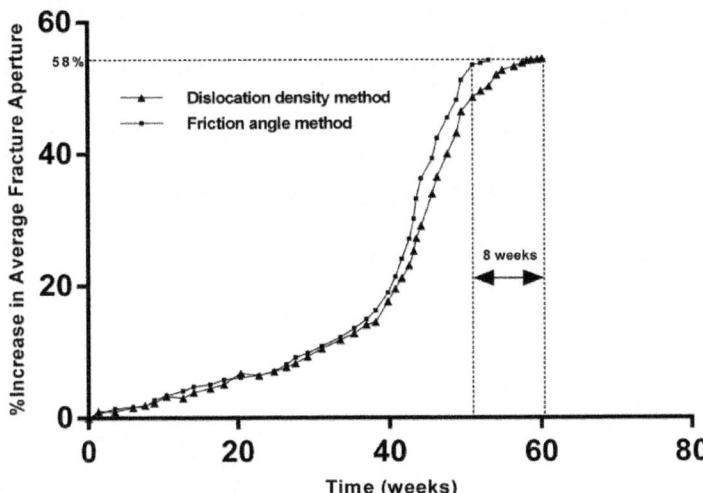

Figure 6. Comparison of Average aperture increase between the current approach and the previous study.

The locations of the dilation events during the stimulation period are shown in Fig. 7. As shown in this figure, after 40 weeks of stimulation about half of the reservoir is affected by the shear dilation and after 52 weeks of injection shear dilation happened in almost all parts of the reservoir.

Also the reservoir pressure and stress distribution profiles (see Figs. 8 and 9) show that after 40 weeks of stimulation the injected fluid pressure affected almost all of the fractures and that after 52 weeks of injection the pressure is established in all part of the reservoir domain. Similarly the x- and y component of the effective stress decreased significantly over the entire reservoir domain towards the end of the stimulation period.

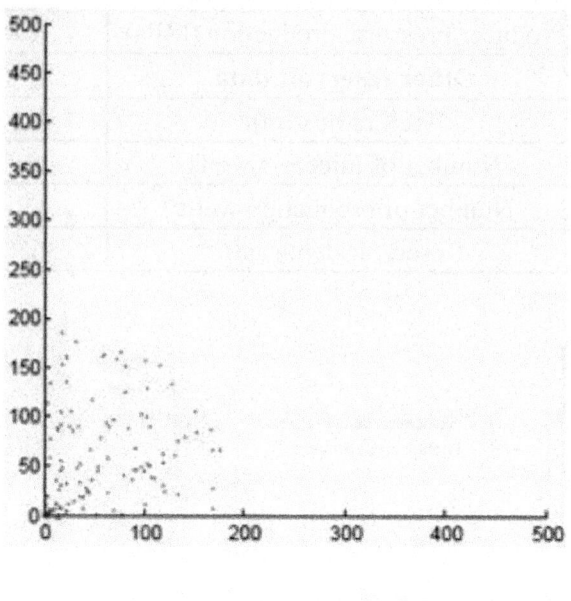

(a)

(b)

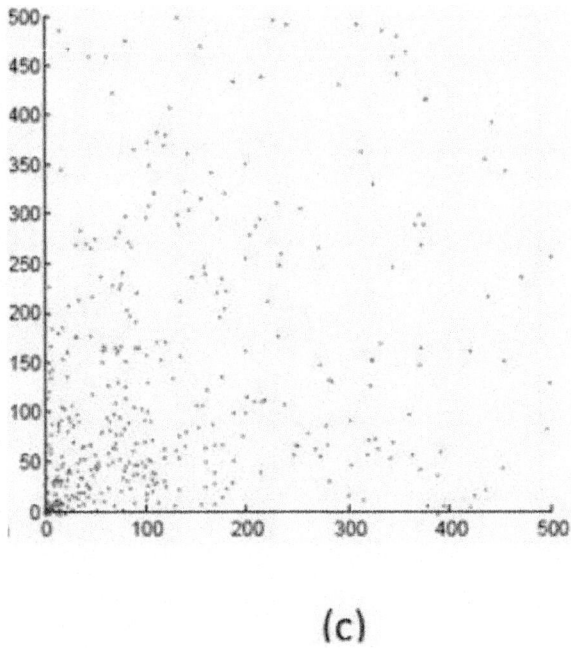

(c)

Figure 7. Location of the dilation events marked by the dots after (a) 1 week (b) 40 weeks and (c) 52 weeks of stimulation with σH = 78.9 MPa and σh = 53.3 MPa, Pinj = 51.7 MPa.

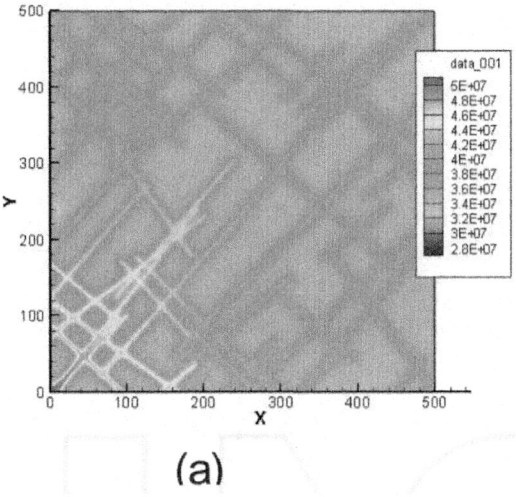

(a)

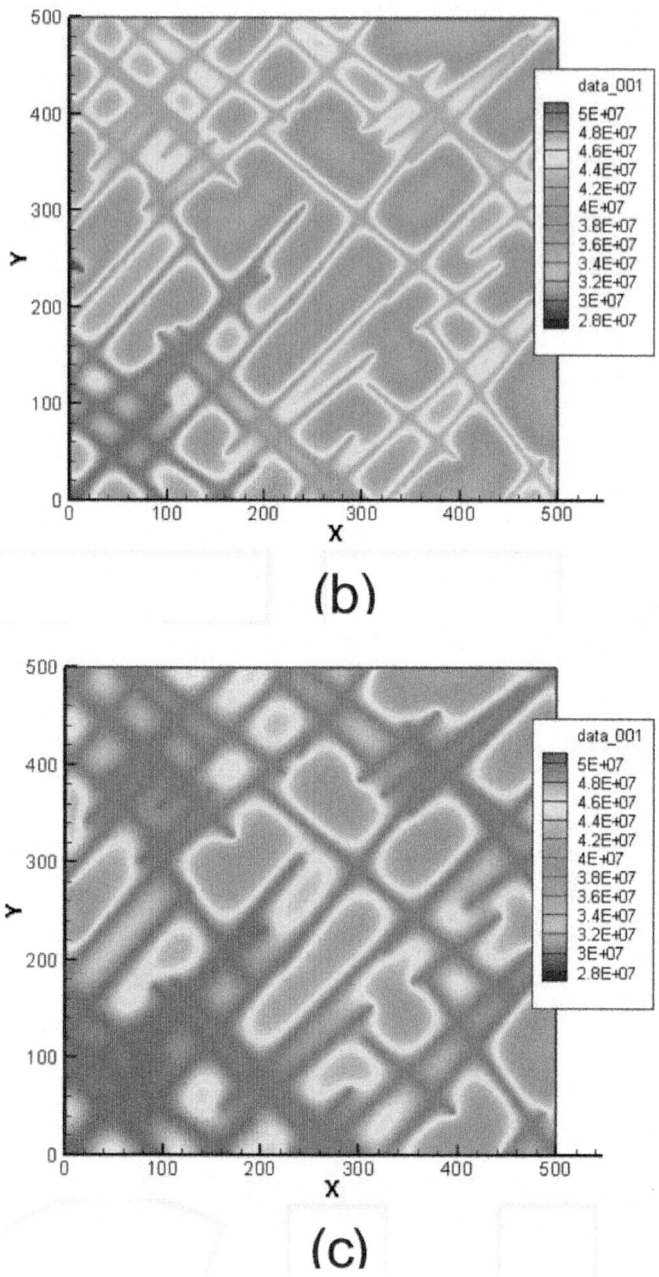

Figure 8. Pore pressure distribution of the fractured reservoir at different stimulation stages: after (a) 1 week, (b) 40 weeks and (c) 52 weeks for a strike slip stress regime with σH = 78.9 MPa and σh = 53.3 MPa, Pinj = 51.7 MPa.

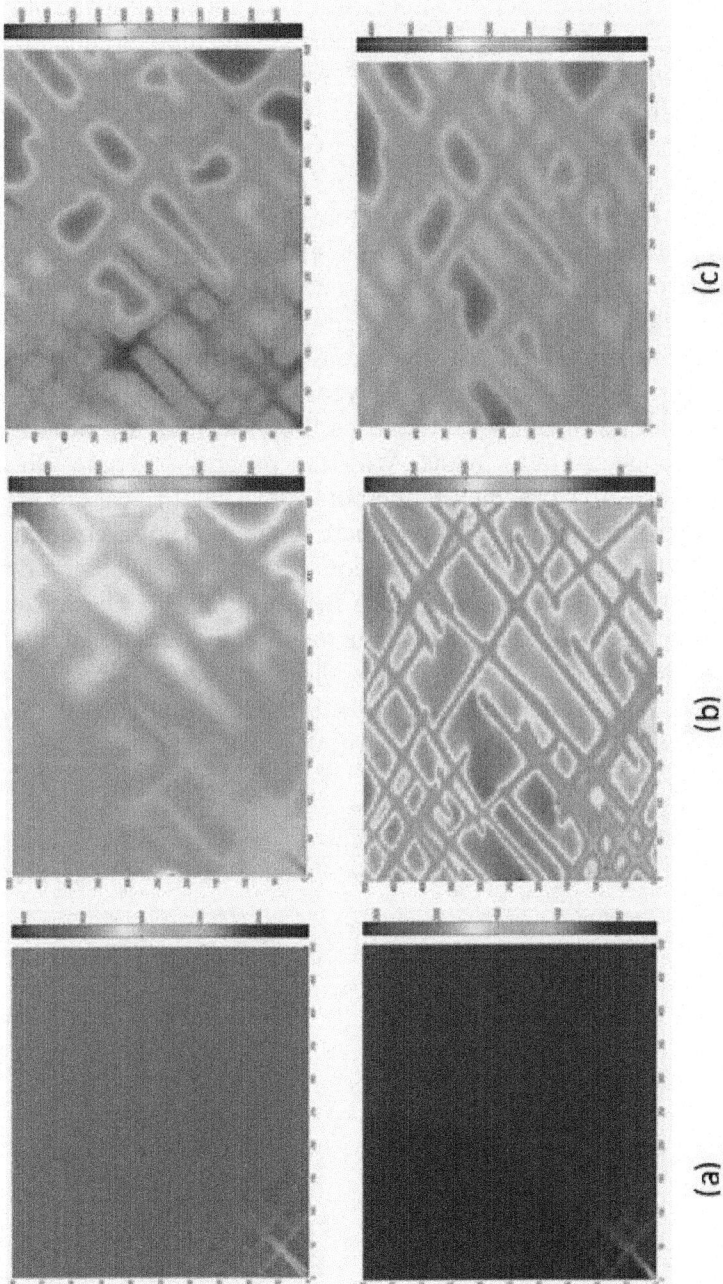

Figure 9. x (top) and y (bottom) components of effective stress after: (a) 1 week, (b) 40 weeks and (c) 52 weeks of stimulation for σH = 78. 9 MPa and σh = 53.3 MPa, P$_{inj}$ = 51.7 MPa.

After the stimulation period a numerical experiment is carried out to assess the produced matrix temperature for 14 years of cold fluid circulation. Because of the low fluid and rock matrix contact area at the early stage of production, the heat transfer and the resulting thermal drawdown is very low (seeFig 10 a). With the pass of time the fluid sweeps over a large part of the reservoir which increases thermal drawdown. At the end of the 14 years of production the average matrix temperature drops from 200 to 150°C which is quite low (drop of 50$^{\circ C}$) compared to previous studies (drop of 80°C over the production period of 14 years as in [29]) under the same reservoir conditions. Also in Fig. 10 (bottom) the Log10 RMS fluid velocity profile after 1 year, 10 years and 14 years of production are presented. From the results it can be observed that during the early production period (1 year) high pore pressure is primarily built up around the injection well and the flow of fluid is primarily through major inter-connected flow paths. With the progress of time the injection pressure advances towards the production well. After 14 years of production, the fluid sweeps through a significant part of the reservoir. Also the x- and y components of effective stress distribution of the Soultz geothermal reservoir during different stages of production are shown in Fig. 11. These results show that by the end of 14 years of production the effective stresses throughout the reservoir are significantly reduced, thus allowing most fractures to open and conduct fluid. The reduction in the effective stresses is caused by the cold circulating fluid as well as thermal drawdown.

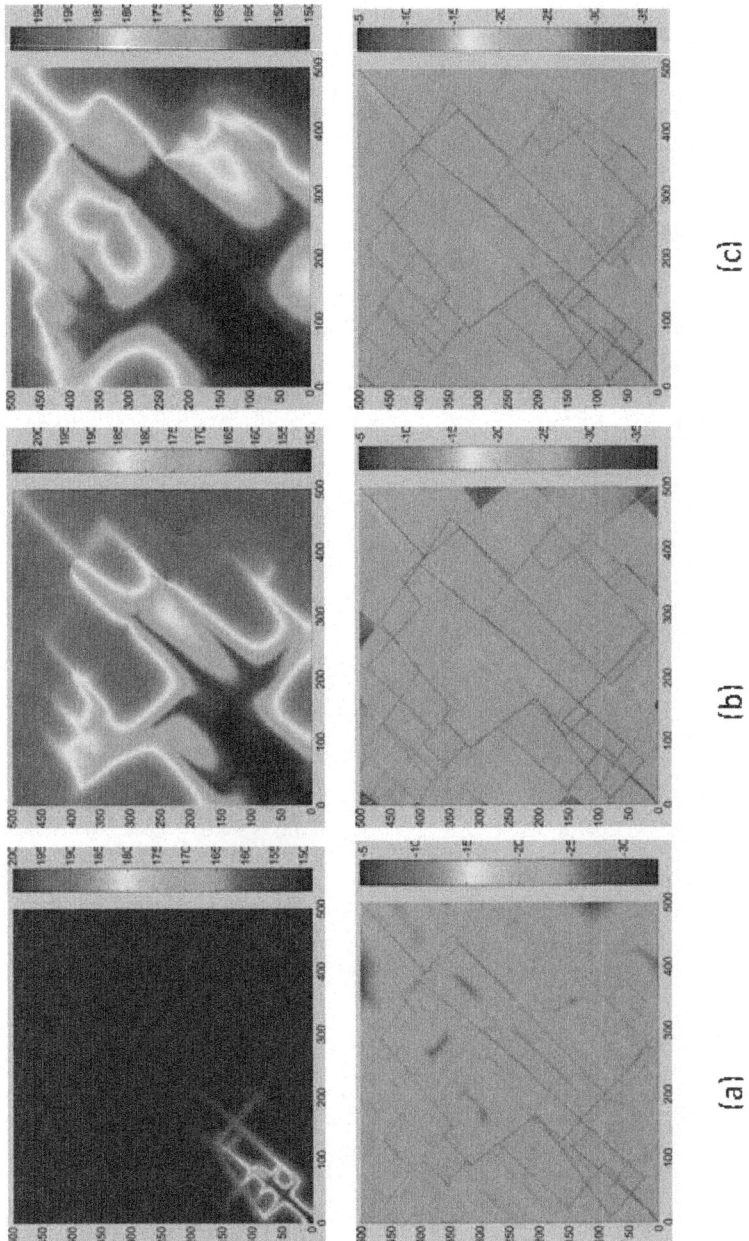

Figure 10. Reservoir temperature profile (top) and Log10RMS fluid velocity profile (bottom) after (a) 1 year (b) 10 years and (c) 14 years of production with σH = 78.9 MPa and σh = 53.3 MPa, P$_{inj}$=44.8 MPa and P$_{prod}$=31 MPa.

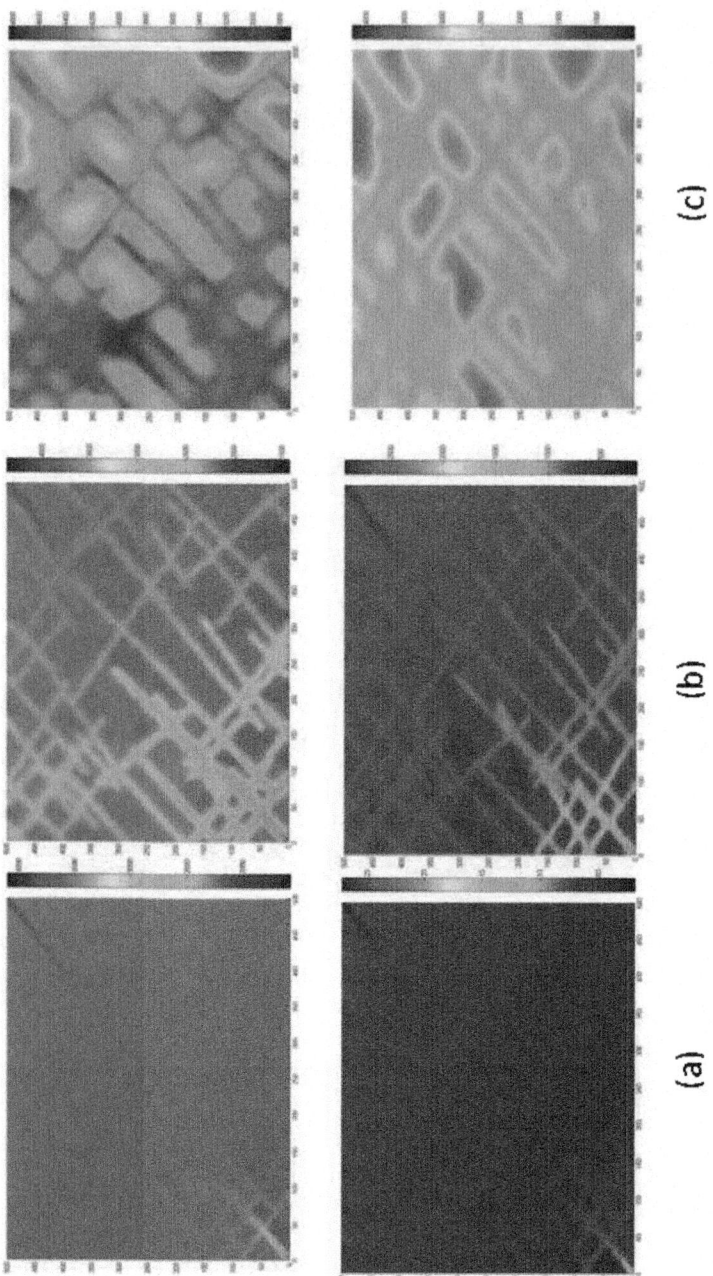

Figure 11. x (top) and y (bottom) component of effective stress after (a) 1 year (b) 10 years and (c) 14 years of production with σH = 78.9 MPa and σh = 53.3 MPa, P_{inj} =44.8 MPa and P_{prod} =31 MPa.

CONCLUSIONS

In this paper, a roughness induced shear displacement model in a poro-thermoelastic environment combined with an advanced computational technique is used to study the effects of induced fluid pressure and thermal stresses (cooling effect) on reservoir permeability and consequent increase in hot water production. It has been shown that surface roughness induced shear displacement provides a more realistic prediction of residual fracture aperture. These results agree well with the experience of existing EGS trials around the world. An average increase in aperture due to fluid induced shear dilation has been found to be lower and time required to obtain a maximum stimulated volume is greater. Results of this study are in consistent with that of previous studies: for every geothermal system there exists an optimum injection schedule (injection pressure and duration). Any further increases in stimulation effort, i.e. stimulation time for a given stimulation pressure, does not provide additional permeability enhancement.

REFERENCES

1. Roshan, H. and S.S. Rahman, Effects of Ion Advection and Thermal Convection on Pore Pressure Changes in High Permeable Chemically Active Shale Formations. Petroleum Science and Technology, 2013. 31(7): p. 727-737.

2. Lockner, D.A., J.B. Walsh, and J.D. Byerlee, Changes in seismic velocity and attenuation during deformation of granite. Journal of Geophysical Research, 1977. 82(33): p. 5374-5378.

3. Hast, N., Limits of stress measurements in the Earth's crust. Rock mechanics, 1979. 11(3): p. 143-150.

4. Solberg, P., D. Lockner, and J.D. Byerlee, Hydraulic fracturing in granite under geothermal conditions. International Journal of Rock Mechanics and Mining Sciences & Geomechanics Abstracts, 1980. 17(1): p. 25-33.

5. Rahman, M.K., M.M. Hossain, and S.S. Rahman, An analytical method for mixed-mode propagation of pressurized fractures in remotely compressed rocks. International Journal of Fracture, 2000. 103(3): p. 243-258.

6. Kotousov, A., L. Bortolan Neto, and S. Rahman, Theoretical model for roughness induced opening of cracks subjected to compression and shear loading. International Journal of Fracture, 2011. 172(1): p. 9-18.

7. Heidinger, P., J. Dornstädter, and A. Fabritius, HDR economic modelling: HDRec software. Geothermics, 2006. 35(5–6): p. 683-710.

8. Blumenthal, M., et al., Hydraulic model of the deep reservoir quantifying the multi-well tracer test.. EHDRA Scientific Conference, Soultz-sous-Forets, 2007.

9. Barton, N. and V. Choubey, The shear strength of rock joints in theory and practice. Rock mechanics, 1977. 10(1-2): p. 1-54.

10. Barton, N., S. Bandis, and K. Bakhtar, Strength, deformation and conductivity coupling of rock joints. International Journal of Rock Mechanics and Mining Sciences & Geomechanics Abstracts, 1985. 22(3): p. 121-140.

11. Bandis, S., Experimental studies of scale effects on shear strength and deformation of rock joints. PhD Thesis, 1980.

12. Piggott, A.R. and D. Elsworth, A Hydromechanical Representation of Rock Fractures, 1991, A.A. Balkema. Permission to Distribute - American Rock Mechanics Association.

13. Olsson, W.A. and S.R. Brown, Hydromechanical response of a fracture undergoing compression and shear. International Journal of Rock Mechanics and Mining Sciences & Geomechanics Abstracts, 1993. 30(7): p. 845-851.

14. Willis-Richards, J., K. Watanabe, and H. Takahashi, Progress toward a stochastic rock mechanics model of engineered geothermal systems. J. Geophys. Res., 1996. 101(B8): p. 17481-17496.

15. Rahman, M.K., M.M. Hossain, and S.S. Rahman, A shear-dilation-based model for evaluation of hydraulically stimulated naturally fractured reservoirs. International Journal for Numerical and Analytical Methods in Geomechanics, 2002. 26(5): p. 469-497.

16. Zhang, X., R.G. Jeffrey, and E. Detournay, Propagation of a hydraulic fracture parallel to a free surface. International Journal for Numerical and Analytical Methods in Geomechanics, 2005. 29(13): p. 1317-1340.

17. Gangi, A.F., Variation of whole and fractured porous rock permeability with confining pressure. International Journal of Rock Mechanics and Mining Sciences & Geomechanics Abstracts, 1978. 15(5): p. 249-257.

18. Snow, D.T., Anisotropie Permeability of Fractured Media. Water Resources Research, 1969. 5(6): p. 1273-1289.

19. Long, J.C.S., et al., Porous media equivalents for networks of discontinuous fractures. Water Resources Research, 1982. 18(3): p. 645-658.

20. Baumgartner, J., P.L. Moore, and A. Gtrard, Drilling of Hot and Fractured Granite at Soultz-sous-Forgts (France). Proceedings of the

World Geothermal Congress, Florence, Italy,International Geothermal Association,, 1995. 4: p. 2657-2663.

21. Rasmussen, T.C., J. Yeh, and D. Evans, Effect of variable fracture permeability/matrix permeability ratios on three-dimensional fractured rock hydraulic conductivity. Proceedings of the Conference on Geostatistical, Sensitivity, and Uncertainty Methods for Ground-Water Flow and Radionuclide Transport Modeling, San Francisco, California, September 1987, B. E. Buxton, Batelle Press, Columbus, OH, 1987, 337, 1987.

22. Lough, M.F., S.H. Lee, and J. Kamath, A New Method To Calculate Effective Permeability of Gridblocks Used in the Simulation of Naturally Fractured Reservoirs. SPE Reservoir Engineering, 1997. 12(3): p. 219-224.

23. Chen, M., M. Bai, and J.C. Roegiers, Permeability Tensors of Anisotropic Fracture Networks. Mathematical Geology, 1999. 31(4): p. 335-373.

24. Lee, S.H., M.F. Lough, and C.L. Jensen, Hierarchical modeling of flow in naturally fractured formations with multiple length scales. Water Resources Research, 2001. 37(3): p. 443-455.

25. Teimoori, A., et al., Effective Permeability Calculation Using Boundary Element Method in Naturally Fractured Reservoirs. Petroleum Science and Technology, 2005. 23(5-6): p. 693-709.

26. Fahad, M., S.S. Rahman, and Y. Cinar, A Numerical and Experimental Procedure to Estimate Grid Based Effective Permeability Tensor for Geothermal Reservoirs. Geothermal Resources Council Transactions, 2011.

27. Rasmussen, M.L. and F. Civan, Full, Short-, and Long-Time Analytical Solutions for Hindered Matrix-Fracture Transfer Models of Naturally Fractured Petroleum Reservoirs, in SPE Production and Operations Symposium2003, Society of Petroleum Engineers: Oklahoma City, Oklahoma.

28. Bathe, K.J., Finite element procedures. 1996: Prentice Hall.

29. Koh, J., H. Roshan, and S.S. Rahman, A numerical study on the long term thermo-poroelastic effects of cold water injection into naturally fractured geothermal reservoirs. Computers and Geotechnics, 2011. 38(5): p. 669-682.

30. Gholizadeh Doonechaly, N., S.S. Rahman, and A. Kotousov, An Innovative Stimulation Technology for Permeability Enhancement in Enhanced Geothermal System--Fully Coupled Thermo-Poroelastic

Numerical Approach. 36th Geothermal Resources Council Transactions, 2012.

31. Gholizadeh Doonechaly, N. and S.S. Rahman, 3D hybrid tectono-stochastic modeling of naturally fractured reservoir: Application of finite element method and stochastic simulation technique. Tectonophysics, 2012. 541–543(0): p. 43-56.

32. Genter, A., et al., Contribution of the exploration of deep crystalline fractured reservoir of Soultz to the knowledge of enhanced geothermal systems (EGS). Comptes Rendus Geoscience, 2010. 342(7–8): p. 502-516.

Chapter 6

RESERVOIR PROCESSES RELATED TO EXPLOITATION IN LOS AZUFRES (MÉXICO) GEOTHERMAL FIELD INDICATED BY GEOCHEMICAL AND PRODUCTION MONITORING DATA

Víctor Manuel Arellano[1], Rosa María Barragán[1], Miguel Ramírez[2], Siomara López[1], Alfonso Aragón[1], Adriana Paredes[1], Emigdio Casimiro[3], and Lisette Reyes[3]

[1]Instituto de Investigaciones Eléctricas, Gerencia de Geotermia, Cuernavaca, México

[2]Comisión Federal de Electricidad, Gerencia de Proyectos Geotermoeléctricos, Morelia, México

[3]Comisión Federal de Electricidad, Residencia Los Azufres, Campamento Agua Fría, México

ABSTRACT

A combined analysis of geochemical and production data of 39 wells of the Los Azufres (Mexico) geothermal field (227.4 MWe) over time was developed to investigate the exploitation-related processes for 2003-2011. In the south zone, important effects of reinjection were observed through Cl increases in some wells (up to 8000 mg/kg) while in wells with significant boiling, Cl has decreased. In most of the north zone wells, the variations in gas data indicated boiling and condensation of a highly gas-depleted brine, which seems to consist of reinjection fluids. It is suggested that this process maintains the production in the zone relatively stable. The main reservoir exploitation-related processes found were: 1) production of reinjection returns; for this, it was possible to distinguish a) wells that produce liquid and steam from injection, and b) wells that produce steam from injection and sometimes condensed steam from injection; 2) boiling: two types of boiling were identified: a) boiling with steam gain, and b) boiling with steam loss. The results indicated that an effective reservoir recharge occurs since very moderate production declining rates were found.

INTRODUCTION

The Los Azufres geothermal field is an intensely-fractured, two-phase, volcanic hydrothermal system located in the northern portion of the Mexican Volcanic Belt, in the state of Michoacán, at an average elevation of 2800 m.a.s.l. (Figure 1). At present, the installed capacity of the field is 227.4 MWe [1]. Based on the occurrence of natural manifestations, the field was divided into two zones: north and south, although both zones are supplied by the same deep aquifer. Reservoir engineering and geochemical conceptual models of the Los Azufres system were established [2] [3] which constitute the reference conditions that allow characteristic changes in parameters over time to be related to exploitation. The reservoir's response to exploitation for 1982-2002 was studied based on a systematic analysis of chemical, isotopic, and production data from 20 production wells [4]. The installed capacity of the field was 88 MWe during that time. Subsequently, since 2003, four additional 25 MWe flash plants were brought on line, bringing the total installed capacity to 188 MWe which operated till first months of 2015. At present, (August 2015) the capacity has increased to 227.4 MWe and then higher rates in fluids extraction are required to supply units. In order to support decisions on field polices regarding sustainable exploitation, the objective of this work was to investigate the exploitation-related processes through the analysis of geochemical and production data of 39 wells. The Los Azufres geothermal field is operated by the Comisión Federal de Electricidad (CFE). In Figure 1 the locations of the field and the wells are given. All of the injection wells are located on the west side of the field. Injectors Az-3, 15, 52, and 61 are located in the northern production area, whereas Az-7A and 8 are in the south.

METHODOLOGY

In order to obtain the thermodynamic characteristics of reservoir fluids and to investigate the dominant processes occurring because of exploitation, a method based on the analysis of the following patterns of behavior of production and geochemical indicators over time was used [4]:

- Well mass flow-rates, well-bottom pressures, enthalpies, and temperatures. Well-bottom thermodynamic conditions (pressures, temperatures, enthalpies and quality of steam) were obtained through WELFLO simulator [5]. Input data consisted of wellhead production data and well geometry. WELFLO is a simulator of the mass and energy flows in the well that considers multiphase, one-dimension and steady state conditions and is suitable for vertical wells with variable diameter. Input data consists of geometry of the well, mass flow rates, and wellhead pressures and enthalpies.

- In order to identify reservoir processes, a method based on the comparison of the total discharge enthalpy (H_{TD}) with other enthalpy estimations obtained through different approaches [6], was used. Thus, besides the reservoir temperatures given by well simulation, the following temperature estimations were obtained.

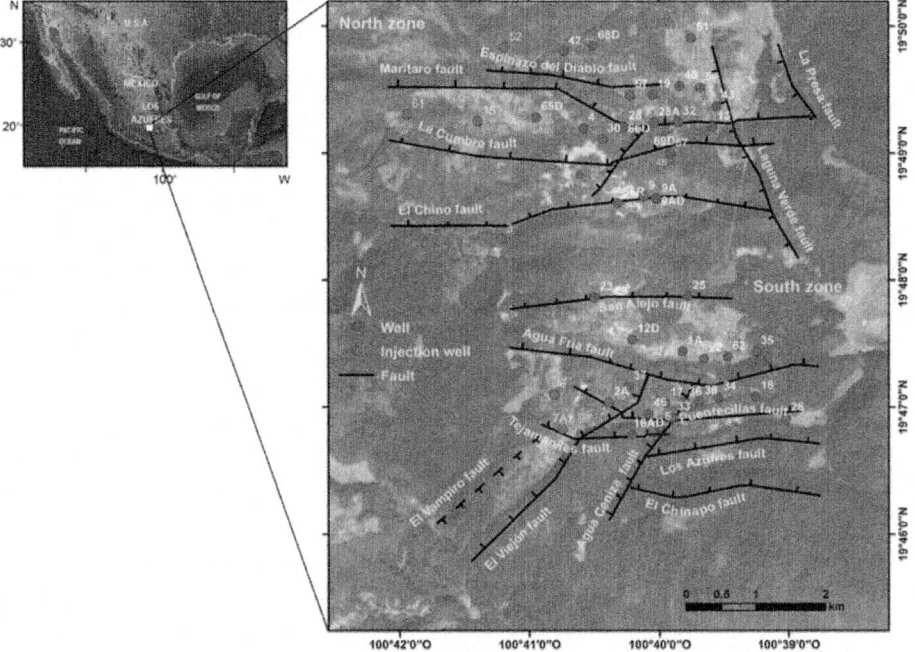

Figure 1. Location of the Los Azufres geothermal field and locations of wells.

For two-phase wells the cationic Na/K [7] and the silica [8] geothermometers were estimated while for the steam wells, the FT-HSH2 gas geothermometer [9] [10] was used. From the cationic and silica reservoir temperatures, the enthalpies ($H_{Na/K}$, H_{Si}) were interpolated from the steam tables considering the occurrence of single saturated liquid in the reservoir. Simulation temperatures reflect actual temperatures at the entrance of wells, they usually compare fairly with gas and silica temperatures. Currently, cation geothermometers such as the Na/K and Na-K- Ca are considered "slow response" because they are based on the kinetics of the water-rock interactions which re-equilibrates relatively slowly and keep the memory of previously conditions. When considering "slow response" cation geothermometers one of the exceptions is the K/Mg which re-equilibrates fast. In contrast, silica geothermometers are considered "fast response" because they are based on silica solubility, which depends on temperature. The general guidelines to interpret the enthalpies comparison tendencies are as follows.

- Fraction of steam entering the well. This parameter can be obtained either from simulation of wells or can be estimated through gas equilibria methods [3] [9] [11]. In this work both methods provided similar results.

- Chloride concentrations in total discharge and in separated water. The total discharge fluid is defined as that ascending from the well to the wellhead before steam separation due to de-pressurization. To calculate the total discharge concentrations of solutes, first, the steam removed (y) during separation is obtained through an enthalpy (H) balance, according to:

$$H_{TD} = y \cdot H_{S,ST} + (1-y) \cdot H_{L,ST} \tag{1}$$

where y is the steam fraction and subscripts TD, S, L and ST stand for total discharge, steam, liquid and separation temperature. From (1):

$$y = \left(H_{TD} - H_{L,ST}\right) / \left(H_{S,ST} - H_{L,ST}\right) \tag{2}$$

Subsequently the total discharge Cl (as well as any other chemical species) concentration is obtained through a mass balance:

$$Cl_{TD} = y \cdot Cl_{S,ST} + (1-y) \cdot Cl_{L,ST} \tag{3}$$

- Total discharge and reservoir CO_2 concentrations. These concentrations were obtained through the SCE- XVAP program [3] [12] with steam data as input.

- Reservoir volumetric liquid saturation (S_L). This parameter is closely related to the production of the wells and was estimated according to the following expression [13]:

$$S_L = (1-y) \cdot V_L / \left\{ (1-y) \cdot V_L + y \cdot V_V \right\} \tag{4}$$

where V_L and V_V are the specific volumes of liquid and steam at reservoir temperature respectively; y is the fraction of steam entering the well.

- $\delta^{18}O$ and δD composition of total discharge fluids. This composition is calculated by using equations 2 and 3 and considering the partition coefficients, which depend on temperature, for every separation step that had occurred previous to sample collection [14].

Chemical, isotopic and production monitoring data were provided by the Residencia Los Azufres, CFE. The study included 39 wells (Figure 1): Az-4,

5, 9, 9A, 9AD, 13, 19, 28, 28A, 30, 32, 41, 43, 45, 48, 51, 56R, 65D, 66D, 67, and 69D from the north and Az-1A, 2A, 6, 16AD, 17, 18, 22, 23, 25, 26, 33, 34, 35, 36, 37, 38, 46, and 62 from the south zone of the field.

FLUID PRODUCTION/INJECTION

Fluid production and injection rates have varied depending on the operation of the generating units over time. Figure 2 shows the production and injection data as a function of time for the field. Fluid extraction started in 1978 (Figure 2) and by December 2011 an amount of 441,744,661 tons of fluids were produced. Of these, 254,376,326 tons (57.6%) were from the south, and 187,368,335 tons (42.4%) from the north. Between 1982 and 2011, an amount of 132,016,271 tons (30%) of produced fluids had been re-injected to the reservoir. Production increased in the south zone in 1987, when the 50 MW power plant came on line [4]. In 2003, fluid production in the north zone increased due to installation of the 100 MW eadditional capacity.

RESULTS

The results obtained allowed identification of the main processes that have occurred and are occurring in different zones of the Los Azufres reservoir. Two main processes were identified as follows. 1) Production of reinjection returns. According to the type of fluid arriving to the well, two sub-processes can occur: a) wells that receive liquid and steam from reinjection, and b) wells that receive steam originated from the boiling of reinjection fluids and/or wells that receive condensed steam originated from boiling of reinjection fluids. 2) Boiling. Depending on the boiling rate (increasing or decreasing rates), boiling can be classified as: a) boiling with steam gain or increasing boiling rate and b) boiling with steam loss.

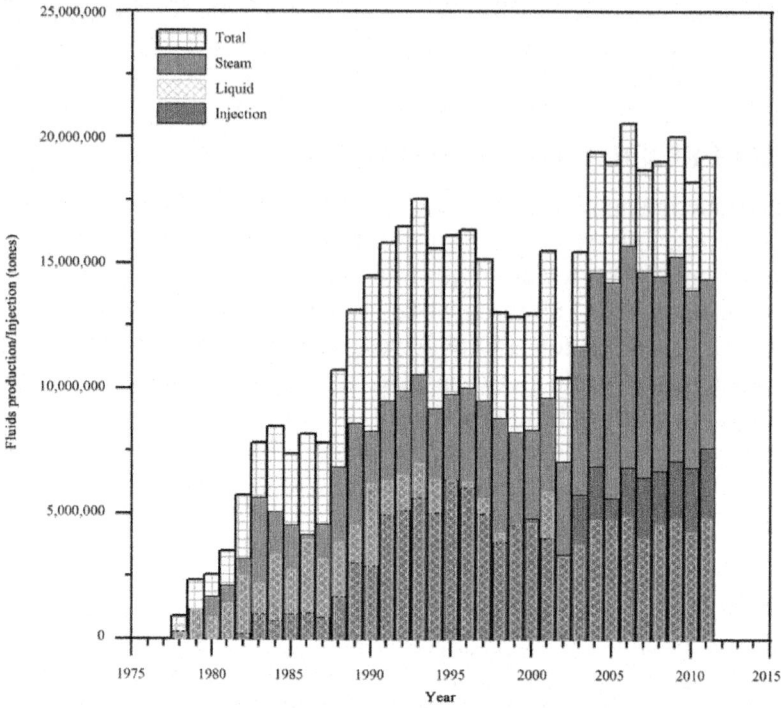

Figure 2. Production and reinjection fluids in Los Azufres geothermal field.

Reservoir Processes

Production of Reinjection Returns (IR's)

In order to illustrate this process, the well Az-33 was selected. This well is 683 m deep, with 81 m of slotted liner that receives (IR's) in liquid phase. In Figure 3 time series of mass flow rates, well bottom pressure, well bottom enthalpy, chlorides, wellhead pressure, comparison of total discharge enthalpy and enthalpy estimations by geothermometers, CO_2, and reservoir liquid saturation for the well Az-33 are given. As seen in Figure 3(A) large variations in mass flow rates between 1988 and 2003 are seen, which are due mostly to variations in water production. Since 2004 the production of well showed a decreasing trend over time while up to middle 2007 the well became dried due to the shutting out of one of the injectors, the well Az-7R. Since then, the well produced small amounts of liquid and the decreasing tendency changed to a more stable production up to 2011 and even an increase in production was noticed during 2008. However it appears that during 2009 and 2010 the liquid production vanished; (B) as in the cases where the wells receive IR's, the well bottom pressure changes abruptly due to the recharge entry, since 2004 the pressure decreases with the same tendency found for mass flow rates and (C) the well bottom enthalpy increases and decreases depending on the entry of liquid recharge; (D) the total discharge chlorides increase and (G) the total discharge CO_2 increases. The pattern for the enthalpies comparison in (F) shows mixing of equilibrated liquid with steam produced by boiling away the well, it is also noticed that between 1996 to 2003 the Na/K enthalpy is slightly lower than the silica enthalpy which is characteristic of the entry of reinjection water. In Figure 4(A) it is seen that the well Az-33 has received reinjection fluids some years through the increase in chlorides and, as the injection rates are not constant over time, the behavior of the well shows intermittent changes in Figures 3(A)-(H) and Figure 4(A) and Figure 4(B). In Figure 4(A) the enthalpy-chloride data of well Az-33 are found to the right hand side of the characteristic boiling line which constitutes the reference for the field [4]. This is due to the chloride enrichment of the fluids discharged because of production of reinjection returns. Chloride and other solutes are highly concentrated in injection fluids since they are evaporated at ambient conditions before injection.

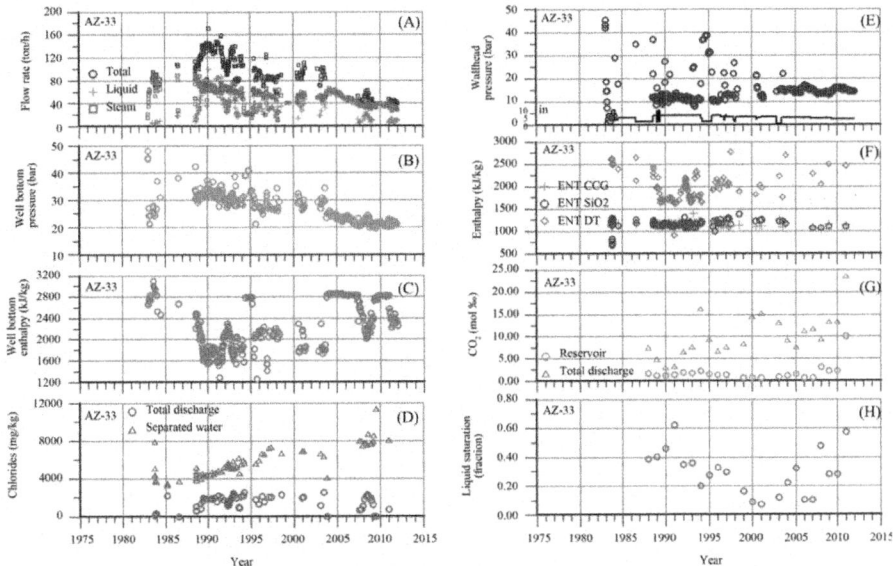

Figure 3. Time series of (A) mass flow rates, (B) well bottom pressure, (C) well bottom enthalpy, (D) chlorides, (E) wellhead pressure, (F) comparison of total discharge enthalpy and enthalpy estimations by geothermometers, (G) CO_2, and (H) reservoir liquid saturation of well Az-33.

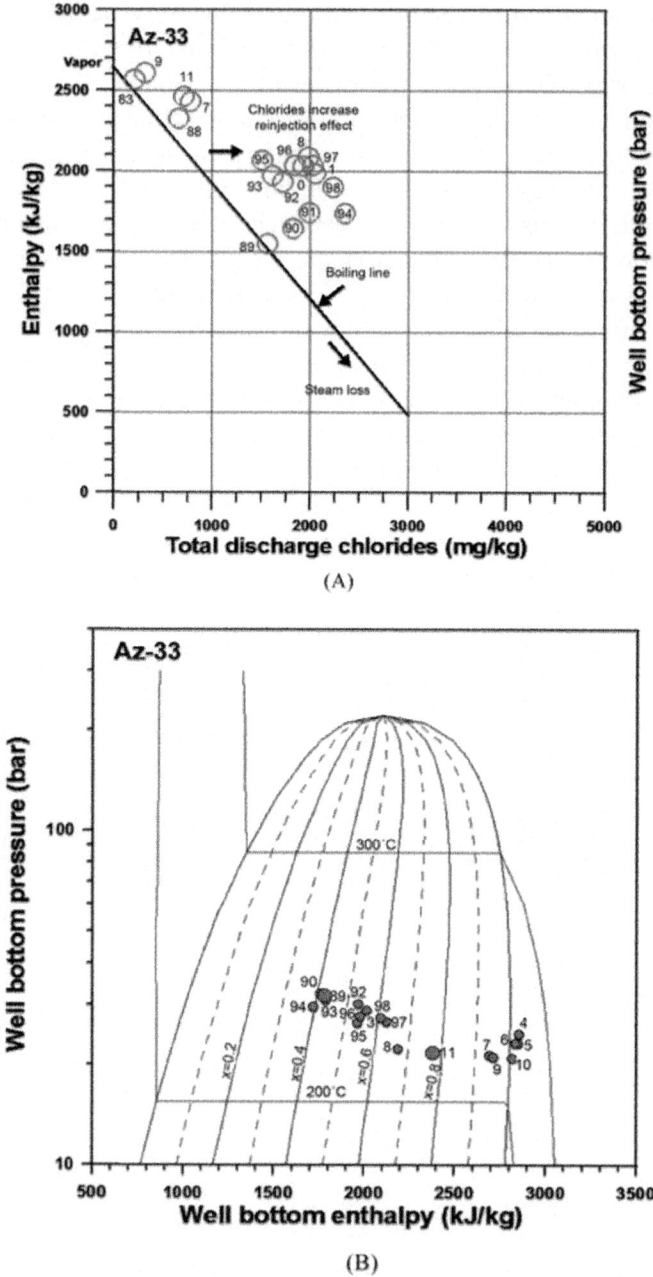

Figure 4. (A) Enthalpy versus chlorides and (B) well bottom pressure versus well bottom enthalpy for well Az-33.

The well Az-33 as other wells, shows important recovery in liquid saturation since 2004-2005 (Figure 3(H)), this can be related to the operation of the injection well Az-7AR which replaced the original injection well Az-7R. As was mentioned, the well Az-33 produced two-phase fluids but became steam producer when the injector well Az-7R was closed in 2003. The well Az-33 started producing liquid intermittently in 2007 (Figure 3(A)). Results from tracer tests [15] indicate that part of the fluids injected in well Az-8R are recovered as both liquid and steam in well Az-33. However the well Az-33 develops a moderate and sometimes important boiling process depending on reinjection rates, which can be seen in Figure 3(C) andFigure 3(F) through the enthalpy variations.

The production of injection returns is shown only by some wells of the south zone such as Az-1A, 2, 2A, 16D, 36 and 46, where injection plays an important role. From these, the wells Az-36 and 46 became dried when the well Az-7R was shut off. The well Az-7R was replaced by the well 7AR, which seemed not to provide enough recharge to the wells Az-33, 36 and 46 to change their flow regime to two-phase. In wells that receive reinjection returns in liquid phase, chlorides concentrations in separated water are as high as 5500 - 8000 (mg/kg) contrasting with initial concentrations which were recorded as between 2500 - 4000 (mg/kg).

The production of steam from the boiling of reinjection fluids and sometimes condensed steam from the boiling of injection fluids, was also identified to occur in some wells of the field. When injection fluids consist of recycled produced fluids, the effects of reinjection in two-phase production wells are routinely noticed on the salts increase in the discharged liquid [4] [14] [16] -[18] as was illustrated for well Az-33. However, injected fluids in contact with reservoir rocks are heated and eventually evaporated by boiling, then either as steam or as condensed steam they flow to the production zones of wells. These types of reinjection returns could not be recognized in production wells by the increase in the salinity of the liquid discharged, since condensed steam is both salts and isotope-depleted. Thus, in two-phase wells that produce evaporated or condensed reinjection returns, a decrease in salinity is observed. In order to recognize the presence of injection returns consisting of steam or condensed steam gas and isotope data should be analyzed. Basically, when production of steam from reinjection occurs, the following characteristics are seen in the produced fluids: 1) N_2/Ar molar ratios lower than that for air saturated water (38); 2) low gas and CO_2 (<5‰) concentrations and 3) relative δD and $\delta^{18}O$ depletion.

In the south zone the production of evaporated or condensed reinjection returns is seenin wells Az-1A, 23 and 25 while in the wells Az-22, 35 and

62, the production is intermittent depending on reinjection rates. In contrast, mainly after 2005, in the north zone a number of wells produce steam or condensed steam from the boiling of reinjection fluids: Az-4, 9, 19, 28, 28A, 45, 51, 57, 66D, 67 and 69D; while wells Az-5, 13, 32, 42 and 48 pro- duce such returns in an intermittent way.

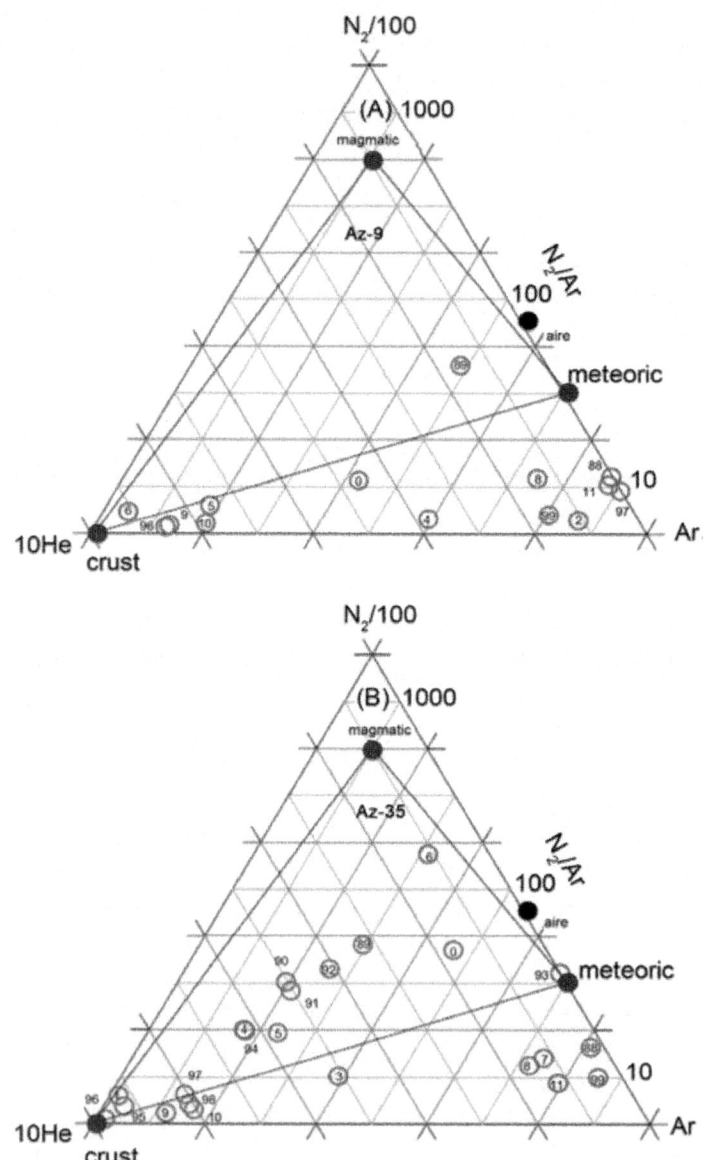

Figure 5. The N_2-He-Ar ternary diagrams for (A) well Az-9 (two-phase) and (B) Az-35 (steam) [21].

In Figure 5 the N_2-He-Ar ternary diagrams for (A) well Az-9 (two-phase) and (B) Az-35 (steam) are given, as it is seen both wells receive evaporated or condensed injection when data plot below the He corner-meteoric line. Well Az-9 produces mostly evaporated or condensed reinjection fluids (Figure 5(A)), which explains the production of slightly lighter fluids in 2010 in Figure 6(A) [18] ($\delta^{18}O$ of −3.5‰, δD of −63.5‰) considering the isotopic composition taken as reference ($\delta^{18}O$ of −2.4‰, δD of −61‰) [3]. The well Az-28 has also become isotopically lighter regarding its reference composition ($\delta^{18}O$ of −2.51‰, δD of −61.8‰) because of the production of condensed reinjection fluids. This is seen in Figure 6(A) where 2010 point is very close to the lighter end of the fitting line [19]. In contrast, according to Figure 5(B) the steam well Az-35 (south zone) in 2010 and 2011, as intermittently occurs, received steam from reinjection. This causes the deuterium enrichment of well Az-35 regarding the fitting line in Figure 6(B), considering that deuterium partitions preferably in steam at temperatures above 220°C [20]. In Figure 6(B) the isotopic compositions of two-phase wells Az-1A, 23 and 25, which produce condensed reinjection fluids, are the more depleted in the south zone because of this effect.

Boiling

The boiling processes in wells are a natural response to exploitation, due to pressure drops induced by fluids extraction. Depending on rates of boiling (increase/decrease) in wells this process has two sub-processes as follows. 1) Boiling with steam gain (increasing boiling rate) and 2) boiling with steam loss (decreasing boiling rate).

The boiling with steam gain process was noticed in some wells of Los Azufres such as Az-5, 13 and 28 from the north zone and Az-18 and 26 from the south zone, among others [4] [21]. From these, the Az-5, 13 and 18 became dried after being first two-phase producers. In order to illustrate the case of boiling with gaining steam the well Az-28 was selected. In Figure 7 the behavior of (A) production mass flow rates, (B) well bottom pressure and (C) enthalpy, (D) Cl, (E) wellhead pressure, (F) enthalpies comparison patterns (G) CO_2 and (H) volumetric liquid saturation over time, is shown. As seen, the liquid production in well Az-28 increased from 1991 to 1993 but after 1995 it decreased to minimum values at the end of 2003. The well became dried from 2006 to 2010 and then started producing liquid again. The steam production showed an increasing tendency from 1991 to the end of 2004, after that and up to 2011 the steam production is rather stable with a slight declining tendency. These variations in production sometimes are due to changes in operating conditions controlled by the orifice plate (E). On average, the pressure declining was estimated as 0.7 bar/year and the enthalpy increased 38.7 (kJ/kg)/year; Cl in total discharge decreased 58 (mg/kg)/year while CO_2 increased 0.03 (‰mol)/year. The enthalpies comparison patterns (F) indicates the mixture of equilibrated liquid with steam formed by boiling far from the well which is due to the development of an important boiling process in this zone of the field. This is shown in Figure 8(B) through the evolution of the thermodynamic conditions that shows production of saturated or even superheated steam during 2006-2010. The 2011 data on this diagram shows that enthalpy has decreased and the two-phase mixture entering the well has a steam fraction of 0.75.

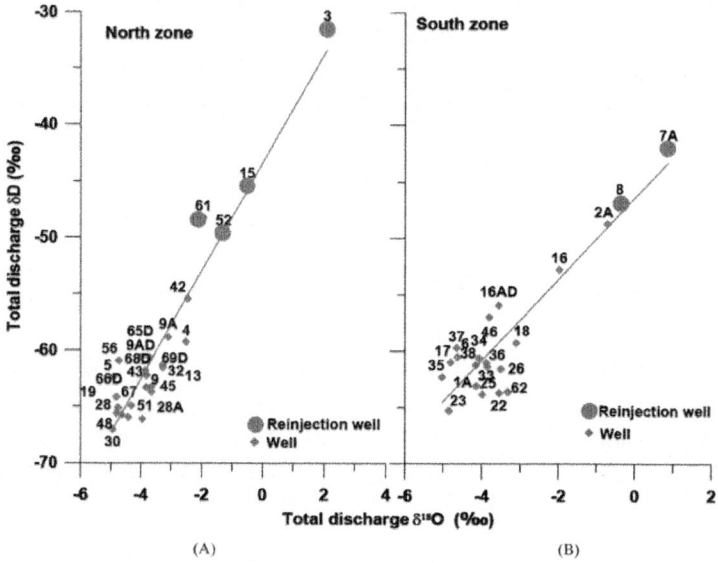

Figure 6. δD versus $δ^{18}O$ compositions of (A) North zone and (B) South zone wells according to 2010 data [18].

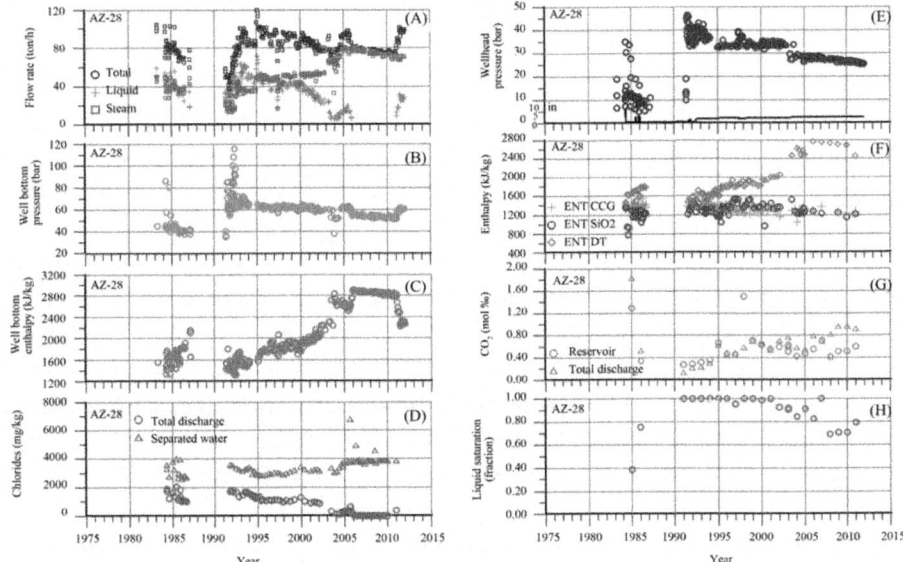

Figure 7. Time series of (A) mass flow rates, (B) well bottom pressure, (C) well bottom enthalpy, (D) chlorides, (E) wellhead pressure, (F) comparison of total discharge enthalpy and enthalpy estimations by geothermometers, (G) CO_2, and (H) reservoir liquid saturation of well Az-28.

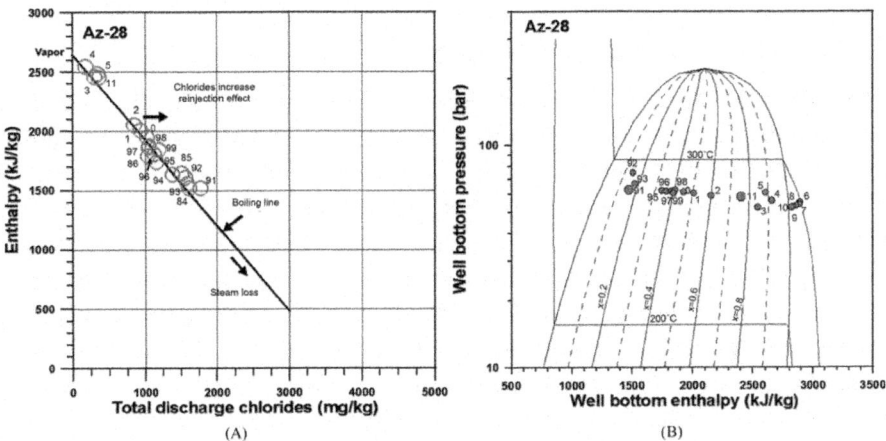

Figure 8. (A) Enthalpy versus chlorides and (B) well bottom pressure versus well bottom enthalpy for well Az-28.

In Figure 8(A) it is seen that this well plots on the characteristic boiling line of the field and no Cl increases are seen related to production of production of returns of reinjection consisting of liquid. However the N_2/Ar ratio is lower than that for air-saturated water and indicates the production of de-gasified fluid while a tracer test [22] concluded connectivity between well Az-28 and the reinjection well Az-15R.

The boiling with steam loss process consists in the decreasing of boiling rate in a well and was identified to occur in very few wells of the field, such as Az-23 and 28A [21] and for short periods of time, in some other wells. Considering the well Az-23, in Figure 9(A) it is seen a decreasing trend for the enthalpy between 2007 and 2010, just above the boiling line, while chloride increases. Also in Figure 9(B) the well Az-23 shows that the well bottom enthalpy decreased between 2006 and 2010. In Figure 10, time series of (A) mass flow rates produced; (B) the total discharge enthalpy and estimated silica and Na/K enthalpies; (C) CO_2 and (D) liquid saturation at reservoir for well Az-23 are given. As seen in Figure 10(A) the mass flow rate shows variations with a decreasing trend since 2008 up to 2010 while it stabilizes in 2011. The patterns of enthalpies comparison provides $H_{TD} = H_{Na/K} > H_{Si}$ indicating the mixing (close to the well) of equilibrated liquid with less-temperature more diluted water probably related to the entry of condensed reinjection returns. This process is the responsible of the decreasing enthalpy trend. As the fluid entering the well comes from reinjection it is CO_2-depleted regarding reservoir fluids, thus CO_2 shows a decreasing trend over time in Figure 10(C). Well Az-23 has a very high liquid saturation as seen in Figure 10(D).

2005-2011 Reservoir Temperatures

The occurrence of the described reservoir exploitation-related processes has changed the distributions of temperature through time as follows. In Figure 11 the temperature distributions of the Los Azufres reservoir for 2005 and 2011 are given. The more important changes include temperature variations in the north zone, where with respect to 2005, the 300°C contour is located toward the east. In 2011 the temperature iso-contours are aligned in an approximately N-S direction in the center of the north zone with a decreasing trend toward the west, where injection wells are located. In 2011 the 290°C contour is lying along both zones of the field while minimum temperatures are located in the west of the south zone where injection wells are located.

CONCLUSIONS

The analysis of chemical and production data for the Los Azufres geothermal field allowed the main reservoir processes related to exploitation to be identified. According to the results, after 27 years of commercial exploitation of the field, very moderate declining in production parameters was estimated which is due to an efficient artificial and natural recharge to the reservoir.

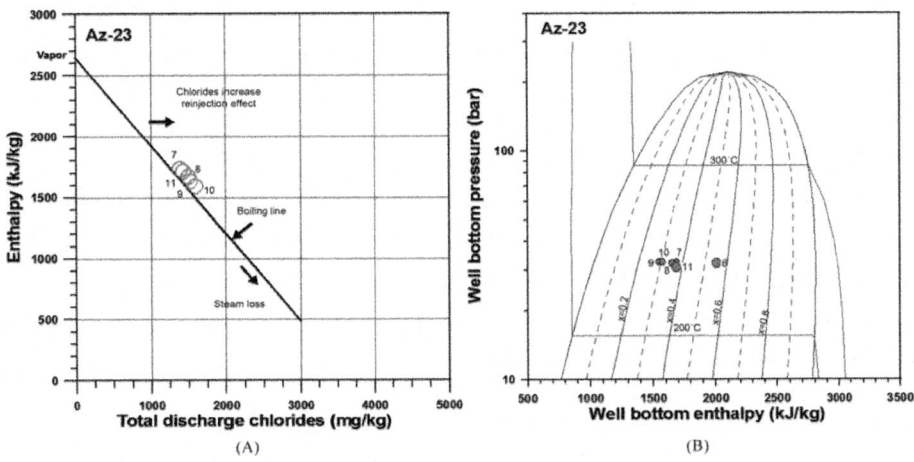

Figure 9. (A) Enthalpy versus chlorides and (B) well bottom pressure versus well bottom enthalpy for well Az-23.

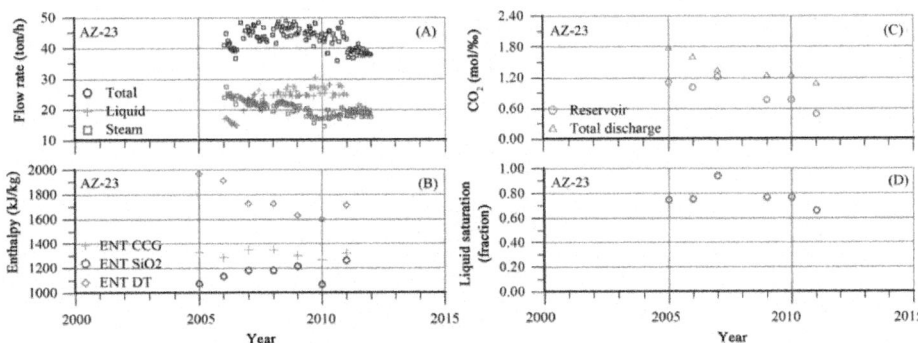

Figure 10. Time series of (A) mass flow rates, (B) comparison of total discharge enthalpy and enthalpy estimations by geothermometers, (C) CO_2, and (D) reservoir liquid saturation of well Az-23.

The extraction and reinjection of fluids from and to the reservoir have induced the occurrence of mainly two physical processes: 1) the production of returns from reinjection and 2) boiling. The more important process was the production of reinjection returns either as liquid or steam, and the production of steam and sometimes condensed steam from boiling of injection fluids. Then, for the first time it is reported that when returns from injection consist of condensed steam, salinity decreases in two-phase wells and isotopic compositions become relatively-depleted. In order to recognize the production of such reinjection returns, the analysis of gas data in particular the N_2/Ar molar ratio is necessary. The other process induced by exploitation was boiling, which can take place either with gaining steam, in which rates of boiling over time tend to increase, or with steam loss, in which the boiling rates over time tend to decrease.

It is recommended to continue the monitoring of geochemical and production data to follow the evolution of the reservoir due to exploitation and support decisions focused on extending the resource lifetime.

ACKNOWLEDGEMENTS

Authors acknowledge the authorities of Gerencia de Proyectos Geotermoeléctricos (Comisión Federal de

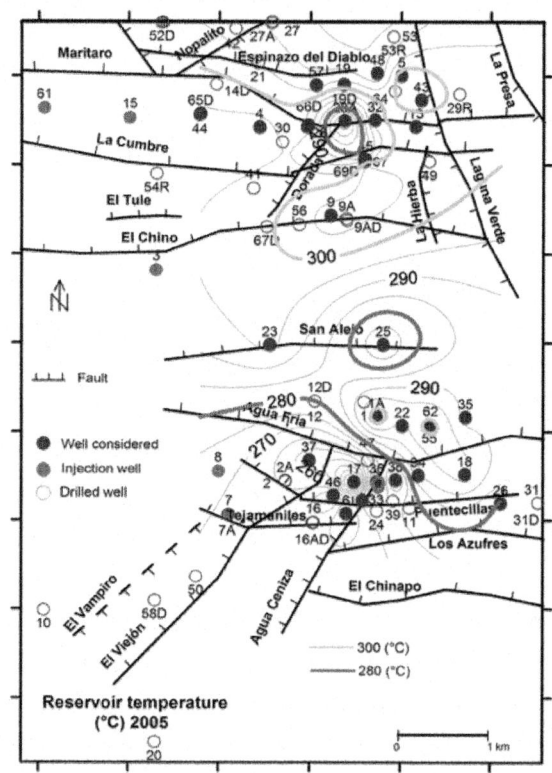

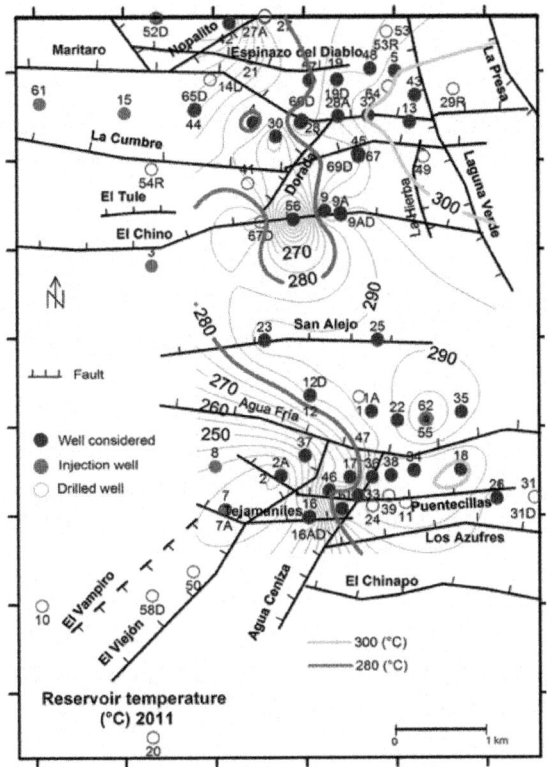

Figure 11. 2005 and 2011 distributions of reservoir temperature in the Los Azufres geothermal field [21].

Electricidad, CFE) for authorizing this publication. Recognition is given to the technical staff of the Residencia Los Azufres (CFE) who provided support for this work.

REFERENCES

1.   Gutiérrez-Negrín, L. (2015) Mexico: Update of the Country Update, and IRENA's Remap. IGA News, 100, 17-19.

2.   Iglesias, E.R., Arellano, V.M., Garfias, A., Miranda, C. and Aragón, A. (1985) A One-Dimensional Vertical Model of the Los Azufres, México, Geothermal Reservoir in its Natural State. Geothermal Resources Council Transactions, 9, 331-336.

3.   Nieva, D., Verma, M., Santoyo, E., Barragán, R.M., Portugal, E., Ortíz, J. and Quijano, L. (1987) Chemical and Isotopic Evidence of Steam Upflow and Partial Condensation in Los Azufres Reservoir. Proceedings of 12th

Workshop on Geothermal Reservoir Engineering Stanford University, Stanford, CA, 20-22 January 1987, 253-259.

4.   Arellano, V.M., Torres, M.A. and Barragán, R.M. (2005) Thermodynamic Evolution of the Los Azufres, Mexico, Geothermal Reservoir from 1982 to 2002. Geothermics, 34, 592-616. http://dx.doi.org/10.1016/j.geothermics.2005.06.002

5.   Goyal, K.P., Miller, C.W. and Lippmann, M.J. (1980) Effect of Measured Wellhead Parameters and Well Scaling on the Computed Downhole Conditions in Cerro Prieto Wells. Proceedings of 6th Workshop on Geothermal Reservoir Engineering Stanford University, Stanford, CA, 16-18 December 1980, 130-138.

6.   Truesdell, A.H, Lippmann, M., Quijano, J.L. and D'Amore, F. (1995) Chemical and Physical Indicators of Reservoir Processes in Exploited High-Temperature, Liquid-Dominated Geothermal Fields. Proceedings of World Geothermal Congress, Florence, Italy, 18-31 May 1995, 1933-1938.

7.   Nieva, D. and Nieva, R. (1987) Developments in Geothermal Energy in Mexico-Part Twelve: A Cationic Geothermometer for Prospecting of Geothermal Resources. Heat Recovery Systems & CHP, 7, 243-258. http://dx.doi.org/10.1016/0890-4332(87)90138-4

8.   Fournier, R.O. and Potter II, R.W. (1982) A Revised and Expanded Silica (quartz) Geothermometer. Geothermal Resources Council Bulletin, 11, 3-12.

9.   Siega, F.L., Salonga, N.D. and D'Amore, F. (1999) Gas Equilibria Controlling H2S in Different Philippine Geothermal Fields. Proceedings of 20th Annual PNOC-EDC Geothermal Conference, Manila, Philippines, March 10-12, 1999, 29-35.

10. Barragán, R.M., Arellano, V.M., Aragón, A., Torres, R., Reyes, N., López, S. and Pati?o, A. (2013) Estudio Isotópico de Fluidos de Pozos Productores y de Reinyección del Campo Geotérmico Los Azufres, Mich. 2013. Final Report, IIE/11/14154/I01/F, Instituto de Investigaciones Eléctricas for the Comisión Federal de Electricidad, Cuernavaca, Morelos, México. (In Spanish)

11. Arellano, V.M., Torres, M.A., Barragán, R.M., Sandoval, F. and Lozada, R. (2003) Chemical, Isotopic and Production Well Data Analysis for the Los Azufres (Mexico) Geothermal Field. Geothermal Resources Council Transactions, 27, 275-279.

12. Barragán, R.M., Arellano, V.M., Rodríguez, M.H., Pérez, A. and Segovia, N. (2010) Gas Geochemistry Related to Wellhead Production Data to

Investigate Physical Reservoir Phenomena in Geothermal Reservoirs: Application at Cerro Prieto IV (Mexico). Geofluids, 10, 511-524. http://dx.doi.org/10.1111/j.1468-8123.2010.00319.x

13. D'Amore, F. and Truesdell, A.H. (1995) Correlation between Liquid Saturation and Physical Phenomena in Vapor-Dominated Geothermal Reservoir. Proceedings of the World Geothermal Congress, Florence, 18-31 May 1995, 1927-1932.

14. Barragán, R.M., Arellano, V.M., Portugal, E. and Sandoval, F. (2005) Isotopic (δ18O, δD) Patterns in Los Azufres (Mexico) Geothermal Fluids Related to Reservoir Exploitation. Geothermics, 34, 527-547. http://dx.doi.org/10.1016/j.geothermics.2004.12.006

15. Iglesias, E.R., Flores-Armenta, M., Torres, R.J., Ramírez-Montes, M., Reyes-Piccaso, N. and Reyes-Delgado, L. (2011) Estudio con Trazadores de Líquido y Vapor en el área Tejamaniles del Campo Geotérmico de Los Azufres, Mich. Geotermia Revista Mexicana de Geoenergía, 24, 38-41. (In Spanish)

16. Arnórsson, S. (2000) Injection of Waste Geothermal Fluids: Chemical Aspects. Proceedings of the World Geothermal Congress, Kyushu-Tohoku, 28 May-10 June 2000, 3021-3024.

17. Arnórsson, S. and D'Amore, F. (2000) Monitoring of Reservoir Response to Production. In: Arnórsson, S., Ed., Isotopic and Chemical Techniques in Geothermal Exploration Development and Use: Sampling Methods, Data Handling, Interpretation, International Atomic Energy Agency, Vienna, 309-341.

18. Barragán, R.M., Arellano, V.M., Aragón, A., Martínez, J.I., Mendoza, A. and Reyes, L. (2010) Geochemical Data Analysis of Los Azufres Geothermal Fluids (Mexico). In: Birkle, P. and Torres-Alvarado, I., Eds., Proceedings of the Water Rock Interaction, Taylor & Francis Group, London, 137-140.

19. Barragán, R.M., Arellano, V.M., Mendoza, A. and Reyes, L. (2011) Chemical and Isotopic (δ18O, δD) Behavior of Los Azufres (Mexico) Geothermal Fluids Related to Injection as Indicated by 2010 Data. Geothermal Resources Council Transactions, 35, 603-608.

20. Truesdell, A.H., Nathenson, M. and Rye, R.O. (1977) The Effects of Subsurface Boiling and Dilution on the Isotopic Compositions of Yellowstone Thermal Waters. Journal of Geophysical Research, 82, 3694-3704. http://dx.doi.org/10.1029/JB082i026p03694

21. Arellano, V.M., Barragán, R.M., Ramírez, M., López, S., Paredes, A., Aragón, A., Casimiro, E. and Reyes, L. (2015) The Los Azufres (México)

Geothermal Reservoir: Main Processes Related to Exploitation (2003-2011). Proceedings of the World Geothermal Congress, Melbourne, 19-25 April 2015, Paper 14028, 1-10.

22. Iglesias, E.R., Flores-Armenta, M., Quijano, J., Torres, M., Torres, R.J. and Reyes-Piccaso, N. (2008) Estudio con trazadores de líquido y vapor en la zona Marítaro-La Cumbre del campo geotérmico de Los Azufres, Mich. Geotermia Revista Mexicana de Geoenergía, 21, 12-24. (In Spanish)

Chapter 7

DOES THE TETHYS BEGIN TO OPEN AGAIN? LATE CENOZOIC TECTONOMAGMATIC ACTIVIZATION OF THE EURASIA FROM PETROLOGICAL AND GEOMECHANICAL POINTS OF VIEW

Marian Petre[1] and Sharkov[2]

[1]Institute of Geology of Ore Deposits, Petrography, Mineralogy and Geochemistry (IGEM), Russia

[2]Russian Academy of Sciences, Moscow, Russia

INTRODUCTION

Alpine-Himalayan-Indonesian Mobile Belt, lasted from Gibraltar to Indonesia, is about 16,000 km long and from 500 to 1500 km width. It is the fourth generation of the Mediterranean mobile belt (Khain, 1984). Its predecessors were Neoproterozoic Baikalian, Caledonian, and Hercynian belts. Now it is represented by the major modern geological structure of the Eaurasia, a huge belt of the Late Cenozoic tectonomagmatic activization (Trans-Eurasian Belt, TEB), which stretches out through the whole continent practically from the Atlantic to the Western Pacific (Fig. 1). TEB has been formed after the closure of the Mesozoic Tethys and is marked by mountain building and appearance of riftogenic structures, numerous Cenozoic basaltic plateaus, and chain of subduction-related andesite-latite volcanic arcs, which trace suture zones of continental plates collision. Two large amagmatic geoblocks (North-Eurasian and Indian) lie on each side of the TEB.

The TEB is the excellent testing area for solving of such problems as interaction of plume head and crustal roof above it and interaction of a mantle superplume head (or asthenospheric rise) with shallow continental lithosphere above it. The goal of this work is to show that the present-day tectonomagmatic activity within the TEB can be interpreted that a new ocean has begun to open here. The work is divided into two parts. The first part discusses the interaction of a mantle plume head with its roof, composed by continental crust as an example of the Syria region, and the second deals with the interaction of the mantle superplume head with lithosphere above it under conditions of continental plate collision.

So, classical case of new ocean opening – Red Sea Rift – is, probably, not a sole case of such process. Likely, the TEB represents an alternative situation when a new ocean is opened under condition of large collision zone. Instead of breakup, a system of large gradually growing caverns are developed here, which begin to divide a body of Eurasia supercontinent starting from its western part. The eastern part of the TEB is characterized only by numerous riftogenic structures yet, which have a chance to regenerate in zones of oceanic spreading.

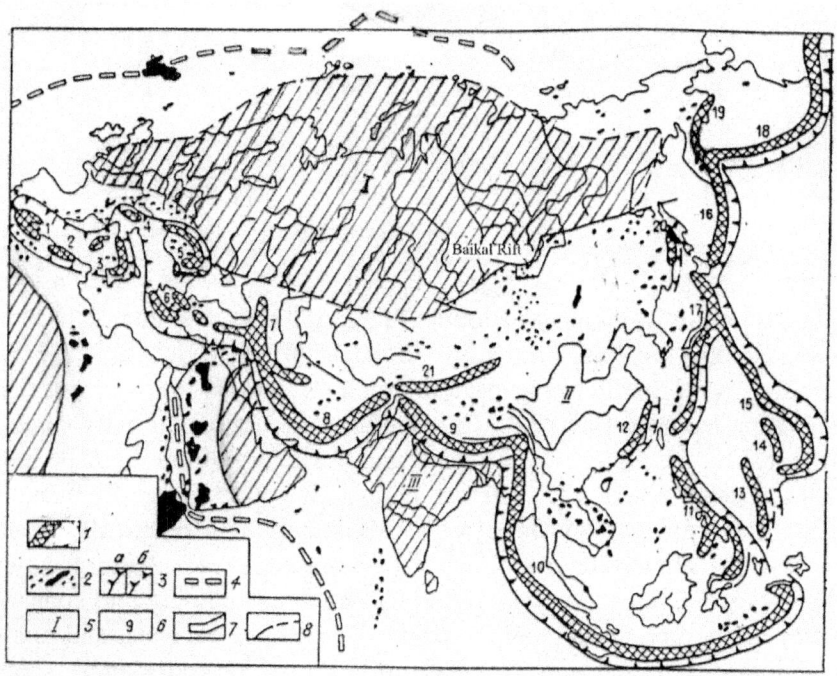

Figure 1. Major Cenozoic Eurasian geoblocks. 1 - Late Cenozoic subduction-related andesite arc; 2 - Areas of the Late Cenozoic flood basalts; 3 - Subduction zones: a - established, b - suggested; 4 - Zones of oceanic spreading; 5 - Geoblocks: I - North-Eurasian; II – Trans-Eurasian Belt; III - Indian: 6 - Baikal Rift: 7 - Boundaries of the Late Cenozoic tectonomagmatic activization expansion on the Eurasian continent.

INTERACTION MANTLE PLUME HEAD WITH CONTINENTAL CRUST: EVIDENCE FROM THE NORTH OF ARABIAN PLATE

One of the best testing region for discussing this problem is an area of the modern within-plate tectonomagmatic activity is the north of the Arabian plate, in Syria (Fig. 2). This region is located on the north-eastern periphery of the Red Sea Rift – newly-formed zone of oceanic spreading; tectonomagmatic activity has begun here in the Late Cenozoic, about 26-25 Ma, and goes on now (Sharkov, 2000). The uniqueness of the situation is that there are widely manifested both powerful processes of intraplate magmatism caused by melting of mantle plume material, and the processes of intraplate crustal deformations associated with the formation of the Palmiride fold-thrusted structure, as well as unique intracontinental Levant (Dead-Sea) transform fault (Kopp, Leonov, 2000). This gives possibility of a substantive discussion of the influence of strains in the crust to the processes in the roof of a mantle plume head.

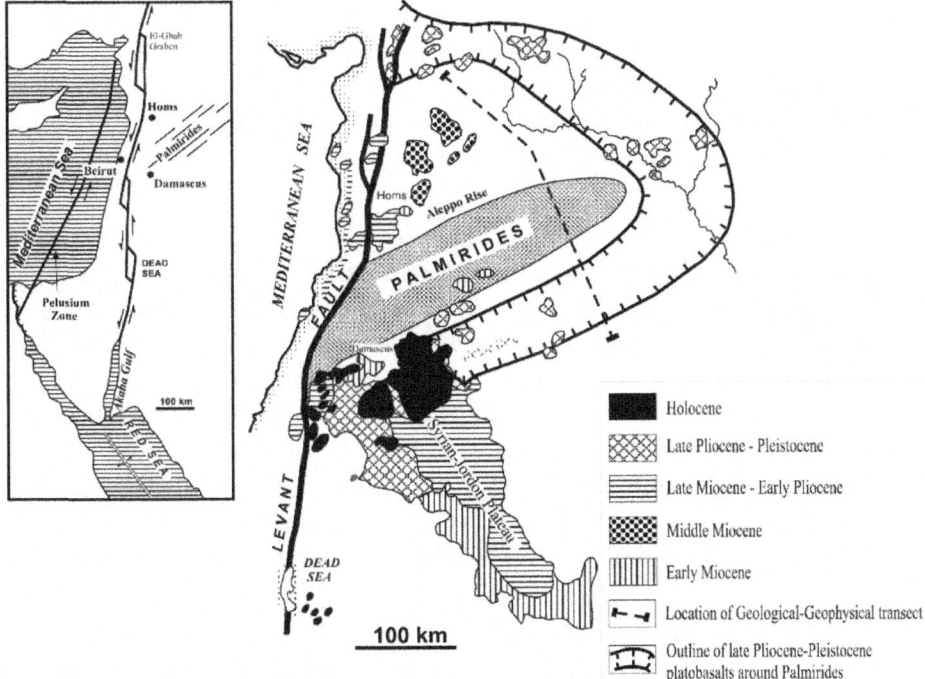

Figure 2. Distribution of the Late Cenozoic basaltic volcanism in Syria.

Such a situation takes place in the north of Arabian plate (Syria). The feature of the region is widespread Late Cenozoic basaltic volcanism, which gives evidence about the presence of mantle plume here. This magmatism has begun about 26 Ma and lasted till Historical time. Our detailed studying of K-Ar isotope ages of basaltic plateaus showed, that the most ancient eruptions were located on the place of the modern Palmyrides – large nappe-folded within-plate structure with roots, dipping to NW direction. Palmyrides appeared in the Middle Miocene and after that front of basaltic volcanism began gradually moving to the north according to their development. Eruptions to the south of Palmyrides occurred practically uninterruptedly at the same time.

Geological Setting

The Arabian Peninsula belongs geologically to formerly single ancient African-Arabian craton, which was broken out into some plates. Two such plates, Arabian and Sinai, divided by large left-lateral Levant Fault (Dead Sea Transform), occur in the region. The most manifestations of the Late Cenozoic basaltic volcanism are located in the Arabian plate, i.e. to the east from Levant Fault; there is practically no essential volcanic activity on the Sinai plate. The major tectonic elements of the Arabian plate, where basaltic volcanism mainly occurred, are stable platform structures – rises Aleppo (on north) and Rutba (on south), divided by within-plate Palmyride Fold Zone (PFZ, Palmyrides). These rises are formed by platform cover assuming 5-6 km thick, which overlaps the Precambrian crystalline basement uncovered by bore holes in Jordan and Syria.

Palmyrides is a modern zone of deformed platform cover, occurred between these stable structures. It is represented by the east-degenerated zone of nappe-folded deformations (Fig. 2). The PFZ appeared in the Middle Miocene, 12-14 Ma ago, practically simultaneously with the Dead Sea transform; moderate seismicity gives evidences that it continues of own development now. Formation of Palmyrides was linked with braking of western edge of the Arabian plate on the place of S-like curve on the Lebanon territory under its moving to the north along the Levant Fault (Kopp and Leonov, 2000). In other words, Palmyrides was formed as a result of the crust's compression, which compensated the northern displacement of the Arabian plate during the late Cenozoic.

According to geophysical and geological data along the 450-km transect in the central Syria (Fig. 2), structure of the earth's crust beneath the eastern Palmyrides is very specific (Al-Saad et al., 1992). Widen upwards plate-like trans-crustal anomaly of 20-30 km thick, composed by rocks of increased density, occur here (Fig. 3). The anomaly plunges to north direction under moderate (30-40°) angles beneath Aleppo Rise, and its upper continuation coincides with the frontal part of Palmyrides on the place of transition to

Rutba rise. In essence, this anomaly represents "roots" of the Palmyrides, body of packet intense deformed rocks, forced (subducted) into lithosphere under influence of north-direction subhorizontal motion of the Arabian plate. Relationships of the Palmyrides "roots" and the mantle were not established yet, because seismic data were obtained only for the upper crust; accordingly, the Moho discontinuity on Fig. 3 is shown arbitrarily.

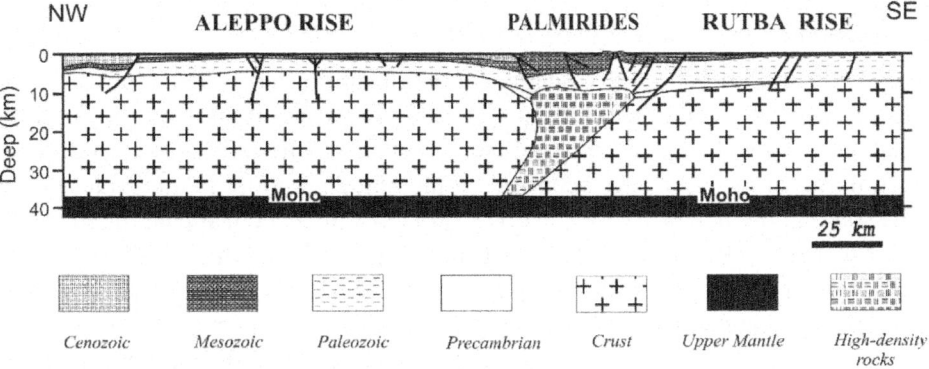

Figure 3. Structure of the Central Syria crust according to gravimetric data along geological-geophysical transect (modified after Al-Saad et al., 1992).

So, Palmyrides was formed as a result of the crust's compression, compensated the northern displacement of the Arabian plate which has begun in the Middle-Late Miocene. Geophysical data (Al-Saad et al., 1992) evidence that beneath the Palmyrides some kind of roots occur: widen upward plate like transcrustal anomaly of 20-30 km thick, composed by rocks of increased density. These "roots" are represented by plate-like body of pact deformed rocks, forced (subducted) into lithosphere under influence of north direction subhorizontal motion of Arabian Plate.

Dynamics of Basaltic Magmatism Development in Connection to Tectonic Processes

The second feature of the region is wide spread occurrence of the Late Cenozoic subaerial basaltic magmatism, represented mainly by high-Ti (TiO_2 ~1.8-3.7 wt.%) subalkaline lavas (Fe-Ti alkali basalts, basanites, hawaiites, etc.) plus rare transitional/tholeiitic basalts and basaltic andesites. All these rocks have geochemical features typical for within-plate plume-related magmas (Sharkov, 2000; Lustrino, Sharkov, 2006). The basaltic magmatism appeared at the end of Oligocene and lasted till the Historical time (Sharkov et al., 1994, 1998). On the basis of incompatible trace element content, the volcanic activity in Syria has been divided into two stages: the first lasted from ~26 to ~5 Ma and the second from ~5 Ma to recent. Indeed, the Syrian lavas show incompatible trace element content increasing with decreasing age from ~26 to ~5 Ma followed by an abrupt decrease to low values roughly at the Miocene-Pliocene boundary; lavas of the second stage show the same variation with age. This temporal shift in composition is related to major tectonic re-organization of the region occurred during Late Miocene, which finally led to appearance of the north continuation of the Levant Transform and intensification of the Palmirides formation (Rukieh et al., 2005; Trifonov et al., 2011).

During all this period volcanic centers have not had a stable position and systematically shifted in the region. Isotopic K-Ar dating of numerous basaltic plateaus showed that the most ancient eruptions (the end of the Oligocene-beginning of the Miocene) occurred on the place of modern Palmirides and to the south of them (Fig. 2). PFZ itself did not exist then yet; it appeared only at the Middle Miocene, and in process of its development front of basaltic volcanism gradually moved to the north and north-west. Quaternary volcanism forms elongated arc which bounded the Palmirides from the north. However, to the south of the Palmirides eruptions have lasted practically uninterrupted from the end of Oligocene till Historical time. There is no consensus about origin of the North Arabian magmatism (Lustrino and Sharkov, 2006 and references in). Some geologists and petrologists suggest that it is related to a mantle plume, however, other investigators suppose that a combination of plume source and asthenospheric mantle occurred here. Most likely that a mantle plume presently lies under shallow lithosphere, having triggered secondary plumes from the lithosphere-asthenosphere boundary, similar to what Gautheron et al. (2005) suggest concerning Cenozoic basaltic magmatism of Europe. In all cases, we have a source body (mantle plume or protuberance of asthenospheric material) beneath the north Arabia plate, which looks like plume head and henceforth it will be named so. Absence of essential evolution of the mantle-derived magmas probably indicates that over this period the mantle plume head in the process of its extension beneath the region has constantly been supported by fresh material.

Deep-Seated Processes

Alkali basalts of the region often contain mantle xenoliths, mainly spinel lherzolites (Sharkov et al., 1996) and, correspondingly, cannot arrive from the transitional within-crustal chambers, because in such case xenoliths should be obligatory sink to their bottom. So, xenoliths arrived directly from the mantle and represent fragments of cooled upper margin of the plume heads above magma-generation zones, captured by magmas in their way. At that, processes of melting occurred due to adiabatic decompression not in the whole plume head, but localized into protuberances on the its surface, which could reach rather shallow levels – 25-30 km (Sharkov and Bindeman, 1990). This is in good agreement with the fact that there are no lower crustal xenoliths in the basalts – only mantle and upper crustal ones. Lower crustal xenoliths (garnet granulites and garnet gabbroids) were found only in Cretaceous diathermes of kimberlite-like rocks in the Coastal Ridge (Sinai plate), to the west of Levant Transform, where Cenozoic volcanism is absent, and structure of Pre-Cenozoic lithosphere cover was not disturbed (Sharkov et al., 1993).

From such point of view, distribution of the basaltic plateaus in space could reflect some local uplifts on the surface of the extended plume head at the moment of the lava plateaus formation, and so migration of the volcanic centers is reflected dynamics of such rises shift. From this follows that some interrelation occurs between displacement of magmatic activity in region due to the Palmyrides development and the gradual penetration (subduction) of the Palmyrides "roots" into the plume head, which led to cessation of the melting in this place. At the same time, it has led to displacement of plastic heated mantle material to the north and formation there of new rises on the mantle plume head with appearance of new magmatic systems in another places (Fig. 4).

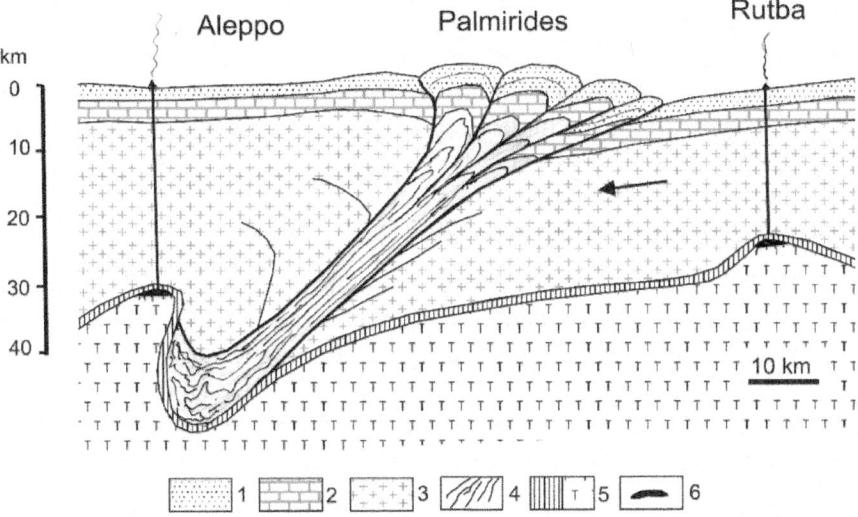

Figure 4. Scheme, illustrating character of interaction of earth crust and mantle plume head in process of the Palmyrides development1 – crust of the Arabian lithospheric plate 2 – cooled rim of mantle plume head; 3 – mantle plume matter; 4 – ancient litho-spheric mantle; 5 –zones of generation of magmatic melt, feeding basaltic plateaus.

Petrological data evidence that basalts were generated into protuberances on the mantle plume head as a result of stress decompression. From such point of view, clear correlation between appearance of basaltic plateaus and Palmyrides development suggests that it was linked with gradually penetration of the Palmyrides "roots" into the plume head which led to moving the plastic mantle material to the north and formation there of new rises with appearance of new magmatic system.

Thereby, the arc of the Quaternary basaltic volcanism (Fig. 3), probably, represents a projection of the present-day Palmyrides plate to the surface. On this basis it is possible to estimate a velocity of forced flow of the mantle heated material here, which was about 14-15 km per million years (14-15 mm/year). It is only one and half mm/year lower than the northdirected velocity motion of the Arabian plate itself, obtained by GPS method which is 20-24 mm/year (Reilinger et al., 1997).

INTERACTION OF SUPERPLUME HEAD WITH CONTI-NENTAL LITHOSPHERE: EVIDENCE FROM ALPINE BELT

As it seen from fig. 1, the Trans-Eurasian Belt consists from two segments (halves): western, represented by Alpine Belt, and eastern, located in Central and Eastern Asia. The most strongly processes of interaction of superplume head and continental lithosphere are manifested within Alpine Belt and to a lesser extent on the eastern continuation of the TEB, where numerous Late Cenozoic basaltic plateaus indicate presence of the mantle plumes, especially beneath Baikal and Eastern Asia rift systems.

Geological Features of Alpine Belt

This belt has the most complicated structure within the Alpine segment, where a system of andesite-latite volcanic arcs and back-arc basins (Alboran, Tyrrhenian, Aegian, and Pannonian depression) are observed. In spite of differences in morphology of these structures, they have a number of common features. Fold-thrusted zones, in a sense of "accretion prisms", evolved along their peripheries to form arc-like mountain ridges: Alps, Carpatian, Gibraltar arc, etc. (Fig. 5).

The TEB is a good example of interaction of superplume head with mobile continental lithosphere, where processes of collision occur now. The main feature of this belt is wide spread of the late Cenozoic volcanism, which has displayed practically coeval on all its length, presuming existence of a superplume (or asthenospheric rise) beneath it. The belt has the most complicated structure within the Alpine segment, where a system of andesite-latite volcanic arcs and back-arc basin, bordering by nappe-folded mountain ridges are observed. In front of these ridges in Western Europe, north-west Africa and Arabia, coeval rift systems and flood basaltic volcanism often occurs.

Not rarely among nappes are found deepsea sediments of the Tethys, ophiolites, and sometimes even blocks of lower-crustal and upper-mantle rocks, such as Ivrea-Verbano, Ronda, Beni Boussera, etc., which indicate powerful deep-seated processes into lithosphere here. Volcanic andesite-latite arcs are situated at the rears of these structures, repeating their configuration. There are three types of such subduction-related arcs: (1) island arcs (Aegian), (2) "semiland" arcs, occurred partly on continent, partly near it (Alborane and South-Italian, including Calabrian arc and Roman province) and (3) within-continental arcs (Carpathian and Anatoly-Caucasus-Elbursian). Depressions with newly-formed thinned crust of transitional type at expense of removing of lower high-velocity layers (Gize and Pavlenkova, 1988) and even oceanic crust with basaltic volcanism located behind the volcanic arcs. Such crust beneath seas of the Western Mediterranean was evolved in the late Cenozoic on the place of the African plate (Ricou et al., 1986; Ziegler et al., 2006), very likely as a result of back-arc spreading (Sharkov, Svalova, this book).

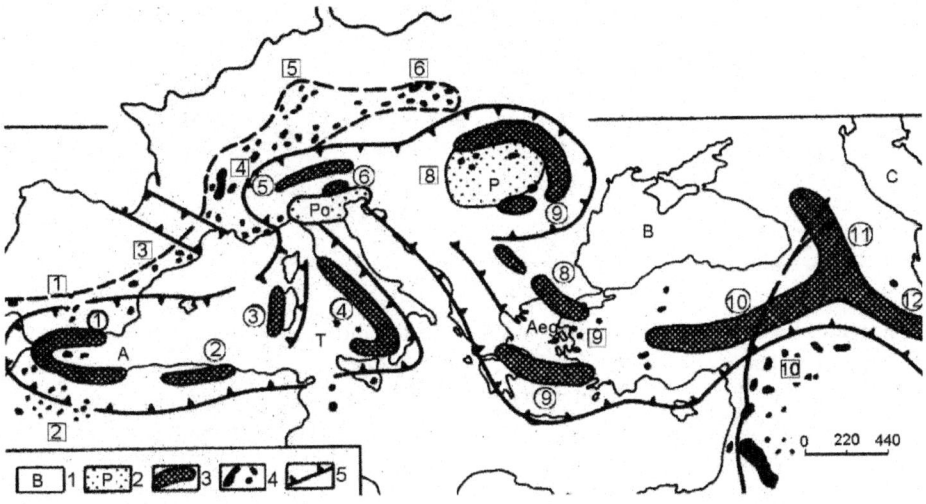

Figure 5. Distribution of the late Cenozoic magmatism within the Alpine Belt1 – back-arc seas (A – Alboran, T – Tyrrhenian; Ae – Aegian) and "downfall" seas (B –Black, C – Caspian); 2 – back-arc sedimentary basins (P- Pannonian, Po – Po valley); 3 – Late Cenozoic andesite-latite volcanic arcs (in circles): 1 – Alboran, 2 – Cabil-Tell, 3 – Sardinian, 4 – South-Italian, 5 – Drava-Insubrian, 6 – Evganey, 7 – Carpatian, 8 – Balkanian, 9 – Aegian, 10-12 – Anatolia-Elbursian (10- Anatolia-Caucasian, 11 – zone of the Modern Caucasus volcanism, 12 – Caucasus-Elbursian); 4 – areas of flood basaltic volcanism (in square): 1 – South Spain and Portugal, 2 – Atlas, 3 – Eastern

Spain, 4 – Central France massif, 5 – Rhine graben, 6 – Czech-Silesian, 7 – Pannonian, 8 – Western Turkey, 9 – northern Arabia; 5 – suture zones of major thrust structures.

Deep-Seated Processes beneath Basins of the Alpine Belt

According to geophysical data, lithosphere beneath the Alpine Belt has very complicated structure and rather different beneath ridges and basins (Hearn, 1999; Artemieva et al., 2006, Kissling et al., 2006, etc.). M. Artemiev (1971) firstly showed that there are two types of basins here. The first type is represented by Tyrrhenian, Alborane, Aegian seas and Pannonian basin which are characterized by positive isostatic anomalies, pointing to excess of mass beneath them and occurrence of basaltic volcanism (Fig. 6). Judging from the magnitude of anomalies, the most intensive is found beneath the youngest back-arc sea – Aegean, and also beneath the Pannonian basin. Obviously, existence of the present-day extended plume heads can explain appearance of such anomalies and their magnitude can evidence about intensity of fresh plume material arrival.

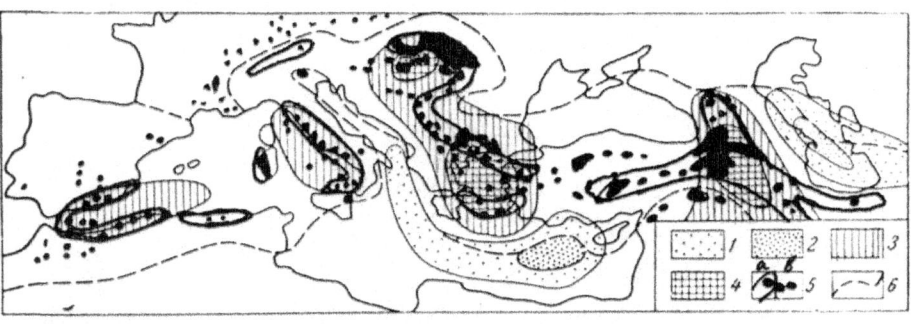

Figure 6. Distribution of main regional isostatic anomalies in connection to areas of Cenozoic volcanism in the Alpine Belt. After M. Artemiev (1971)Regional minimums: 1 – lows intensity, 2 – high intensity; regional maximums: 3 –average intensity, 4 – high intensity; 5 – volcanic areas: a – calc-alkaline rocks, b – basalts; 6 – boundaries of the Alpine Belt.

Some depressions of the Alpine belt (Tyrrhenian, Aegian, Alboran, and Pannonian) are characterized by positive anomalies, which evidence about excess of mass beneath them. Probably, they represent the present-day plume heads, which support basaltic volcanism and lead to onwards displacement of andesitic volcanic arcs in time. Another depressions (Eastern Mediterranean and Caspian Sea), in contrast, are characterized by negative anomalies or neutral and represent deficit of mass beneath them probably considered with descending mantle currents amongst extended plume heads. The Black Sea has no essential isostatic anomaly; deep-seated situation beneath it is stabilized now.

According to geophysical data, the plumes joint together in common layer at the depth 200-250 km, forming of Circum-Mediterranean Common Magmatic Reservoir (Lustrino & Wilson, 2007) which is, probably, extended superplume head.

According to Hearn (1999), Smewing et al. (1991) and others, these plumes join together in a single asthenospheric rise at depths of 200-250 km. It begins in Eastern Atlantic, near Azores, spreads to the east to Europe (Hoernle et al., 1995). Extension of plume heads, judging by geological data, has been directed – beneath Carpathians mantle-plume material and its crustal roof moved to the east (Royden, 1989), beneath Tyrrhenian Sea – to the south-east (Rehault et al., 1987), beneath Alboran Sea – to the west (Lonengran, White, 1997), etc. The material of continental crust above extended plume head was transported to its frontal edge where it was involved into descending mantle flows with formation of subduction zones and appearance of subduction-related volcanic arcs and back-arc basins in their rear (Sharkov, Svalova, this book; Bogatikov et al., 2009).

The second type of the basins is represented by Eastern Mediterranean, including Ionian Sea, as well as Black and Caspian seas. In contrast to the above mentioned basins, large negative isostatic anomalies occur beneath them, indicating deficit of mass, probably linked with descending mantle flows ("cool plumes"). One of the youngest basin (Levantine basin) occurred on the place where about 3-3.5 Ma part of the Eastern Mediterranean was submerged (Geological…, 1994; Emels et al., 1995). Origin of such "cool plumes" obviously linked with appearance of excess of mantle material between extended plume heads.

These seas have passive margins and oceanic crust, covered by sediments of 10-15 km thick. Depressions of Black and Caspian seas form large "downfall", or caldrons which were cut off pre-Pliocene structures of Caucasus and Kopet-Dag. Formation of these seas has, probably, begun in the Cretaceous, however, essential deepening of the basins occurred at the Oligocene-Miocene boundary, and after that gradually shallowing of them took place in the Miocene (Zonenshain and Le Pichon, 1986). New essential deepening of the Black Sea and South-Caspian Deep has begun in the Pliocene-Quaternary; it occurred simultaneously with uprising of Crimea and Caucasus mountains (Grachev, 2000).

According to geophysical data, along margins of such "downfall" basins (for example of north-eastern Black Sea and north of the Eastern

Mediterranean) there are observed powerful local positive gravity anomaly and steep seismoactive zones, traced into the mantle till 60-70 km deep (Zverev, 2002;Shempelev et al., 2001). These zones are analogues to structures which evolved along passive non-volcanic continental margins (Louden, Chian, 1999).

Other Features of the Alpine Belt Structure

The only exception from common situation is a large positive isostatic anomaly beneath the Lesser Caucasus located between "downfall" Black and Caspian seas, which is a northern continuation of anomaly of the Syrian region. Obviously this anomaly also considered with ascending of a mantle plume. There is no depression here, but front of Anatolian-Caucasus- Elbursian volcanic arc sharply displaces to the north, forming a zone of young Caucasus volcanism. Judging on the maximum location, further northwards extension of this plume head could lead to rupture of this arc in two independent parts, how it was occurred in case of modern Calabrian and Betic-Rif arcs (Lonengran and White, 1997).

Simultaneously with formation of depressions, Cenozoic rift systems and flood basaltic volcanism with typical Fe-Ti-affinities (rifts of Central and Western Europe, Atlas, basaltic plateaus of northern Africa and Arabia, etc) was developed before fronts of mountain ridges in Western and Central Europe and north-west Africa (Grachev, 2003; Wilson, Downes, 2006). Judging on geochemical-isotopic data this anorogenic circum-Mediterranean magmatism has common source – so-called Common Magmatic Reservoir (Lustrino, Wilson, 2007). It is, obviously, evidence about existence beneath the region a present-day mantle superplume; Alpine Orogen with complicate combination of mountain ranges and basins is located in its inner part.

Such detailed works are absent yet for the eastern part of the TEB, where powerful processes of mounting building, andesite-latite volcanic arcs, riftogenic structures, basaltic plateaus occur (Fig. 1)and complicate structure of the upper mantle icluding subduction zones was established by seismic tomography (Kulakov et al., 2003). Numerous late Cenozoic basaltic plateaus indicate presence of the mantle plumes, especially beneath Baikal and Eastern Asian rift systems (Grachev, 2000; Stupak et al., 2008). It may suggest that beneath of the eastern part of the TEB a mantle superplume occurs also. From the south this part of the TEB is confined by chains of subduction-related South-Afganian, Kunlun and Birma-Indonesian andesite-latite volcanic arcs. Chain of Neogene granites, which traced the Himalayan collision suture, probably derived as a result of shear heating during active displacement of rocks (Harrison et al., 1997).

Western and Eastern superplumes are divided by mountain system the Hindu Kush - Pamir - Karakorum and northern ledge of the Indian plate (Indian syntax). However, because tectonomagmatic processes in western and eastern parts of the TEB occurred and developed practically simultaneously, their appearance was likely arose from the same reason – collision of Eurasia with Gondwana. From this view they can be jointed in the single tectonomagmatic system which is comprised of two halves.

Thus, according to geological, geophysical and petrological data, it is assumed that beneath the superplume an elongated superplume or combination of two superplumes occur. Relief of the belt surface is very complicated, with numerous protuberances (plumes). Above the largest of extended plume heads (in Alpine belt) back-arc depressions occurred with new-formed oceanic crust (Western Mediterranean seas on place of ancient African plate, part of the Pannonian basin crust) and rift systems (Red-Sea, Central-European, Baikal, Eastern-Asian, etc.). Formation of the "downfall-type" basins with passive margins was probably linked with places of descending mantle flows ("cool plumes"). Probably, their oceanic crust survived since the Mesozoic Tethys, because the modern basaltic magmatism is absent there in contrast to back-arc basins of the Western Mediterranean.

DISCUSSION

Judging by the presented geological, geophysical and petrological data, beneath the TEB a huge sublatitudinal mantle superplume or two superplumes, generated secondary plumes, exist now. Existence of such plastic heated "basement" under condition of collisional tangential tensions obviously makes easier passing of different deformational processes into superposed cool rigid shallow lithosphere. The latter was broken apart around extended plumes heads, which led both to depressions formation and to uploading of crustal material around their peripheries to form mountain ridges, etc. (Sharkov, Svalova, this book). Some lithoplastines could reach depths ~200 km (Laubscher, 1988). Partitioning of this superplume head on secondary plumes, as the discussed above on example of the Arabian plate, could also explain its mechanical interaction with shallow lithosphere. This interaction is resulting from appearance of lithoplastines and subduction zones, which are established by both geophysical methods and corresponding volcanism. Lithoplastines and subduction zones penetrated into superplume head and pressed out its plastic matter to higher levels, leading to appearance of independent secondary mantle plumes, in which heads melting processes occur. Heads of the largest plumes, when they arrived to their floatage level, began to extend into lithosphere lead to appearance of continental rifts and/or large depressions with newly formed oceanic crust. The Alpine back-arc basins and the rifts, surround mountains of the Alpine Belt, are among such structures as well as Baikal and East-Asian rift systems.

It was shown above that extension of the secondary plumes heads could influence to the morphology of subduction zones. Good example is the Anatolian-Elbursian subduction-related volcanic arc, which Caucasus part of curved to northern direction follows by northwards of the plume head extension. In future it, probably, will tear at two independent fragments, like Calabrian and Betic-Rif arcs (see section 2.3).

Between ascended mantle plumes domains with descending mantle movements ("cool plumes") occurred, where earth crust sank downwards. They form depressions of "downfall" type with survived ancient oceanic crust, covered by thick sedimentary piles. Along margins of such depressions specific steep-dipping deformation structures occur, where blocks of ancient oceanic lithosphere sink into mantle up to level of their isostatic compensation. Such compensation in the Black Sea, probably, already reached, whereas in Caspian Sea, Ionic Sea and, especially, in the Eastern Mediterranean the process is still continuous.

The TEB on its scale and elongated morphology looks like structure of middle-oceanic ridge. Probably, it is true. The situation, probably, can be explained by such way.

Feeding system of the oceanic-spreading zone (alongated mantle superplume or asthenospheric lens like observed beneath the Mid-Atlantic Ridge (Ritsema, Allen, 2003)), which provided existence of the Mesozoic Tethys, preserves its activity in a considerable degree even be overlapped by continental plates in process of collision; due to more intense motion of Indian plate to north, this system was divided in two halves. As a result, forcing of the plastic mantle material occurs into clearance between coming together Laurasia and Gondwana plates. According to geomechanics (Bobrov and Trubitsin, 2003), after peak of compression a tensile stress must follow. In the case under consideration, it would be expected that these plates have begun to go away with appearance of numerous secondary plumes, which ascended from the surface of the superplume head and bring to rifting, formation of back-arc spreading structures, descending of some crust blocks between the plumes, etc. All these processes can act only under condition of preservation of fresh material supply from the core-mantle boundary.

Further development of the TEB can truly bring to a new dividing of Eurasia and the revival of the Tethys, which repeatedly occurred in geological history. The first steps in this direction became clear in the late Cenozoic in the appearance of the Mediterranean, which gradually extends to east; in the sharp deepening of the Black and Caspian seas and coeval formation of numerous basaltic plateaus and continental rift systems in Europe and Asia. Evidently, classical case of new ocean opening – Red Sea Rift – is not the only one case of such process. Another, more substantial case is discussed above. It is represented by a situation when at the beginning, instead of breakup, a system of large gradually growing caverns was developed, which broke up a body of supercontinent Eurasia in our case.

Thus, the data available show that from the Late Precambrian, due to oscillatory mechanical processes in zone of collision of large continental plates, periodically opening and closure of sublatitude Tethys ocean has occurred. Probably, a new stage of this ocean opening has begun in the late Cenozoic.

CONCLUSIONS

- The major feature of the modern geological structure of Eurasia is a huge belt of the late Cenozoic tectonomagmatic activization (Trans-Eurasian Belt, TEB), which stretches out through the whole continent from the Atlantic till the Western Pacific. It has been formed after closure of the Mesozoic Tethys and marked out by numerous Cenozoic basaltic plateaus, riftogenic structures, and chain of subduction-related andesite-latite volcanic arcs, which have traced suture zone of the continental plates collision. Two large amagmatic geoblocks (North-Eurasian and Indian) lie on each side of TEB.

- Practically coeval wide-spread mantle-derived magmatism within the TEB along its whole length can indicate an existence of mantle superplume beneath it. Relief of its head has very complicated structure due to many protuberances (secondary plumes) which go aside from its surface. This is evident from numerous late Cenozoic basaltic plateaus, continental rift zones (including Central Europeans, Baikal, East Asian, etc.) and appearance of depressions with newly-formed oceanic crust (Western Mediterranean, for example). Parts of the Mesozoic Tethys lithosphere survived on the places of descending mantle flows among ascending plumes in the form of "downfall" seas (Eastern Mediterranean, Black and Caspian seas).

- Formation of such complicated structures of the superplume's roof was probably linked with its interaction with rigid shallow continental lithosphere, which has been subjected to powerful deformation processes related to collision of continent-continent. Numerous lithoplastines and some subduction zones are noted here by seismic tomography. Obviously, they penetrated into the upper part of superplume head (or asthenosphere lens) and pressed out portions of hot plastic material, which further evolved as independent plumes.

- The TEB on its elongated morphology and scale looks like mid-oceanic ridge. Probably, it is true, which suggests that feeding system of the oceanic-spreading zone (alongated mantle superplume or asthenospheric lens), which provided existence of the Mesozoic Tethys, has existed now even been overlapped during collision of continental plates. According to geomechanics, after peak of compression a tensile stress must follow. In the case under consideration, it would be expected that these plates should begin to go away in future as a result of appearance of numerous secondary plumes, ascended from the surface of the superplume head (or asthenospheric lens). Extending of their head leads to continental rifting and to formation of volcanic arc-backarc basins systems, where old continental crust is involved in subduction process gives way to appearance of newly-formed oceanic crust.

- All these processes have developed within the TEB, which is, very likely, is a projection of the superplume head (or asthenospheric lens) to the surface. Judging from theoretical data and the observed features, the further development of the TEB could lead to revival of the Tethys.

REFERENCES

1.   D. Al-Saad, T. Sawaf, A. Gebran, et al. 1992 Crustal structure of central Syria: the intracontinental Palmiride mountain belt. Tectonophysics, 207 345358 .

2.   M. E. Artemiev, 1971 Some peculiarities of deep-seated structure of depressions of Mediterranean type: evidence from data on isostatic gravity anomalies. Bull. Soc. of the Nature Investigators of Moscow. Geol. Dept., 4, 5-10 (in Russian)

3.   I. M. Artemieva, H. Thybo, M. K. Kaban, 2006 Deep Europe today: geophysical synthesis of the upper mantle structure and lithospheric processes over 3.5 Ga. In: European Lithosphere Dynamics. D.G. Gee and R.A. Stephenson (eds). Geol. Soc. London Mem. 32 1142 .

4.   A. M. Bobrov, V. P. Trubitsyn, 2003 Evolution of viscous tensions in mantle and moving continents under process of formation and disintegration supercontinent. Physics of the Earth, N 12 313 .

5.   K. C. Emels, A. Robertson, C. Richer, 1995 Mediterranean Sea.1. Science operator report DSDP Leg.160 1120 .

6.   G. Gautheron, M. Moreira, C. Allegre, 2005 He, Ne and Ar composition of the European lithospheric mantle. Chem. Geology, 1-2: 97-112.

7. Geological structure of the Eastern Mediterranean. 1994 V.A. Krasheninnikov and J.K. Hall (Eds.). Jerusalem.

8. P. Gize, N. I. Pavlenkova, 1988 Structural maps of the Earth's crust of Europe. Physics of the Earth, N 10 3 14 .

9. A. F. Grachev, 2000 (ed.), Neotectonics, geodynamics and seismicity of the Northern Eurasian. PROBEL, Moscow, 487 p. (in Russian).

10. A. F. Grachev, 2003 Final volcanism of Europe and it's geodynamic nature. Physics of the Earth, N 5 11 46 .

11. S. Harangi, H. Downes, I. Seghedi, 2006 Tertiary-Quaternary subduction processes and related nagmatism. In: European Lithosphere Dynamics. D.G. Gee and R.A. Stephenson (eds). Geol. Soc. London Mem. 32 167 190 .

12. T. M. Harrison, O. M. Lovera, M. Grove, 1997 New insights into the origin of two contrasting Himalayan granite belts. Geology 25 899 902 .

13. T. M. Hearn, 1999 Uppermost mantle velocities and anisotropy beneath Europe. Journ. Geophys. Res., 104(B7): 15123-15139.

14. K. Hoernle, Y. S. Zhang, D. Graham, 1995 Seismic and geochemical evidence for large-scale mantle upwelling beneath the eastern Atlantic and western and central Europe. Nature, 374 6517 34 39 .

15. F. Horváth, G. Bada, P. Szafiáan, D. Tari, A. Ádam, S. Cloetingh, 2006 Formation and deformation of the Pannonian Basin: constraints from observational data. In: European Lithosphere Dynamics. D.G. Gee and R.A. Stephenson (eds). Geol. Soc. London Mem. 32 191 206 .

16. E. Kissling, S. M. Schmid, R. Lippitsch, J. Ansorge, B. Fugenschuh, 2006 Lithosphere structure and tectonic evolution of the Alpine arc: new evidence from high-resolution teleseismic tomography. In: European Lithosphere Dynamics. D.G. Gee and R.A. Stephenson (eds). Geol. Soc. London Mem. 32 129 145 .

17. M. L. Kopp, Y. G. Leonov, 2000 Tectonics / in: Outline of geology of Syria. Ed. Yu.G. Leonov. Moscow, Nauka Publ., 7-104 (in Russian with Enlish abstract).

18. I. Y. Kulakov, S. A. Tychkov, N. A. Bushenkova, A. N. Vasilevsky, 2003 Structure and dynamics of the upper mantle of Alpine-Hymalayan folded belt on data of seismic tomography. Russian Geology and Geophysics, 44 566 586 .

19. H. Laubscher, 1988 Material balance in Alpine orogeny. Geol. Soc. Amer. Bull., 100 83 122 .

20. L. Lonergan, N. White, 1997 Origin of the Betic-Rif mountains belt.

Tectonics, 16 504522 .

21. K. E. Louden, D. Chian, 1999 The deep structure of non-volcanic rifted continental margins. Roy. Soc. of London Phil. Trans, 357 767804 .

22. M. Lustrino, M. Wilson, 2007 The circum-Mediterranean Anorogenic Cenozoic Igneous Province. Earth Sci. Rev. 81. 165 .

23. M. Lustrino, E. Sharkov, 2006 Neogene volcanic activity of western Syria and its relationship with Arabian plate kinematics. J. Geodynamics, 42 115139 .

24. T. G. Rehault, E. Moussat, A. Fabri, 1987 Structural evolution of Tyrrhenian back-arc basin // Marine Geology, 74 123150 .

25. R. E. Reilinger, S. C. Mc Clusky, M. B. Oral, et al. 1997 Global Position System measurements of present-day crustal movements in the Arabia-Africa-Eurasia plate collision zone. J. Geophys. Res., 102(B5): 9983-10000.

26. L. E. Ricou, J. Dercourt, J. Geyssant, et al. 1986 Geological constrain on the Alpine evolution of the Mediterranean Tethys. Tectonophysics, 123 83122 .

27. J. Ritsema, R. M. Allen, 2003 The elusive mantle plume. Earth Planet. Sci. Lett., 207 112 .

28. L. H. Royden, 1989 Late Cenozoic tectonics of the Pannonian Basin System. Tectonics, 8 5161 .

29. E. V. Sharkov, 2000 Mesozoic and Cenozoic basaltic magmatism. In: Outline of geology of Syria. Yu.G. Leonov (Ed.). Moscow, Nauka Publ., 177200 (in Russian with English abstract).

30. E. V. Sharkov, G. A. Snyder, L. A. Taylor, E. E. Laz'ko, E. Jerde, S. Hanna, 1996 Geochemical pecularities of the asthenoshere beneath the Arabian plate: Evidence from mantle xenoliths of the Quaternary Tell-Danun Volcano (Syrian-Jordan Plateau, Southern Syria). Geochemistry Intern., 33 (9), 819-835.

31. E. V. Sharkov, I. N. Bindeman, 1991 Petrology of xenolith-bearing basalts from the Baikal rift zone: Tunka, Jida and Vitim areals. Volcanology and Seismology, N 12 443458 .

32. E. V. Sharkov, E. E. Laz'ko, S. Hanna, 1993 Plutonic xenoliths from Nabi Matta explosive center, Northwest Syria. Geochem. Intern., 30 2344 .

33. E. V. Sharkov, V. B. Svalova, 2005 Late Cenozoic geodynamic of Alpine Folded Belt in connection to formation of withincontinental seas (petrologo-geomechanical aspects). Proc. of Higher Educational Establishments. Geology and Exploration, № 1: 3-11 (in Russian).

34. A. G. Shempelev, N. I. Prutsky, I. S. Feldman, S. U. Kukhmazov, 2001 Geologo-geophysical model along profile Tuapse-Armavir. In: Tectonics of the Neogea: general and regional aspects. Proceedings of tectonic meeting. Moscow, GEOS, 316320 .

35. W. Spakman, S. van der Lee, R. van der Hilst, 1993 Travel-time tomography of the European-Mediterranean mantle down to 1400 km. Phys. Earth Planet Inter., 79, 374 .

36. V. G. Trifonov, A. E. Dodonov, E. V. Sharkov, et al. 2011 New data on the Late Cenozoic basaltic volcanism in Syria, applied to its origin. Journal of Volcanology and Geothermal Res., 199. 177192 .

37. M. Wilson, H. Downes, 2006 Tertiary-Quaternary intra-plate magmatism in Europe and its relationship to mantle dynamics. In: European Lithosphere Dynamics. D.G. Gee and R.A. Stephenson (eds). Geol. Soc. London Mem. 32 147166 .

38. P. A. Ziegler, M. E. Schumacher, P. Dezes, J. D. Wees, S. Cloetingh, 2006 Post-Variscan evolution on the lithosphere in the area of the European Cenozoic Rift System. In: European Lithosphere Dynamics. D.G. Gee and R.A. Stephenson (eds). Geol. Soc. London Mem. 32 97112 .

39. L. R. Zonenshain, X. Le Pichon, 1986 Deep basins of the Black Sea and Caspian Sea as remnants of Mesozoic back-arc basin. Tectonophysics, 123 181212 .

40. S. M. Zverev, 2002 Peculiarities of structure of sedimentary pile and basement in the frontal zone of the Cyprian arc (Eastern Mediterranean). Oceanology, 42 416428 .

Chapter 8

ENVIRONMENTAL FLOWS FROM ALTERNATE LAND USES IN THE DELTA, PACIFIC, AND THE SOUTHEASTERN STATES: 1947-2007

Charles B. Moss and Andrew Schmitz

Food and Resource Economics Department, University of Florida, Gainesville, FL, USA

ABSTRACT

Land use policy involves allocating land between production alternatives to meet society's wants and desires. Increase in the affluence in the United States has increased the demand for environmental flows that could be met from public ownership or as joint products of private ownerships. The empirical results of this study indicated that land use patterns remained relatively unchanged between 1947 and 2007. The lack of change suggests that a large part of the demand for environmental services is being as byproducts of other commercial decisions.

INTRODUCTION

This study examines the changes in land use between 1947 and 2007 focusing on the possibility that commercial uses generate significant environmental benefits. In the early twentieth century, land distributions under the Land Ordinance of 1790 as modified by the Homestead Act of 1862 came to a close. While vast tracts in the United States remained under the control of the federal and state governments, private land ownership was limited. Once the distributions from public land had been limited, land in private hands started to accrue rents (i.e., no additional land could be brought into production—increasing the rent accruing land previously in production under Ricardo's model). These increased rents from commercial uses of land increased the opportunity cost of less intensive uses of land. In addition, urban growth increased the demand for the conversion of farmland into residential and other urban uses. Taken together, higher rents from agricultural and the increased demand for urban uses are typically hypothesized to reduce the amount of land generating environmental services.

The increasing affluence in the United States has given rise to demands for the environmental services generated by land. Environmental values were first given voice in the establishment when the National Park Service was established at the dawn of the twentieth century under Theodore Roosevelt. The trend toward environmental values continues under different auspices. Today, states and local communities have developed mechanisms to retire land from private ownership into public use. For example, in Alachua County Florida, Alachua Forever establishes funding for the purchase of environmentally sensitive ground. In California, developers often are required to invest in environmental offsets in order to develop a specific parcel of land. These offsets are intended to produce environmental flows.

The crux is that some land once deeded to the private sector is now being reclaimed by the public sector. Such transactions secure environmental benefits or environmental flows. The mechanism for capturing these environmental flows, however, may be imperfect. While a variety of trends suggest that consumers have become more environmentally aware, it is not obvious that the allocation at the margin is optimal. Unlike the allocation of goods under the price system, consumers under this allocation method do not balance marginal benefits and marginal costs.

THEORY

Consider the possible outputs from a particular parcel of land (s)

$$\{y_{1s}, y_{2s}, \cdots, y_{ns}\} = F(z_{1s}) \tag{1}$$

where y_{is} is the level of output i that can be obtained from parcel s, $F(\cdot)$ is a surface of possible outputs (depicted for three different outputs in Figure 1), and z_{1s} are a bundle of physical characteristics of parcel s. The total value to parcel s becomes

$$y_{1s} R_{1s} + y_{2s} R_{2s} + \cdots y_{ns} R_{ns} = \kappa \tag{2}$$

where the R_{is} is the rent accruing to the use of parcel s to use i. This relationship is depicted in Figure 1 for three outputs. Essentially, if all land was privately owned, the land owner would choose the vector of outputs $\{y_{1s}, y_{2s}, \cdots, y_{ns}\}$ that maximized the total rents given the characteristics of the land parcel.

We consider four different outputs from farmland: agriculture (y_{1s}), forestry (y_{2s}), urban uses (y_{3s}), and environmental flows (y_{4s}). The total value from all the parcels of land in a region can be written as

$$\sum_{s=1}^{S}\left[y_{1s} \times R_1\left(\{y_{1r}\}_{r=1}^{S}, z_2\right) + y_{2s} \times R_2\left(\{y_{2r}\}_{r=1}^{S}, z_3\right) + y_{3s} \times R_3\left(\{y_{3r}\}_{r=1}^{S}, z_4\right) + y_{4s} \times R_4\left(\{y_{4r}\}_{r=1}^{S}, z_5\right) \right]$$

$$\{y_{1s}, y_{2s}, y_{3s}, y_{4s}\} = F\left(z_{1s}\right)$$

(3)

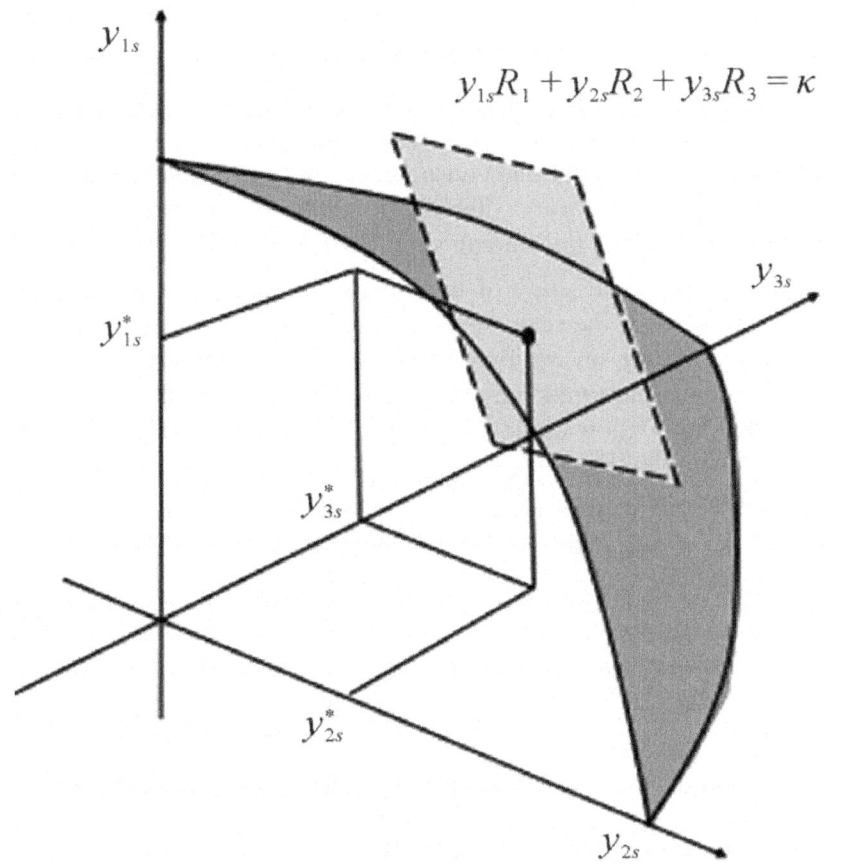

Figure 1. Production possibilities frontier for outputs for land.

where the rents on each activity i for parcel s is a function of the amount of each output supplied from other parcels and a set of exogenous variables (i.e., z_2 are a set of variables that affect the profitability of agriculture such as food prices and the cost of production). These rents follow the complementary slackness conditions from general equilibrium solutions [1]

$$\left[y_1^D - \sum_{s=1}^{S} y_{1s} \right] \times R_1\left(\{y_{1r}\}_{r=1}^{S}, z_2 \right) \leq 0$$

$$\left[y_2^D - \sum_{s=1}^{S} y_{2s} \right] \times R_2\left(\{y_{2r}\}_{r=1}^{S}, z_3 \right) \leq 0$$

$$\left[y_3^D - \sum_{s=1}^{S} y_{3s} \right] \times R_3\left(\{y_{3r}\}_{r=1}^{S}, z_4 \right) \leq 0$$

$$\left[y_4^D - \sum_{s=1}^{S} y_{4s} \right] \times R_4\left(\{y_{4r}\}_{r=1}^{S}, z_5 \right) \leq 0$$

$$(4)$$

where y_i^D is the quantity of output i demanded by consumers. Given these equilibrium conditions, a positive rental rate implies that the demand for output i equals the amount that consumers wish to consume (or producers can profitably use in production). Alternatively, if the amount that consumers wish to consume is less than the amount supplied, no rents will accrue to this use.

A key issue in the study of land use is the jointness of production. As depicted in Figure 1, the selection of output is continuous. Any parcel of land may produce one or several outputs at the same time. For example, pasture land and commercial forest may provide habitat for wildlife. A more realistic vision of the production surface is presented in Figure 2 where the production surface is a simplex (i.e., the tradeoff between outputs is linear). In this case production typically occurs at one of the corners of the production surface. However, the production of $\{\tilde{y}_{1s}, \tilde{y}_{2s}, \tilde{y}_{3s}\}$ is not on the y_{1s} axes (i.e., $\tilde{y}_{2s} > 0$ and $\tilde{y}_{3s} > 0$). At this point agricultural land could produce forestry outputs and urban uses. This production in a sense is free—maybe this land has sandy ridges that are used for windbreaks.

In the case of forests, commercial forests may generate significant environmental flows such as wildlife habitat. Hence, if we assume $s_2 \subset S$ to be that set of parcels used for commercial forests, we would expect that

$$y_2^D - \sum_{s \in s_2} y_{2s} = 0 \Rightarrow R\left(\{y_{2r}\}_{r=1}^{S}, z_3 \right) \gg 0$$

$$(5)$$

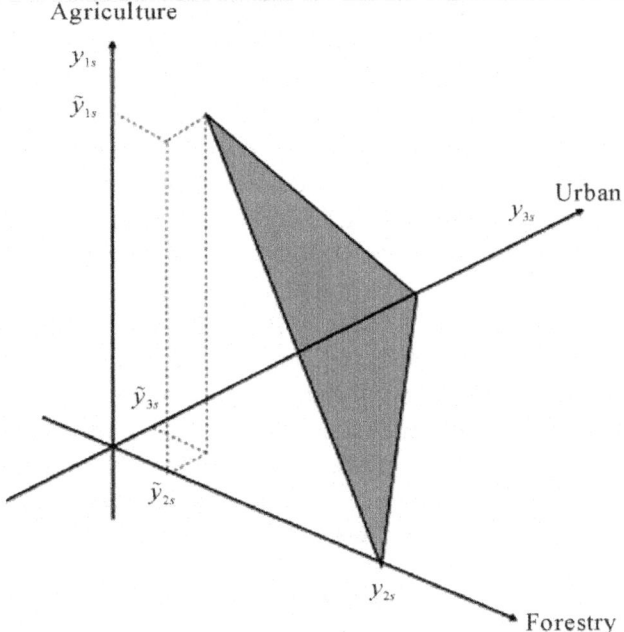

Figure 2. Linear production possibilities frontier for land outputs.

or that the number of parcels planted to commercial forestry exactly meets the demand for products from commercial forestry implying positive rents to commercial forestry. However, in meeting these demands, the forestry sector generates environmental flows

$$\sum_{s \in S_2} y_{4s} \geq 0$$

(6)

The question in analyzing the land use solution is whether this indirect source of environmental services is adequate to satisfy the demand

$$y_4^D - \sum_{s \in S_2} y_{4s} < 0 \Rightarrow R\left(\{y_{4r}\}_{r=1}^S, z_5\right) = 0$$

$$y_4^D - \sum_{s \in S_2} y_{4s} \geq 0 \Rightarrow R\left(\{y_{4r}\}_{r=1}^S, z_5\right) \geq 0.$$

(7)

In the first scenario (where the rents for allocating land to environmental uses is equal to zero), no additional land will be allocated to produce environmental flows.

An underlying argument to support programs that remove land from private ownership is that the market solution has generated insufficient environmental services. The contention is that private market does not allocate enough land to generate the desired level of environmental services. This disequilibrium is typically hypothesized to be driven by a market failure such as the non-exclusionary nature of most environmental services. The argument is then that the appropriate level of environmental services will only be forthcoming if a government or philanthropic entity enters the land market to produce the desired level of services. However, once the values of the environmental flows are separated from the market transaction (i.e., land is no longer allocated across uses based on market rents), it is difficult if not impossible to answer the basic allocation question—is too much or too little land allocated to a specific use? Schmitz, Kennedy, and Hill-Gabriel [2] examine this question in the context of the buyout of sugar land from US Sugar to aid in the restoration of Florida's Everglades. They find that the cost of buying the sugar land exceeds the benefit unless the environmental impact is taken into account. Given this anomaly, they hypothesize the existence of an environmental equivalence (e.g., an amount of environmental benefits to justify the policy action). They find that the sugar buyout cost is far less than the benefits when the environmental benefits are included.

DATA AND METHODOLOGY

In this study, we examine the changes in the uses of land in three significant agricultural regions in the United States—the Delta States (Arkansas, Louisiana, and Mississippi), the Pacific States (California, Oregon, and Washington), and the Southeast (Alabama, Florida, Georgia, and South Carolina). These regions represent areas of environmental concern. For example, in Florida the policy questions include the effect of commercial agriculture on the Everglades as well as the urban encroachment on other environmentally sensitive lands such as wetlands. California's policy questions include the effect of urban growth in environmentally sensitive areas such as Chaparral. These states are also similar in the importance of high-valued agriculture such as fruits and vegetables to agriculture. Finally, these regions have similar levels of public and private forests in each state. The data used in this study are from Nickerson, Borchers, and Carriazo [3] who report the land allocated to cropland, pasture, forests, special uses (including rural transportation, rural, state, and national parks, and military uses), urban uses, and other (including marshes, swamps, bare rock, and other unclassified uses). The detailed data is available online [4]. Table 1 presents the land use for each region.

To analyze the change in land use over time we use the information inequality to measure the amount of statistical information in the distribution of use. Specifically, we define the statistical information index (I) as

$$I(q,p) = \sum_{i=1}^{N} q_i \ln\left(\frac{q_i}{p_i}\right) \quad \ni: \sum_{i=1}^{N} q_i = 1 \text{ and } \sum_{i=1}^{N} p_i = 1$$

(8)

where q_i is a posterior distribution and p_i is a prior distribution. The information index in Equation (8) was popularized by Theil [5]. Later, Theil [6] exemplifies much of this work which focuses on the use of the $I(q,p)$ to measure income inequality. In this application, q_i is the share of income in region i while q_i is the share of population. Mathematically, as the inequality of between income shares and population shares increases the information measure becomes larger.

Table 1. Land use by region in 1000 acres, 1945-2007.

Year	Total Land	Cropland	Pasture	Forest	Special	Urban	Other
Delta States							
1945	93,006	22,192	7215	51,404	3085	649	8461
1949	92,855	24,283	6017	52,715	3507	841	5492
1954	92,855	22,162	8501	51,641	3556	815	6180
1959	92,690	20,808	9358	53,245	3772	1118	4389
1964	92,600	20,238	9433	54,624	3874	1194	3237
1969	92,269	24,558	8433	50,471	3162	1200	4445
1974	92,269	25,054	7449	50,470	3171	1407	4718
1978	92,269	25,950	5777	49,453	3358	1849	5882
1982	92,053	24,978	7390	47,827	3470	2122	6266
1987	92,053	23,888	7307	47,443	3566	2416	7433
1992	91,235	23,739	6357	48,269	3710	2717	6442
1997	91,235	22,031	5534	50,672	3694	3065	6241
2002	91,224	21,046	6246	50,667	4418	2251	6595
2007	91,224	18,230	7209	52,317	4500	2284	6683
Pacific States							
1945	204,883	23,404	56,824	96,546	12,117	1857	14,135
1949	204,699	27,023	60,550	97,160	12,474	2722	4770
1954	204,699	26,243	59,850	89,905	14,745	2085	11,871
1959	204,500	26,134	53,965	89,863	14,621	3218	16,699
1964	204,422	25,451	54,307	89,819	18,948	3681	12,216
1969	204,233	24,302	52,594	89,952	19,521	4117	13,747
1974	204,233	24,786	53,761	89,747	19,663	4621	11,655
1978	204,233	25,378	52,595	85,882	21,106	5215	14,057

1982	204,156	25,403	52,296	85,077	21,890	5815	13,675
1987	204,156	25,236	51,981	80,576	24,974	6755	14,634
1992	203,876	23,928	54,480	79,278	23,282	7377	15,530
1997	203,876	24,367	52,144	76,661	31,228	7903	11,573
2002	203,840	23,949	52,337	78,296	32,344	7124	9791
2007	203,840	22,110	57,040	74,021	36,821	7239	6610
Southeast States							
1945	124,450	26,973	8686	72,994	5212	1279	9306
1949	124,242	27,919	6776	74,926	5696	1876	7049
1954	124,242	24,824	9967	78,114	6618	1858	2861
1959	124,068	21,071	13,939	76,855	7035	2904	2264
1964	123,817	18,880	12,564	78,992	7352	3179	2850
1969	123,581	20,424	10,498	77,061	7232	3323	5043
1974	123,581	20,708	11,341	76,256	7738	4042	3496
1978	123,581	21,150	9285	75,078	8633	5852	3583
1982	123,635	20,338	10,387	73,356	8503	6815	4236
1987	123,635	18,290	10,044	73,500	8630	8373	4798
1992	123,377	18,053	9780	73,434	9245	8042	4823
1997	123,377	17,982	9116	71,938	8984	9136	6220
2002	123,319	14,824	8281	73,661	9099	8707	8748
2007	123,319	12,483	10,288	75,150	9698	8887	6815

However, others have used Equation (8) as an estimation tool. Specifically, Salois, Moss, and Erickson [7] formulate the information inequality as

$$J\left(v,p,y|\phi\right) = \sum_{i=1}^{N} v_i \ln\left(\frac{v_i}{\phi p_i + \left(1-\phi\right)y_i}\right) \tag{9}$$

where v_i is the share of farmland values in state i, p_i is the share of agricultural income in state i, y_i is a measure of the share of urban pressure in state i, and ϕ is a parameter. They then determine the significance of agricultural income versus urban pressure by selecting ϕ to minimize the information inequality ($J(\cdot)$). Hence, the information inequality measures the distribution between two distributions.

We are interested in two applications of the general inequality measure. First, we are interested in the difference in the distribution of land use by land type between states

$$I_{irt} = \sum_{s=1}^{S_r} \left(\frac{Q_{st}q_{sit}}{Q_{rit}} \right) \ln \left(\frac{\frac{Q_{st}q_{sit}}{Q_{rit}}}{Q_{st}} \right)$$

(10)

where Q_{st} is the share of land in state s as a percent of total land in region r, q_{sit} is the share of state s's land used for land use i at time t, and Q_{rit} is the share of land in region r in use i at time t. Essentially, $Q_{st}q_{sit}/Q_{rit}$ is equal to the share of land use i in state s (e.g., across all states). If this share is higher than the overall share of state s's in total land, then that use is concentrated in that state. Alternatively, if this share is lower than the overall share of the state's land in the region, then that land use must be relatively higher in the remaining states in the region. Next, we aggregate the inequality across all states

$$\overline{I}_{rt} = \sum_{i=1}^{N} Q_{st}I_{sit}$$

(11)

where $\overline{I}_{rt}$ is the average use inequality in region r at time t.

Following Moss, Mishra, and Erickson [8], we use the information measure to analyze the change in land use between two periods in time. We assume that the land use i at time t_0 is p_{it_0} while the share of the same land use at time t_1 is q_{it_1}. The change in information between the two periods is then

$$\tilde{I}(t_0,t_1) = \sum_{i=1}^{N} q_{it_1} \ln \left(\frac{q_{it_1}}{p_{it_0}} \right)$$

(12)

Intuitively, if $\tilde{I}(.) \to 0$ there has been little change in land use.

RESULTS

The inequality between land uses in each region is presented in Table 2. The overall inequality of land use is slightly higher in the Pacific States. Most of this inequality is explained by differences in the urban land use. Numerically, 78.8 percent of the urban land use across the Pacific states was in California, but California only accounts for 49.0 percent of the land in this region. This dominance falls slightly to 71.4 percent in 2007 with most of the gains occurring in Washington that increases from 13.3 percent of the urban use in 1945 to 19.3 percent in 2007. Apart from the changes in the urban use, there have been random fluctuations in the special land use. Again, California dominates this use with 71.7 percent of special land use in 1945 declining to 68.9 percent in 2007.

In the Southeastern States, Florida dominates the urban land use with 40.1 percent of all urban use in 1945 compared with 27.9 percent of all land. In addition, the Southeast is different in that the share of urban use has been relatively stable over time. Most of the changes in the differences in land use in the Southeast involve differences in pasture. Again, the difference is Florida which accounted for 46.1 percent of all pastureland in the region in 1945. This increased to 54.0 percent in 2007.

The dispersion of land use in the Delta States is fairly small. The differences occur in the other land use category. Most of this variation can be attributed to Louisiana which accounted for 60.3 percent of this category in 1945 increasing to 65.5 percent in 2007 compared to Louisiana's share of land which was 31.0 percent. Much of this use is attributable to Louisiana's swamps such as the Atchafalaya Basin.

Table 2. Inequality of land use in each region, 1945-2007.

Year	Cropland	Pasture	Forest	Special	Urban	Other	Average
Delta States							
1945	0.01580	0.04892	0.00161	0.00168	0.09724	0.18496	0.02602
1949	0.01522	0.02256	0.00065	0.00594	0.09763	0.16084	0.01643
1954	0.00980	0.02312	0.00025	0.00341	0.06142	0.09049	0.01129
1959	0.01193	0.00065	0.00030	0.00132	0.06250	0.51640	0.02817
1964	0.01236	0.02759	0.00098	0.00482	0.06350	0.64836	0.02978
1969	0.01254	0.00066	0.00016	0.01346	0.03134	0.31669	0.01961
1974	0.01240	0.00131	0.00016	0.01347	0.01307	0.33931	0.02157
1978	0.00984	0.00030	0.00076	0.01842	0.01710	0.32985	0.02523
1982	0.00908	0.00331	0.00089	0.01950	0.02178	0.24643	0.02120
1987	0.01214	0.00382	0.00132	0.01722	0.00732	0.19219	0.02051
1992	0.01318	0.00642	0.00138	0.02464	0.02363	0.25793	0.02453
1997	0.01842	0.00142	0.00423	0.02579	0.02829	0.39166	0.03567
2002	0.01832	0.00280	0.00412	0.04693	0.06406	0.23695	0.02769
2007	0.01732	0.01760	0.00522	0.04938	0.06499	0.26237	0.03113
Pacific States							
1945	0.03458	0.04496	0.00362	0.13846	0.20894	0.33894	0.05159
1949	0.03041	0.02818	0.00271	0.13068	0.22195	0.04883	0.02569
1954	0.03433	0.04381	0.00576	0.14340	0.17196	0.27485	0.04776
1959	0.03632	0.03873	0.00544	0.13497	0.18242	0.16725	0.04343
1964	0.04115	0.03240	0.00530	0.09629	0.20396	0.31661	0.04758
1969	0.05180	0.04427	0.00450	0.10485	0.19805	0.18391	0.04594

1974	0.04974	0.04670	0.00429	0.09991	0.18847	0.17688	0.04419
1978	0.04814	0.05036	0.00700	0.08632	0.18841	0.20276	0.04958
1982	0.04829	0.03462	0.00657	0.08937	0.18764	0.17270	0.04411
1987	0.04634	0.04643	0.00287	0.06901	0.19838	0.09593	0.04056
1992	0.05155	0.03322	0.00551	0.05843	0.18590	0.12732	0.04017
1997	0.05223	0.04012	0.00865	0.10603	0.18142	0.04835	0.04577
2002	0.04502	0.04794	0.00727	0.09775	0.14526	0.12570	0.04702
2007	0.05164	0.03715	0.03178	0.11040	0.14422	0.19577	0.05895
Southeast States							
1945	0.09015	0.10462	0.00284	0.01048	0.04479	0.04558	0.03282
1949	0.07574	0.13066	0.00217	0.02821	0.04108	0.01881	0.02844
1954	0.05715	0.11441	0.00007	0.04217	0.00982	0.12467	0.02590
1959	0.05536	0.15445	0.00082	0.04772	0.01297	0.24549	0.03475
1964	0.02358	0.15979	0.00381	0.02657	0.01894	0.37707	0.03298
1969	0.02477	0.19141	0.00655	0.07173	0.05435	0.14155	0.03587
1974	0.01629	0.17373	0.00620	0.10603	0.08086	0.10446	0.03474
1978	0.01223	0.22659	0.00870	0.09300	0.04086	0.17520	0.03791
1982	0.01577	0.23907	0.00877	0.08652	0.04611	0.09931	0.03977
1987	0.01444	0.21309	0.00938	0.13306	0.05512	0.13060	0.04311
1992	0.01424	0.21631	0.01195	0.14475	0.04356	0.08541	0.04337
1997	0.03012	0.23996	0.01627	0.12904	0.04967	0.10010	0.04973
2002	0.00434	0.20823	0.01839	0.10480	0.08902	0.05079	0.04311
2007	0.01461	0.18497	0.01384	0.12536	0.09123	0.02748	0.04329

Overall the inequality of land use does not show dramatic changes in land use over time. Especially apparent is the lack of reallocation to either the special or other land use categories, the exception being Florida. The land use inequality for the special category use in the Southeastern States increased from 0.01048 in 1945 to 0.12536 in 2007. This change in inequality is associated with an increase from 30.0 percent of special land use in Florida for 1945 to 51.6 percent in 2007. At the same time the other land use category for Florida declined from 32.3 percent of the Southeast in 1947 to 12.8 percent in 2007.

There is more information regarding the changes in land use across regions as depicted in Table 3. The regions have very different allocations to pasture use that persist over time. Interestingly, the Pacific States accounted for 78.1 percent of pasture use in 1945 declining only slightly to 76.5 percent in 2007 compared with the fact that the Pacific States account for 48.6 percent of land in our sample. Interestingly, the inequality in the special use substantially increases over time from 0.02625 in 1945 to 0.12037 in 2007. This growth can be primarily attributed to an increase in the special land use in the Pacific States. In 1945, 59.4 percent of the special land use occurred in the Pacific States. This amount increased to 72.2 percent in 2007.

Finally, turning to the change in land allocation over time, Table 4 presents the change in information for each state and region between 1945-1954 and 1997-2007. In general, the Delta States have seen very little change in the allocation of land between uses while the results for the other two regions are more mixed. The largest change in allocation occurs in Florida. Three categories account for this shift—the amount allocated to forestry has fallen from 64.8 percent to 43.8 percent while the amount of land allocated to special land uses have increased from 6.0 percent to 13.8 percent and the urban use increased from 1.8 percent to 11.1 percent. The next largest change was for South Carolina. Most of South Carolina's change results from a reduction in the use of land for crops, an increase in the land devoted to forests, and a slight increase in the share of land for urban uses. In the Pacific States, the changes in Washington are due to a slight decrease in the share of forestry and a rather significant increase in the special land use and urban uses of land. Similarly, the change in California is due to an increase in the special land use from 9.4 percent of California's land to 20.3 percent. A decline in forests in California is from 42.3 percent of California's land to 32.1 percent. And an increase from 1.7 percent to 5.4 percent of land allocated to urban uses.

IMPLICATIONS AND DISCUSSION

A widely held belief is that the increased affluence in the United States since World War II has increased the demand for environmental services and these are being met through the reallocation of land from commercial uses such as agriculture into reserves, parks and other environmental uses.

Table 3. Inequality between regions, 1945-2007.

Year	Cropland	Pasture	Forest	Special	Urban	Other	Average
1945	0.05505	0.18546	0.00490	0.02625	0.00890	0.00620	0.04577
1949	0.04435	0.25053	0.00592	0.01820	0.01521	0.09184	0.05995
1954	0.03529	0.16342	0.01272	0.02795	0.02240	0.07109	0.04852
1959	0.02669	0.09338	0.01233	0.02155	0.03002	0.13964	0.03744
1964	0.02673	0.10502	0.01460	0.04614	0.02960	0.07436	0.03800
1969	0.05572	0.12883	0.01182	0.06730	0.02985	0.02374	0.04397
1974	0.05586	0.13538	0.01135	0.06410	0.03502	0.03822	0.04592

1978	0.05767	0.17955	0.01339	0.06327	0.05940	0.05551	0.05549
1982	0.05281	0.14077	0.01222	0.06608	0.06464	0.03788	0.04817
1987	0.04995	0.14464	0.01655	0.08276	0.07652	0.03689	0.05273
1992	0.05694	0.17039	0.01837	0.06353	0.05142	0.03538	0.05651
1997	0.04364	0.18284	0.02141	0.11431	0.05778	0.00607	0.06121
2002	0.04675	0.18453	0.02094	0.10363	0.08290	0.01930	0.06242
2007	0.04271	0.16220	0.02943	0.12037	0.08390	0.05809	0.06962

Table 4. Change over time.

State	Inequality	
Arkansas	0.01887	
Louisiana	0.02767	
Mississippi	0.03273	
Delta States		0.01651
California	0.09043	
Oregon	0.04799	
Washington	0.10001	
Pacific States		0.07112
Alabama	0.08107	
Florida	0.18794	
Georgia	0.08064	
South Carolina	0.10028	
Southeast		0.08728
Total		0.05471

For example, the farm bills have included payments for the conservation reserve program (CRP) which removed environmentally sensitive land from production. These CRP payments increase the environmental flows from agricultural lands. In addition programs such as Florida's plan to buyout US Sugar's land between Lake Okeechobee and the Everglades was intended to provide environmental benefits to the Everglades and Florida Bay between the mainland of Florida and the Florida Keys. In addition, several contend that the desire for environmental flows may limit the conversion of land into urban uses. In this study we examine the allocation of land across uses including cropland, pasture, forests, special (which includes the local, state and national parks), urban, and other. In general, changes between these uses have been small over time. There is some evidence of increased demand for environmental flows with the growth in the special category in California and Florida. However, these increases are often created by reductions in forests that also provide certain environmental amenities. In addition, there is little evidence that the desire for increased environmental flows limit the growth of urban land use over time.

The lack of private markets for environmental flows from land in most states suggests that these demands are met through traditional land uses. Forestry provides many of the same environmental flows produced by the transfer of land public ownership (i.e., parks and wildlife reserves). One of the dominant questions remains—of the land transferred from other uses into parks and recreational areas, how much is transferred at full price? Stated slightly differently, how much of the land transferred into environmental uses meets the marginal rental condition posited in the theoretical model?

Agricultural technology also has contributed to significance improvements in environmental stewardship. For example, the adoption of no-till air seeding equipment has greatly reduced soil erosion in the high plains grain growing area of North America [9]. These innovations have not only increased yields and reduced energy costs, but also reduced water use in these areas.

A debate centers on the private stewardship of land and the need for regulation. Evidence suggests that farmers and ranchers have increasingly adopted best management practices that are environmentally friendly. For example, set-backs from waterways increase the land's ability to provide water quality. Ranchers in North Florida may receive the County Alliance for Responsible Environmental Stewardship (CARES) [10] award for responsible management of nitrogen and animal runoff into Florida's springs. Similarly, areas of the private land may be declared wetlands; farmers may then follow prescribed agronomic practices in these areas. In some states, farmers are restricted from controlling wildlife activity on their land—for example elk damage in Minnesota. In return for this restriction, these farmers can apply for compensation when wildlife damage occurs. We have not taken these best management practices into account in our analysis. At one level, these regulations imply government provision of environmental services. This provision could be compensated (as in the case of elk damage) or uncompensated (as in the case of set-backs). Regardless of the mechanism, the level of environmental services is increased. However, since this provision does not typically occur in a market system, it is difficult to determine whether society is made better off by the acquisition. Following Schmitz, Kennedy, and Hill- Gabriel [2], these programs imply an environmental equivalence when imposed on private land uses.

REFERENCES

1. Moss, C.B. and Schmitz, A. (2014) Valuing Carbon Recycling through Ethanol: Zero Prices for Environmental Goods. Theoretical Economics Letters, 4, 235-240. http://dx.doi.org/10.4236/tel.2014.43032

2. Schmitz, A., Kennedy, P.L. and Hill-Gabriel, J. (2012) Restoring the Florida Everglades through a Sugar Land Buyout: Benefits, Costs, and Legal Challenges. Environmental Economics, 3, 74-85.

3. Nickerson, C., Ebel, R., Borchers, A. and Carriazo, F. (2011) Major Uses of Land in the United States, 2007. United States Department of Agriculture, Economic Research Service; Economic Information Bulletin No. 89.

4. Borchers, A. (2014) Major Land Uses. http://www.ers.usda.gov/data-products/major-land-uses.aspx#25979

5. Theil, H. (1967) Economics and Information Theory. North-Holland Publishing Company, Amsterdam.

6. Theil, H. (1989) The Development of International Inequality 1960-1985. Journal of Econometrics, 42, 145-155.http://dx.doi.org/10.1016/0304-4076(89)90082-1

7. Salois, M., Moss, C.B. and Erickson, K. (2012) Farm Income, Population and Farmland Prices: A Relative Information Approach. European Review of Agricultural Economics, 39, 289-307. http://dx.doi.org/10.1093/erae/jbr032

8. Moss, C.B., Mishra, A.K. and Erickson, K. (2007) Next Year on the US Farmland Market: An Information Approach. Applied Economics, 39, 581-585. http://dx.doi.org/10.1080/00036840500447831

9. Schmitz, A. and Moss, C.B. (2014) Mechanized Agriculture: Labor Displacement and Machine Adoption. Symposium in Honor of Wallace Huffman, Ames, 1-2 August 2014.

10. Florida Farm Bureau (2014) Farmers CARE about Florida's Natural Resources. Florida Agriculture, 74, 10-11.

Chapter 9

COMPRESSIVE MECHANICAL PROPERTIES AND MICROMECHANICAL CHARACTERISTICS OF WARM AND ICE-RICH FROZEN SILT

Yugui Yang[1,2], Feng Gao[2], and Yuanming Lai[3]

[1]State Key Laboratory for Geomechanics and Deep Underground Engineering, China University of Mining and Technology, Xuzhou, Jiangsu 221008, China

[2]School of Mechanics and Civil Engineering, China University of Mining and Technology, Jiangsu 221116, China

[3]State Key Laboratory of Frozen Soil Engineering, Cold and Arid Regions Environmental and Engineering Research Institute, Chinese Academy of Sciences, Lanzhou 730000, China

ABSTRACT

It is recognized experimentally that the compressibility of warm and ice-rich frozen soil is remarkable under loading, which will cause a significant deformation and affect the stability of infrastructure constructed in cold region. In this paper, the real-time computerized tomography tests of warm and ice-rich frozen silt were carried out. The microstructure characteristics in the process of loading were studied, and the macromechanical behaviors were obtained at the same time. The test results showed that the stress-strain curve of warm and ice-rich frozen silt is sensitive to temperature; the peak stress was greatly enhanced with the decrease of temperature, and the section area increases with the increase of axial strain; the water content mainly decreases with the increase of axial strain at $-1°C$; the change of water content is not obvious at $-2°C$ in the loading process. The density damage changes little at first and then increases with the further increase of axial strain.

INTRODUCTION

About 20% continental land in China is covered by permafrost. In permafrost regions, mechanical properties have been one of the most extensively studied aspects in many engineering problems. The mechanical behavior of frozen soil is crucial to the stability of construction of embankment engineering, such as highways, railways, and other engineering activities in permafrost regions [1–5]. Moreover, with the climate getting warm, the engineering construction in permafrost regions will encounter a lot of problems in the embankment stability of warm and ice-rich frozen soil. Therefore in the evaluation of embankment stability in permafrost regions, the possible influences of warm and ice-rich frozen soil have to be considered [6–8]. In order to explore the mechanical properties of frozen soil and make the design of frozen soil engineering more scientific and reasonable, a series of studies in warm and ice-rich frozen soils have been carried out. Considering the effect of water contents and temperature, [6] studied the strength characteristic of warm and ice-rich frozen clay by carrying out uniaxial compression test. The test results indicated that the type of stress-strain curves of warm and ice-rich frozen clays is strain-hardening, and the form of sample failure is plastic. Reference [9] carried out experimental studies on the compressible behavior of warm and ice-rich frozen clay and found that the frozen clay is essentially sensitive to both load and temperature. Reference [10] studied the dynamic strength characteristic of ice-rich frozen clay under various temperatures and confining pressures. The creep behaviors of warm and ice-rich frozen soils were also studied by researchers. Reference [11] investigated uniaxial creep tests of frozen clay with various water contents (40%, 80%, and 120%) at warm temperature (−0.3°C, −0.5°C, and −1.0°C) and found that the strain rate decreased with the increasing in time. Reference [12] carried out a series of investigations on the creep deformation of warm and ice-rich frozen clay with various water contents at different temperatures.

The studies mentioned above are the latest developments in warm and ice-rich frozen silt. It can be seen that the macromechanical characteristic of warm and ice-rich frozen has been acquired; however, there are few studies concerning the micromechanical characteristic of warm and ice-rich frozen soil. The mechanical properties of frozen soil are essentially governed by the constituent grains' properties and their structures. The microstructure of ice-rich frozen soil is shown in Figure 1.

Grain properties include its size, shape, crushability, and so on, while the frozen soil structure can be represented by void ratio, fabric tensor, the orientation of grain's axes, and so on. It is necessary to know both the macromechanical property and the microstructure evolution of frozen soils during deformation process. Recent development of computer ability has made it possible to handle such microscopic properties. The development of noninvasive imaging allied with computed tomography has begun to allow the study of inner structure propagation of frozen soil and the measurement of localized change of water content. CT is a quantitative measurement technology based on digital techniques, which is used to detect and describe the sectional characteristics of tested materials, and has been applied to investigations of frozen soil [13].

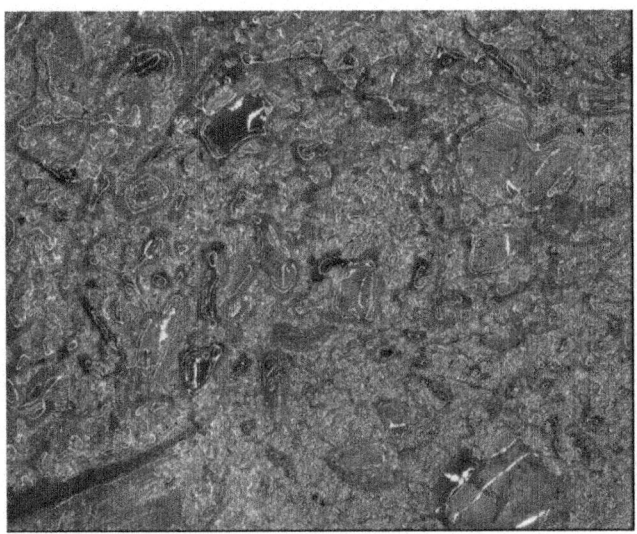

Figure 1: The electron microscope scanning image.

In order to explore the deformation and failure behavior of warm and ice-rich frozen soil, it is necessary to carry out the studies on micromechanical microstructure. In this paper, uniaxial compression tests of warm and ice-rich frozen silt were conducted, and the real-time computerized tomography tests of the microstructure propagation law in the whole deformation process have been completed. The stress-strain curve and CT image of cross section have been analyzed, and the area of cross section, volumetric water content, and density damage propagations are obtained. The results of this study may help elucidate why warm and ice-rich frozen soil is so subject to damage and displays such a contrasting temperature behavior, which is critical to designing possible solutions to reduce their damage in cold region engineering.

TESTING METHOD AND PROCESS

Sample and Equipment

In this paper, the soil used in test was taken from Qinghai-Tibet Railway constructions' site and particle size distribution in Figure 2. The specimens were prepared as cylinders with 6.18 cm in diameter and 12.5 cm in height. The water content and the average dry density of the specimens tested were 40.0% and 1.43 g/cm³, respectively. The preparations of specimens had been given a detailed introduction by [14]. After the preparation of the specimen ends, the specimen was placed into the loading equipment.

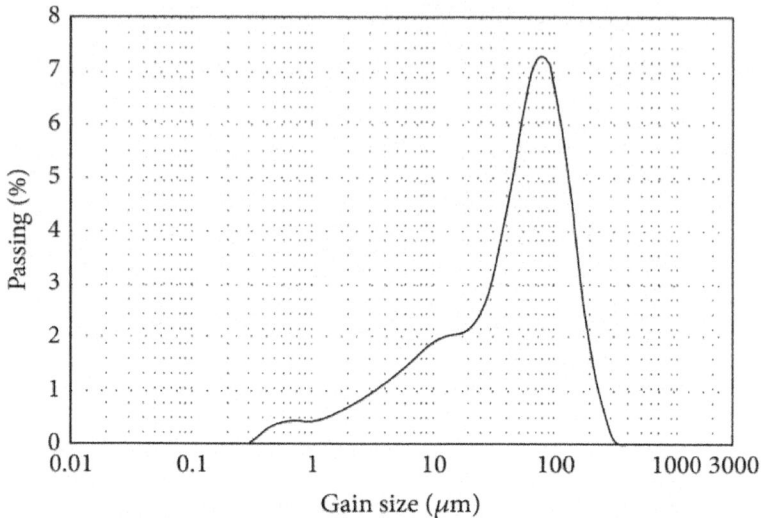

Figure 2: Particle size distribution of silt.

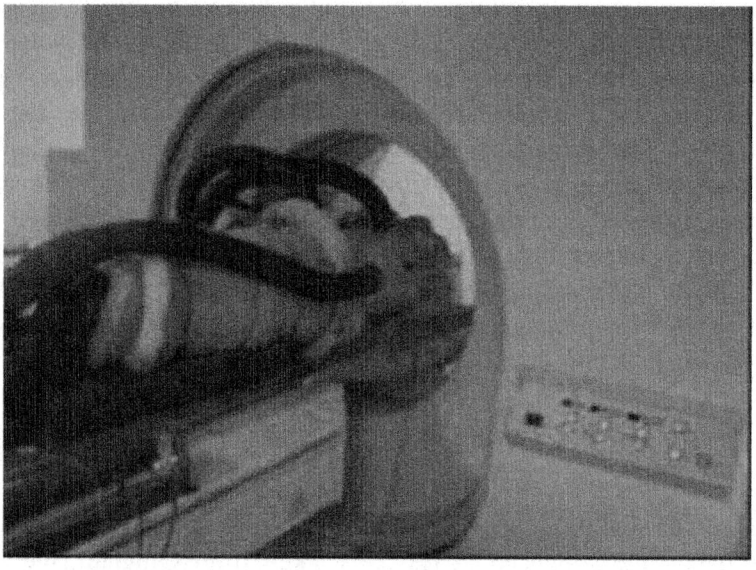

(a)

(b)

Figure 3: (a) The computerized tomography Siemens SOMATOM-PLUS X-ray and (b) the loading equipment matching of computerized tomography for frozen soil (after reference [15]).

The computerized tomography equipment adopted in this study is the spiral scanning Siemens SOMATOM-PLUS X-ray (see Figure 3). This test system includes 4 parts: the cooling bath cyclic system, the loading system, computerized tomography machine, and data acquisition system. The special loading apparatus is winded with cold bathing pipes. The alcohol is used as the cold bath medium. The range of temperature for the digital temperature control system is −20~25°C and we had a temperature accuracy of ±0.1°C.

Test Method and Process

After the specimens were placed into the loading equipment, the loading equipment was wrapped with foam rubber sponge, which was used to keep the temperature stability of loading apparatus and frozen soil. Then, the temperature was adjusted to the specified temperature by the digital temperature control system. The sample was loaded at a rate of 0.3 mm/min. When the axial strains were 0, 1, 3, 6, 10, and 15%, the specimen location was adjusted to proper scanning stratum position, as shown in Figure 4; then it was scanned by the computerized tomography method at specific horizon of sample of frozen soil. The force value, section area, and CT value were recorded by the computer system which corresponds to the loading system.

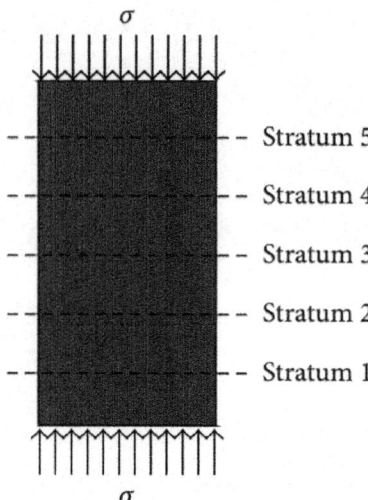

Figure 4: The scanning stratum positions of specimen.

ANALYSIS OF THE TEST RESULTS

We review observations on behavior in the elastoplastic deformation that throw light on the evolution of microstructure and thus on the mechanism of failure. We shall now be concerned with the complete progression of events as the whole stress-strain or force-displacement curve. The stress is computed by taking the average of all forces and dividing by appropriate areas. The strain in the axial direction is computed using the current specimen length and the original specimen length. In order to accurately study the stress-strain characteristic of warm and ice-rich frozen silt, the section area evolution of frozen soil sample should be researched. The cross section deformation subjected to loading is studied using CT scanning. From Figure 5, it can be seen that there is a marked change in section area of warm and ice-rich frozen silt in the loading process. The section areas increase with the increase of axial strain. The maximum amount of cross section area increases about 20%.

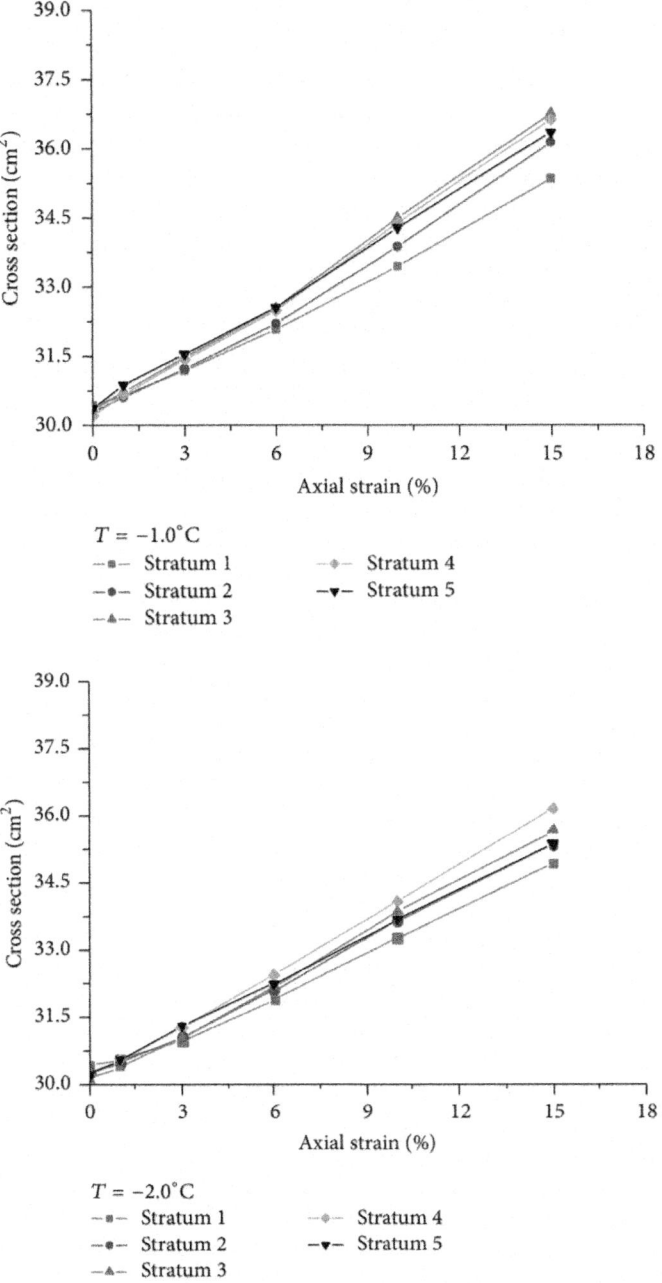

Figure 5: Cross section propagation of warm and ice-rich frozen silt.

The stress-strain curves of warm and ice-rich frozen silt are shown in Figure 6. It can be seen that there is significant difference between nominal stress and true stress in the compression process. The true stress is greatly larger than nominal stress with increasing in axial strain. The result also indicates that the calculation of stress should take the change of section area into account. The stress-strain behavior of warm and ice-rich frozen silt approximately experiences two stages: the initial linear elastic stage, where stress increases linearly with the increasing of axial strain, and there is little plastic strain in this stage, and the plastic stage, where the plastic deformation is dominating in the further loading process, and the elastic deformation is relatively subordinate.

The dry density and water content can reflect changes in the microstructure of the frozen silt associated with the deformation process that involves pores or microcrack. Inelastic density change is designated as dilation or compaction depending on whether there is volume increase or decrease, respectively. Warm and ice-rich frozen silt may be considered as elastoplastic material, of which the compressibility is remarkable under loading.

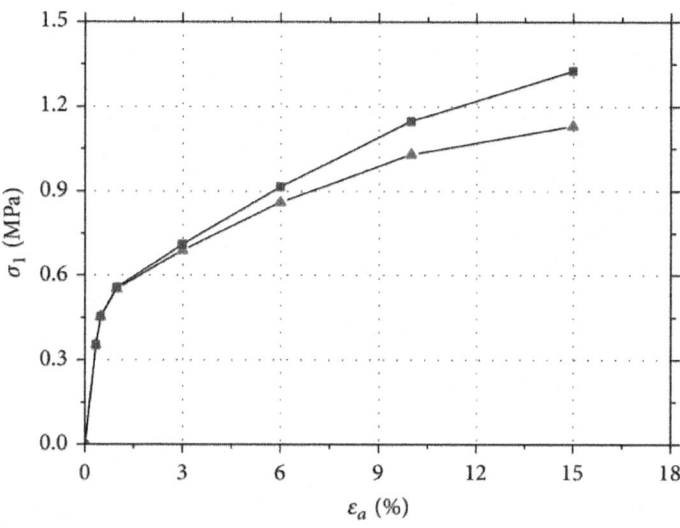

$T = -1.0°C$

$--▲--$ True stress
$--■--$ Nominal stress

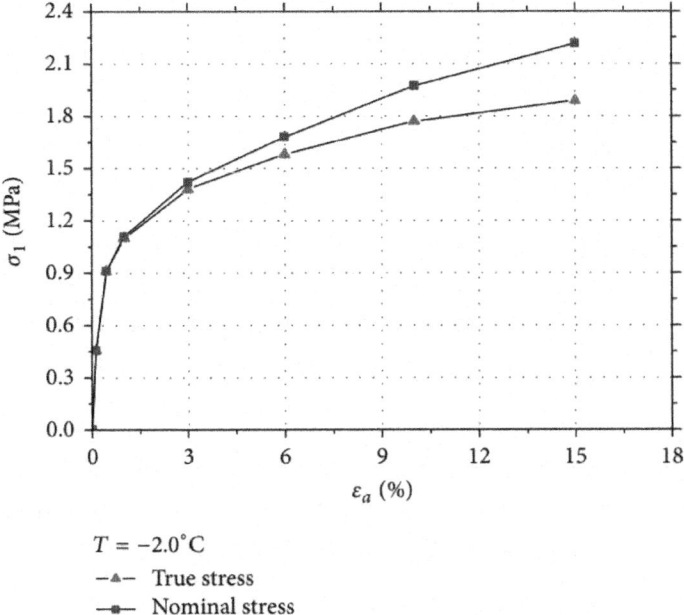

$T = -2.0°C$

−▲− True stress

−■− Nominal stress

Figure 6: Stress-strain curves of warm and ice-rich frozen silt.

Considering that the plastic material has no obvious crack in the loading process, the accurate damage propagation cannot be directly obtained from the visual image. The accurate damage propagation can be identified by the corresponding CT values of scanned slices. It is more significant to investigate the change of regularity of CT values in the loading process. The following equation shows the relationship between CT value and X-ray absorption coefficient of test material [16]:

$$H = \frac{\mu - \mu_W}{\mu_W} \times 1000, \tag{1}$$

where μ is the X-ray absorption coefficient per unit tested material mass; μ_w is the X-ray absorption coefficient of water. H is the CT value of tested material.

The X-ray absorption law complies with addition principle. The X-ray absorption coefficient of frozen soil can be expressed as the following equation [16]:

$$\mu = W_V \times \mu_W + \gamma_d \times \mu_S, \tag{2}$$

where W_v is volumetric water content; γ_d is dry density; μ_w and μ_s are the

X-ray absorption coefficients of water and soil particle

Table 1 contains the CT values of scanned slices corresponding to strains 0, 1, 3, 6, 10, and 15%. It is noted that the CT value of every scanned slice has certain regularities with the damage propagation of frozen silt. The CT values increase with increasing axial strain; with further increase of axial strain, the CT values decrease.

Table 1: Distribution regularities of CT values of scanned slices.

Strain (%)	CT value at −1°C.	CT value at −1°C.	CT value at −2°C.	CT value at −2°C.
	Entire section area	Major section area	Entire section area	Major section area
0	1105.5	1223.2	1099.7	1219.5
1	1105.7	1224.8	1100.7	1222.1
3	1105.3	1226.6	1099.1	1223.0
6	1103.8	1227.8	1097.6	1224.6
10	1100.1	1226.0	1092.3	1221.8
15	1092.7	1218.5	1083.2	1210.9

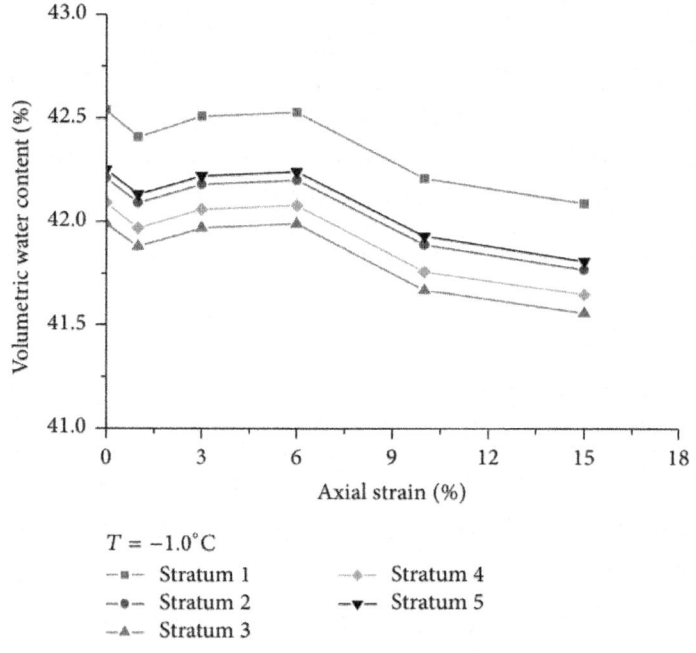

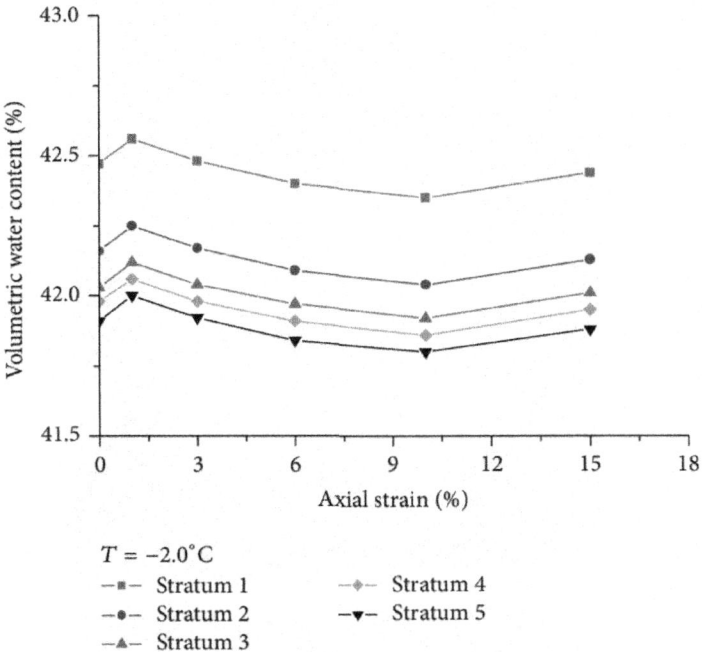

Figure 7: Distributions of volumetric water contents for warm and ice-rich frozen silt.

Figure 7 depicts the regularity of water content obtained from the seismic tomography. The water content of warm and ice-rich frozen silt at −1°C decreases with the increase of axial strain at first; then the water content increases. With the further increase of axial strain, the water content decreases. The water content of warm and ice-rich frozen silt at −2°C changes little in the loading process.

Figure 8 shows CT images of warm and ice-rich frozen silt. It is evident that the edge of specimen at $\varepsilon_a = 0$ is slightly smooth; however, once the strain increases to $\varepsilon_a = 15\%$, the edge becomes rough and has many burrs, which means that the structure of frozen soil has been damaged. The visual observation of CT image cannot represent the accurate change of inner structure. In order to better describe propagation of inner structure, the density damage has been defined by the following equation [16]:

$$\Omega = \left[1 - \frac{\gamma_d}{2.8 \times (1 - w_v \times B)} \right] \times 100\%,$$

(3)

in which γ_d is the dry density of soil particle; $\gamma_d = 2.8$ g/cm3 ; w_v is volumetric water content; B is expansion efficient

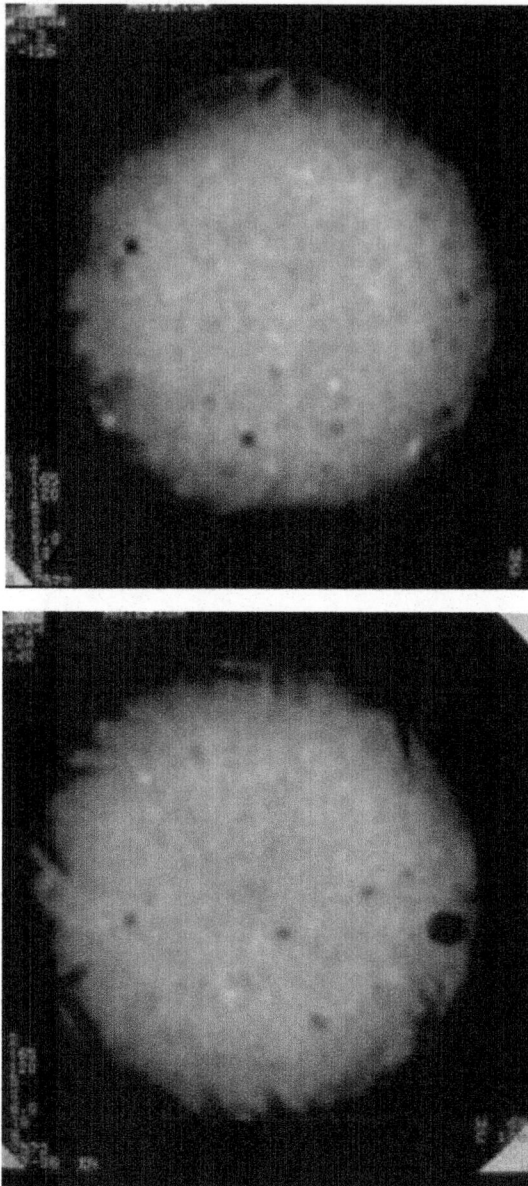

Figure 8: CT images with $\varepsilon_a = 0$ and $\varepsilon_a = 15\%$ of warm and ice-rich frozen silt.

Based on the analysis of water content, the density damage of warm and ice-rich frozen silt at various loading stages has been given in Figure 9.

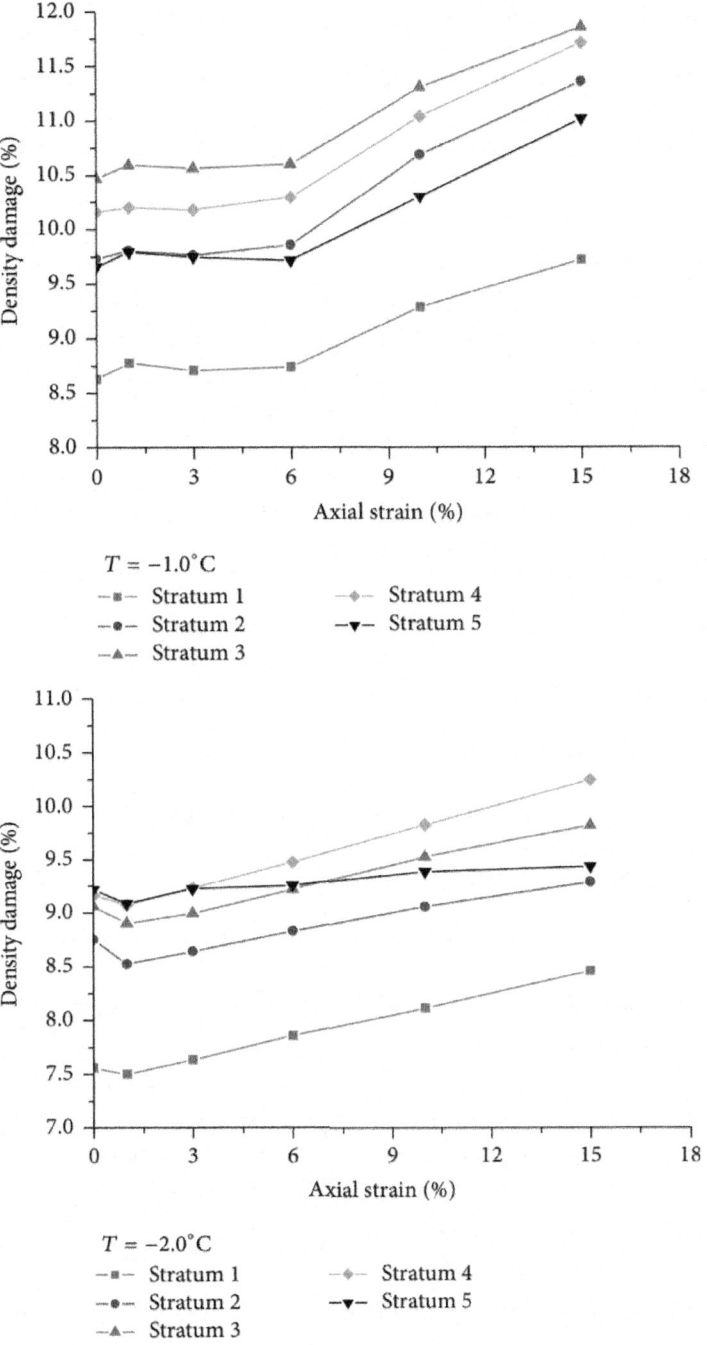

Figure 9: Distributions of density damage of warm and ice-rich frozen silt.

The result shows that the initial density damage at different slices of sample is varying, but the change tendency is similar. Generally, the density damage changes little at first; then it increases with the further increase of axial strain. The nonlinear characteristic of density damage curve is obvious at −1°C; however, the damage curves show almost linear increase with increasing in axial strain at −2°C.

CONCLUSIONS

The nondestructive techniques computerized tomography is employed to assess the inner structure propagation of the warm and ice-rich frozen silt. The peak stress of warm and ice-rich frozen silt, which is sensitive to temperature, can be greatly enhanced with temperature decreasing from −1°C to −2°C. The true stress of warm and ice-rich frozen silt is obviously larger than nominal stress, so the mechanical analysis of warm and ice-rich frozen soil should take the variation of section area into account. With the increase of axial strain, the section area increases, and the maximum amount of cross section area increases about 20%; the water content mainly decreases with the increase of axial strain at −1°C, and the change of water content is not obvious at −2°C in the loading process. The density damage changes little at first; then it increases with the further increase of axial strain.

ACKNOWLEDGMENTS

This research was supported by the National Natural Science Foundation of China (51204161) and the Fundamental Research Funds for the Central Universities (2012QNA57).

REFERENCES

1.   V. R. Parameswaran and S. J. Jones, "Triaxial testing of frozen sand.," Journal of Glaciology, vol. 27, no. 95, pp. 147–155, 1981.

2.   A. M. Fish, "Strength of frozen soil under a combined stress state," in Proceedings of 6th International Symposium on Ground Freezing, vol. 1, pp. 135–145, A.A. Balkema, Rotterdam, The Netherlands, 1991.

3.   G. X. Cui, "Mechanics of frozen soil for deep alluvium- a new field of frozen soil mechanics," Journal of Glaciology and Geocryology, vol. 20, no. 2, pp. 97–100, 1998.

4.   W. Ma, G.-D. Cheng, Q.-B. Wu, and D.-Y. Wang, "Application on idea of dynamic design in Qinghai-Tibet Railway construction," Cold Regions Science and Technology, vol. 41, no. 3, pp. 165–173, 2005.

5. Y. Lai, L. Jin, and X. Chang, "Yield criterion and elasto-plastic damage constitutive model for frozen sandy soil," International Journal of Plasticity, vol. 25, no. 6, pp. 1177–1205, 2009.

6. W. Ma, G. Feng, Q. Wu, and J. Wu, "Analyses of temperature fields under the embankment with crushed-rock structures along the Qinghai-Tibet Railway," Cold Regions Science and Technology, vol. 53, no. 3, pp. 259–270, 2008.

7. Y. Lai, S. Li, J. Qi, Z. Gao, and X. Chang, "Strength distributions of warm frozen clay and its stochastic damage constitutive model," Cold Regions Science and Technology, vol. 53, no. 2, pp. 200–215, 2008.

8. S. W. Liu and J. M. Zhang, "Review on physic mechanical properties of warm frozen soil," Journal of Glaciology and Geocryology, vol. 34, no. 1, pp. 120–129, 2012.

9. Y. H. Qin, J. M. Zhang, B. Zheng, and X. Ma, "Experimental study for the compressible behavior of warm and ice-rich frozen soil under the embankment of Qinghai-Tibet Railroad," Cold Regions Science and Technology, vol. 57, no. 2-3, pp. 148–153, 2009.

10. Z. H. Gao, J. Shi, S. J. Zhang, and L. J. Luo, "Experimental study of the dynamic strength characteristics and residual strain of ice-rich frozen soil," Journal of Glaciology and Geocryology, vol. 31, no. 6, pp. 1143–1149, 2009.

11. X.-J. Ma, J.-M. Zhang, X.-X. Chang, B. Zheng, and M.-Y. Zhang, "Experimental study on creep of warm and ice-rich frozen soil," Chinese Journal of Geotechnical Engineering, vol. 29, no. 6, pp. 848–852, 2007.

12. B. Zheng, J. M. Zhang, X. J. Ma, et al., "Study on compression deformation of warm and ice-enriched frozen soil," Chinese Journal of Rock Mechanics and Engineering, vol. 28, supplement 1, pp. 3063–3068, 2009.

13. Z. W. Wu, W. Ma, Y. B. Pu, et al., "CT analysis on structure of frozen soil in creep process," CT Theory and Applications, vol. 4, no. 3, pp. 31–34, 1995.

14. Y. Yang, Y. Lai, and X. Chang, "Experimental and theoretical studies on the creep behavior of warm ice-rich frozen sand," Cold Regions Science and Technology, vol. 63, no. 1-2, pp. 61–67, 2010.

15. J. F. Zheng, Research on Meso-Structure Change of Frozen Soil under the Action of Load, Cold and Arid Regions Environmental and Engineering Research Institute of the Chinese Academy of Sciences, Lanzhou, China, 2009.

16. X. R. Ge, Macromechanical and Micromechanical Experiment Research of Damage for Geomaterials, Science Press, 2004.

Chapter 10

GENETIC AND ENVIRONMENTAL CONTROLS ON NITROUS OXIDE ACCUMULATION IN LAKES

Jatta Saarenheimo[1], Antti J. Rissanen[1], Lauri Arvola[2], Hannu Nykänen[1], Moritz F. Lehmann[3], and Marja Tiirola[1]

[1]Department of Biological and Environmental Science, University of Jyväskylä, 40014, Jyväskylä, Finland

[2]Lammi Biological Station, University of Helsinki, 16900, Lammi, Finland

[3]Department for Environmental Science, University of Basel, CH-4058, Basel, Switzerland

ABSTRACT

We studied potential links between environmental factors, nitrous oxide (N_2O) accumulation, and genetic indicators of nitrite and N_2O reducing bacteria in 12 boreal lakes. Denitrifying bacteria were investigated by quantifying genes encoding nitrite and N_2O reductases (*nirS*/*nirK* and *nosZ*, respectively, including the two phylogenetically distinct clades $nosZ_I$ and $nosZ_{II}$) in lake sediments. Summertime N_2O accumulation and hypolimnetic nitrate concentrations were positively correlated both at the inter-lake scale and within a depth transect of an individual lake (Lake Vanajavesi). The variability in the individual *nirS*, *nirK*, $nosZ_I$, and $nosZ_{II}$ gene abundances was high (up to tenfold) among the lakes, which allowed us to study the expected links between the ecosystem's *nir*-vs-*nos* gene inventories and N_2O accumulation. Inter-lake variation in N_2O accumulation was indeed connected to the relative abundance of nitrite versus N_2O reductase genes, i.e. the (*nirS*+*nirK*)/$nosZ_I$ gene ratio. In addition, the ratios of (*nirS*+*nirK*)/$nosZ_I$ at the inter-lake scale and (*nirS*+*nirK*)/$nosZ_{I+II}$ within Lake Vanajavesi correlated positively with nitrate availability. The results suggest that ambient nitrate concentration can be an important modulator of the N_2O accumulation in lake ecosystems, either directly by increasing the overall rate of denitrification or indirectly by controlling the balance of nitrite versus N_2O reductase carrying organisms.

INTRODUCTION

Nitrous oxide (N_2O) is an important greenhouse gas and the single most important ozone destroying chemical [1]. N_2O in the biosphere is produced as an intermediate molecule in denitrification or nitrifier-denitrification, or as a by-product during nitrification or dissimilatory nitrate reduction to ammonium (DNRA) [2, 3]. The denitrification pathway includes four enzymatically catalyzed reductive steps: nitrate reduction (*nar*), nitrite reduction (*nir*), nitric oxide reduction (*nor*), and nitrous oxide reduction (*nos*) [4]. Reduction of nitrite, where the first gaseous form of fixed nitrogen (N) (i.e. NO) is produced, is catalyzed by two analogous genes: *nirK* and *nirS* genes encoding a copper nitrite reductase and a cytochrome cd1-nitrite reductase, respectively [4]. These two genes prevail in different organisms and their differential distributions in nature seem to be modulated by the redoxconditions, with *nirS*being preferentially expressed under low dissolved oxygen conditions [5, 6]. Recent studies have also revealed that *nosZ* genes encoding N_2O reductase actually belong to two phylogenetically distinct clades [7, 8], here referred to as *nosZ*$_I$ and *nosZ*$_{II}$, which need to be analyzed by separate PCR primer sets. As with *nir* genes, the relative importance of *nos*genes seems to systematically differ between habitats and with environmental conditions [8], yet the exact controls that modulate their relative abundance in nature are uncertain. Some denitrifiers are lacking the *nosZ* gene completely and perform the truncated denitrification pathway, where N_2O is produced as an end-product [9]. In fact, genome sequencing showed that one third of the cultivated denitrifying bacteria lack the *nosZ* gene [10].

Since denitrifier community structure is likely to have an effect on net N_2O production and emission [11, 12], denitrifier communities have been studied through the analysis of sequence variation and/or the abundance of *nirS*, *nirK*, and *nosZ* genes in many ecosystems [13, 14, 15, 16, 17]. High availability of nitrate and nitrite has been shown to be conducive to N_2O accumulation [18, 19], fostering the increase the $N_2O/(N_2O+N_2)$ ratio in the gaseous denitrification products [20, 21]. Such correlations may simply indicate nitrate-induced enhancement of denitrification rates (and thus N2O accumulation), but they may also be the result of microbial community adaptation. Philippot et al. [22], for example, demonstrated that the relative abundance of the *nosZ* gene was a strong predictor of the $N_2O/(N_2O+N_2)$ production ratio.

In soils, microbially produced N_2O is likely lost to the atmosphere by turbulent diffusive escape. In contrast, in aquatic environments, the diffusivity of gases is much slower ($K_z$values on the order of 10^{-5} to 10^{-6} cm^2 s^{-1}, [23]), reducing diffusive loss rates and improving the N_2O availability for *nosZ* carrying bacteria. More complete denitrification and lower N_2O/N_2 gas emission ratios should, therefore, be expected for the aquatic versus soil environments. Still, lake ecosystems have shown to be important sites of N_2O emissions [19,24], and, as in soils, N_2O production and accumulation in lakes appears to be dependent on the ambient nitrate and oxygen concentrations [25, 26, 24, 27]. Although the importance of lacustrine N_2O production is well recognized [19, 26], and albeit the fact that benthic denitrifier community structure has been studied in some lakes [28, 29], it is not known whether variations in the accumulation of N_2O are mostly directly dependent on the environmental conditions, or whether they rather are indirectly constrained by the denitrifying community structure. With some recent exceptions [7, 8] the role of the $nosZ_{II}$ clade remained mostly unconsidered in this context.

Here, we evaluated genetic and environmental factors that likely modulate N_2O production and accumulation in lake ecosystems, especially focusing on the benthic abundance of $nirS$, $nirK$, $nosZ_I$, and $nosZ_{II}$ genes during the summertime N_2O accumulation period. Anticipating close links between nitrate concentrations and the N_2O accumulation, we hypothesized 1) that high hypolimnetic nitrate concentrations would decrease the relative abundance of the $nosZ$ genes (i.e., increase the nir/nos ratio) within lacustrine sediments, and 2) that higher nir/nos ratios would lead to enhanced N_2O accumulation. The linkage between benthic denitrification gene frequency and N_2O accumulation was assessed in an inter-lake study of 12 boreal lakes in southern Finland, pooling the lakes into two groups based on their hypolimnetic nitrate concentrations (high-NO_3^--lakes and low-NO_3^--lakes). In addition, denitrification gene abundance and N_2O accumulation was investigated along a littoral-to-pelagic transect in a large stratified lake (Vanajavesi) with relatively high hypolimnetic nitrate levels (24.0−44.9 µmol l^{-1}).

RESULTS

Comparison of Denitrification Genes in High- Versus Low-Nitrate Lakes

Considerable inter-lake variation was observed with regards to the nitrate (0.4−79.1 µmol l^{-1}), ammonium (0.6−61.4 µmol l^{-1}), and oxygen (1.9−333.4 µmol l^{-1}) concentrations (S1 Table). The lakes were classified into two groups based on their nitrate concentration, which generally reflected land use in the catchment area: high-NO_3^--lakes comprised lakes mostly with extensive agricultural activity in their catchment area and one urban lake (Jyväsjärvi), while low-NO_3^--lakes included lakes mostly with little agricultural land in their catchment area. Other environmental parameters did not differ significantly between the two groups (Table 1).

Table 1. Environmental parameters (mean and SE) for high-NO$_3^-$-lakes (n = 6) and low-NO$_3^-$-lakes (n = 6), and results of a t-test or Mann-Whitney U-test* comparing the oxygen, nitrate, ammonium and phosphate concentrations, temperature, catchment field area (ha), averaged N$_2$O$_{excess}$ concentrations, and maximum observed N$_{2excess}$ concentrations between the two lake groups.

	O$_2$ (μmol l^{-1})	NO$_3$ (μmol l^{-1})	NH$_4^+$ (μmol l^{-1})	PO$_4^-$ (μmol l^{-1})	T (C°)	Field area (ha)	N$_2$O$_{excess}$ (μmol m^{-3})	N$_{2excess}$ (μmol l^{-1})
High-nitrate lakes								
Mean	101.15	39.30	15.48	0.15	11.41	37990	18.14	5.55
(±SE)	(±51.30)	(±8.76)	(±8.55)	(±0.02)	(±1.49)	(±35870)	(±4.97)	(±0.79)
Low-nitrate lakes								
Mean	77.72	0.64	34.67	0.08	15.78	500	1.36	1.12
(±SE)	(±35.77)	(±0.11)	(±17.68)	(±0.03)	(±2.42)	(±245)	(±1.50)	(±0.47)
Pairwise test results								
High vs. low nitrate	high = low	high > low	high = low	high = low	high = low	high > low	high > low	high > low
P	ns	0.007	Ns	ns	ns	0.012*	0.001	0.001

doi:10.1371/journal.pone.0121201.t001

Throughout the studied lakes, the abundances of $nirS$, $nirK$, $nosZ_I$, and $nosZ_{II}$ relative to 16S rRNA genes varied between 0.6–12.9% (Table 2), and the gene copy numbers ranged between 4.8 and 580 per ng of DNA (S2 Table). The ratio of $nirS/nirK$ ranged between 0.5–2.0 (average 1.0), and the ratio of $nosZ_I/nosZ_{II}$ varied between 0.5–5.7 (average 1.9). Neither environmental factors (oxygen, temperature, nitrate concentration) nor N_2O accumulation showed any significant correlation with the gene abundance, gene copy numbers, or with $nirS/nirK$ or $nosZ_I/nosZ_{II}$ gene ratios (Pearson correlations, p values >0.05). The relative proportion of the previously unaccounted $nosZ_{II}$ gene was of a similar magnitude as that of $nosZ_I$, but showed a markedly higher inter-lake variability (Table 2). Although not statistically significant, $nosZ_I$ and $nosZ_{II}$ seemed slightly more abundant in the low-NO_3^- group of lakes, while $nirS$ and $nirK$ seemed less abundant, (Fig. 1A). The $(nirS+nirK)/nosZ_I$ ratio was higher in high-NO_3^--lakes than in low-NO_3^--lakes (Fig. 1B). In addition, the $(nirS+nirK)/nosZ_I$ gene ratio correlated positively with the estimated net N_2O production, as well as with nitrate and phosphate concentrations (Table 3.). As for $(nirS+nirK)/nosZ_{II}$ and $(nirS+nirK)/(nosZ_I+nosZ_{II})$, we also observed a tendency for higher ratios in the high-NO_3^--lakes compared to low-NO_3^- lakes (Fig. 1B). However, correlation between nitrate and $(nirS+nirK)/(nosZ_I+nosZ_{II})$ was only weakly significant (p = 0.06) (Table 3).

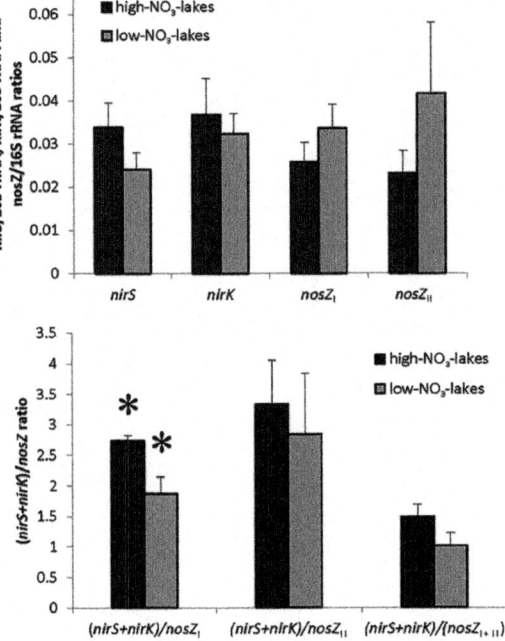

Figure 1. Abundance of $nirS$, $nirK$, $nosZ_I$, and $nosZ_{II}$ genes relative to the amount of 16S rRNA genes (A), and ratios of nir and nos genes in sediments of lakes with

high and low nitrate concentrations (high-NO3⁻-lakes and low-NO3⁻-lakes) (B). $_*$ = significantly different between the two lake groups (Mann-Whitney U-test, p = 0.006).

Table 2. Copy numbers (mean ±**SE**) of *nirS*, *nirK*, $nosZ_I$, and $nosZ_{II}$ gene amplicons as percentages of 16S rRNA gene copy numbers (nd, no data).

Inter-lake comparison	Denitrification gene (% of 16S rRNA gene)				High-nitrate/ Low-nitrate
	nirS	*nirK*	$nosZ_I$	$nosZ_{II}$	
Pääjärvi	5.5	7.9	4.7	2.1	High
(±SE)	(±0.25)	(±0.26)	(±0.34)	(±0.07)	
Mommilanjärvi	4.5	2.3	2.5	2.4	High
(±SE)	(±0.28)	(±0.26)	(±0.24)	(±0.18)	
Ormajärvi	3.6	3.9	2.4	2.1	High
(±SE)	(±0.09)	(±0.34)	(±0.08)	(±0.02)	
Vanajavesi	2.9	3.2	2.4	1.3	High
(±SE)	(±0.31)	(±0.28)	(±0.32)	(±0.14)	
Jyväsjärvi	3.0	3.4	2.5	5.2	High
(±SE)	(±0.08)	(±0.24)	(±0.15)	(±0.22)	
Suolijärvi	0.9	1.3	0.9	1.5	High
(±SE)	(±0.38)	(±0.21)	(±0.19)	(±0.10)	
Ekojärvi	1.2	2.5	2.1	2.9	Low
(±SE)	(±0.07)	(±0.31)	(±0.23)	(±0.19)	
Kataloistenjärvi	4.4	5.2	3.9	1.2	Low
(±SE)	(±0.20)	(±0.4)	(±0.32)	(±0.24)	
Teuronjärvi	2.5	3.0	2.8	1.3	Low
(±SE)	(±0.28)	(±0.28)	(±0.29)	(±0.04)	
Kyynäröjärvi	2.5	3.5	2.9	2.5	Low
(±SE)	(±0.25)	(±0.26)	(±0.09)	(±0.07)	
Kastanajärvi	1.9	1.4	6.1	12.9	Low
(±SE)	(±0.04)	(±0.05)	(±0.25)	(±0.25)	
Lehee	2.0	3.8	2.3	4.1	Low
(±SE)	(±0.21)	(±0.22)	(±0.12)	(±0.26)	
Intra-lake depth transect					
Vanajavesi2	4.9	2.4	2.1	1.8	
(±SE)	(±0.28)	(±0.20)	(±0.28)	(±0.08)	
Vanajavesi3	6.2	3.5	3.4	2.4	
(±SE)	(±0.33)	(±0.32)	(±0.43)	(±0.09)	
Vanajavesi4	4.8	3.6	3.4	1.7	
(±SE)	(±0.26)	(±0.32)	(±0.36)	(±0.05)	
Vanajavesi5	3.3	4.2	2.9	1.4	
(±SE)	(±0.29)	(±0.30)	(±0.22)	(±0.10)	
Vanajavesi6	2.4	4.2	2.4	0.6	
(±SE)	(±0.23)	(±0.27)	(±0.26)	(±0.13)	
Vanajavesi7	2.9	3.2	2.4	1.3	
(±SE)	(±0.31)	(±0.28)	(±0.32)	(±0.14)	
Vanajavesi8	3.1	2.2	1.1	nd	
(±SE)	(±0.20)	(±0.19)	(±0.14)		

doi:10.1371/journal.pone.0121201.t002

Table 3. Correlations of functional gene ratios and accumulated N_2O and N_2 gas concentrations with environmental parameters in the inter-lake dataset. Correlation coefficients with $0.01 < p < 0.05$ and $p < 0.01$ are written in normal text and bold, respectively.

	Gene ratios			Gas accumulation measurements		
	$(nirS+nirK)/nosZ_I$	$(nirS+nirK)/nosZ_{II}$	$(nirS+nirK)/(nosZ_{I+II})$	N_2O_{excess} (µmol m^{-3})	N_2O production (µmol N m^{-2} d^{-1})	$N_{2excess}$ (µmol l^{-1})
O_2 (µmol l^{-1})	-	-	-	-	-	-
NO_3^- (µmol l^{-1})	**0.78**	-	(0.55)*	0.66	0.74	0.58
NH_4^+ (µmol l^{-1})	-	-	-	-	-	-
PO_4^- (µmol l^{-1})	0.67	-	-	-	-	-
T (C°)	0.61	-	-	-	-	-
N_2O_{excess} (µmol m^{-3})	-	-	-	1	0.96	**0.80**

* Marginally significant (p = 0.06)

doi:10.1371/journal.pone.0121201.t003

N₂O and N₂ Accumulation in High- Versus Low-Nitrate Lakes

During the summer sampling (late July), most of the study lakes were oversaturated with respect to N_2O (i.e. the depth-integrated mean N_2O_{excess} was >0). N_2O_{excess} in the water column varied between 0.9–37.1 nmol l^{-1} (11–337% oversaturation). The highest N_2O_{excess} concentrations were observed either in near-bottom waters of the lakes or, in the case of stratified lakes (five lakes were stratified with regards to oxygen and displayed an anoxic hypolimnion), at the oxic-anoxic interface within the water column (S1 Fig.). Maximum $N_{2excess}$ concentrations measured using membrane inlet mass spectrometry (MIMS) were generally slightly higher than the equilibrium concentration at given temperatures (<2% oversaturation). $N_{2excess}$ was significantly higher in high-NO_3^- lakes than in low-NO_3^- lakes (Table 1) and correlated with nitrate concentrations (Table 3). Moreover, the depth-integrated N_2O_{excess} concentrations (0–20.3 μmol m^{-3}) and net N_2O production rates (0–11.2 μmol N m^{-2} d^{-1}) estimated from the N_2O concentration profiles were significantly higher in high-NO_3^--lakes than in low-NO_3^--lakes (Table 1), and both correlated with NO_3^- concentration (Table 3). Maximum $N_{2excess}$ concentrations were found to correlate with the depth-integrated N_2O_{excess} concentration (Table 3).

Denitrification Genes and N₂O Accumulation in Lake Vanajavesi

In Lake Vanajavesi, hypolimnetic temperature and oxygen concentrations were tightly correlated, indicating the effect of thermal water column stratification on the vertical distribution of dissolved oxygen (correlation r = -0.98 and p = 0.000). Sampling sites 1–3 (water depths 2–6 m) were fully aerated, sites 4–6 (water depths 8–12 m) displayed lower oxygen concentrations, and the two deepest sampling sites (water depths 14 and 16 m) were anoxic at the bottom of the hypolimnion (S3 Table). Nitrate concentrations (24.0–44.9 μmol l^{-1}) were consistently high at all sampling sites, whereas ammonium (1.1–57.4 μmol l^{-1}) and phosphate (0.03–0.7 μmol l^{-1}) concentrations displayed strong variability between strongly oxygen-depleted and oxygen-replete conditions (S3 Table).

The relative abundances of *nir*S, *nir*K, *nosZ*$_I$, and *nosZ*$_{II}$ genes in Lake Vanajavesi varied between 0.6 and 6.2% of the total 16S rRNA genes (Table 2), with *nosZ*$_I$ or *nosZ*$_{II}$ being the least abundant of the denitrifying genes at all sites. In contrast to observation at the inter-lake scale (where nitrate concentrations were generally lower and more variable), we observed a strong positive correlation between nitrate concentrations and the (*nirS*+*nirK*)/*nosZ*$_{I+II}$ ratio (r = 0.98 and p = 0.001) (Fig. 2). The correlation with either *nosZ*$_I$ or *nosZ*$_{II}$ only was not significant (p > 0.05).

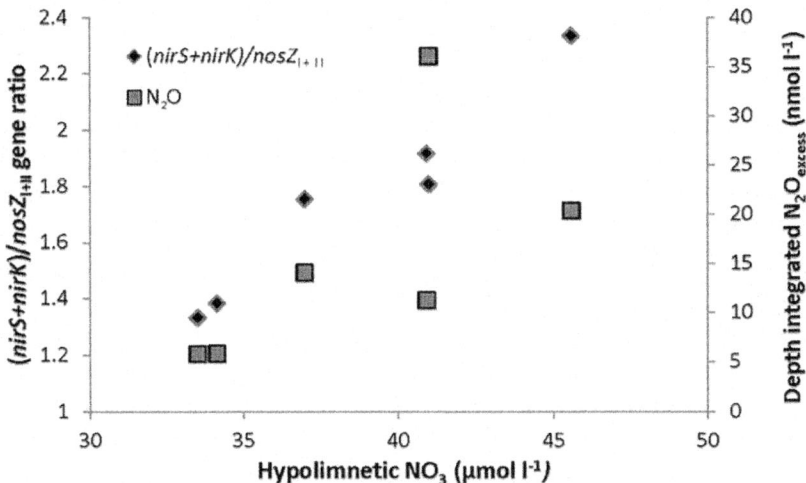

Figure 2. Relationship between hypolimnetic nitrate concentration and the sedimentary (0–2cm) $(nirS+nirK)/nosZ_{I+II}$ gene ratio (r = 0.98 and p = 0.001), and depth-integrated N_2O_{excess} (r = 0.89 and p = 0.02) in Lake Vanajavesi.

At all sampling sites, essentially the entire water column was oversaturated with respect to equilibrium N_2O concentrations (S2 Fig.). The N_2O profiles of Sites 1, 2, and 3 indicated a homogenized water column, with an equal degree of oversaturation throughout. At the deeper Sites 4, 5, and 6, a markedly higher N_2O oversaturation was observed at the bottom of the lake. The degree of N_2O oversaturation was even higher at the oxic-anoxic interface in the water column of Sites 7 and 8 (S2 Fig.). Depth-integrated N_2O_{excess} varied between 5.7–36.0 nmol l^{-1} (62–337% oversaturation) and correlated positively with the nitrate concentration in Lake Vanajavesi (r = 0.89 and p = 0.02) (Fig. 2). A negative correlation was observed with respect to the oxygen concentration (r = -0.90 and p = 0.002) and temperature (r = -0.95 and p < 0.001).

DISCUSSION

To our knowledge, this is the first study combining N_2O measurements and molecular analyses of denitrification genes in lake ecosystems. This is also the first time that the abundance of *nirS* and *nirK* genes together with both clades of *nosZ* genes were investigated in freshwater sediments. The total *nir/nos* ratio was above 1:1 in nearly all study lakes, indicating that the microbial community had a higher potential to produce N_2O than to reduce it. This implies that the accumulation of N_2O is linked to genetic factors.

All the studied denitrification genes (*nir* and *nos* variants) were present in the lake sediments, although their abundance largely varied among the lakes and along the Vanajavesi transect. The qPCR results also revealed that $nosZ_{II}$ genes are as frequent as the canonical $nosZ_I$ genes in the freshwater sediments, which emphasizes the need to further study the ecology of $nosZ_{II}$ encoding organisms in future studies. The relatively high abundance of individual *nirS*, *nirK*, $nosZ_I$, and $nosZ_{II}$ genes highlights the important biogeochemical role of denitrification in boreal lake sediments. For comparison, the abundance of individual denitrification genes *nirS*, *nirK*, and *nosZ* have previously been found to range between 0.5 and 6.8% of the 16S rRNA gene abundance in various soil and sediment samples [14, 21, 30, 31]. Bioavailability of copper (Cu) and iron (Fe) can control the expression and activity of nitrite and nitrous oxide reductases. While *nirK* and *nosZ* are copper-containing reductases, *nirS* is an iron containing cd1-type reductase. Possible Cu limitation may lead to *nirS* dominance and, thus, to increased N_2O accumulation. Unfortunately, data on Fe and Cu concentrations were not available, and we cannot fully exclude Cu versus Fe limitation as a controlling factor in N_2O accumulation in the study lakes. Yet, the equal abundance of *nirS* and *nirK* genes does not suggest any adaptation of the microbial community to Cu limitation.

Data on the *nirS/nirK* gene ratios in lakes are rare. The only study we know of in this context is by Martins et al. [31], who reported that *nirS* genes were more abundant than *nirK* genes in sediments of freshwater lakes on the Azores. In contrast, the average *nirS/nirK* gene ratio observed in this study was 1:1. Different from the subtropical lakes studied by Martins et al. [31], boreal lakes experience seasonal variations in redox and other physico-chemical conditions, which may increase the diversity of ecological niches and prevent certain microbial ecotypes from dominating an ecosystem. Since the distribution of *nirS* and *nirK*genes is phylogenetically scattered [10], the ratio of these two evolutionarily separate, but functionally equivalent, nitrite reductase gene types does not necessarily reflect the dominance of one taxonomical group over another as a function of environmental conditions. Instead, the relatively strong variability in the *nirS* and *nirK* gene ratio between the existing studies highlights the need to quantify both genes when studying the factors affecting N_2O accumulation. Although the *nir/nos* ratio at the DNA level does not necessarily correspond to the respective ratios at the level of mRNA transcripts or enzyme molecules on short-term time scales, it may indicate longer-term genetic adaptation, which was the focus of this study.

When comparing lakes at different spatial scales and between various geographical regions, denitrification rates have shown a clear positive correlation with nitrate availability [32]. This correlation was further corroborated by the observed co-variation of NO_3^- and $N_{2excess}$ in the lakes studied here. Our study also showed the linkage between NO_3^- concentration and N_2O accumulation, which is in agreement with previous work in boreal lakes [18]. Based on previously published N_2 production rates for five of the lakes in this study [29, 32] (unpublished results), the N_2O production rates reported here correspond to 0.2–1.7% of the total gaseous N production ($N_2O/(N_2+N_2O)$ ratio). These values fall within the range of previously reported estimates (0.1–4.1%) for freshwater systems [33]. Besides total denitrification rates, it is the balance between nitrite reduction and N_2O reduction which controls the build-up of N_2O. This balance has been shown to be sensitive to changes in redox conditions [34]; however, the role of longer-term nitrate availability in modulating this balance is uncertain. NO_3^- is generally the preferred electron acceptor for the denitrifying community when compared to N_2O (except for some $nosZ_{II}$ carrying organisms, see the discussion below). Hence, when the competition for nitrate is tighter, reduction of N_2O becomes a more feasible trait for the heterotrophic micro-organisms [35].

At the inter-lake scale, $nir/nosZ_I$ ratios correlated with the nitrate concentrations and N_2O_{excess}. These correlations suggest that the denitrifying communities were adapted to varying nitrate levels within the lake and that they control the ratio of N_2O production versus reduction. Moreover, both at the inter-lake scale and within the Lake Vanajavesi transect the combined nir/nos ratio (i.e. $[nirS+nirK]/nosZ_{I+II}$) correlated with ambient nitrate. In contrast, the $nir/nosZ_I$ ratio did not display any statistically significant correlation with (the less variant) nitrate concentration in Lake Vanajavesi. This apparent difference with regards to the role of $nosZ_I$ and $nosZ_{II}$ may be related to the known genetics of $nosZ_{II}$ carrying organisms. The N_2O reductase $nosZ_I$ has only been found for *Alpha*-, *Beta*-, and *Gammaproteobacteria* and some archaea, whereas $nosZ_{II}$ reductases may be common in a wider range of bacterial and archaeal phyla [7, 8]. While most of the typical $nosZ_I$-harboring microbes have the complete set of denitrification genes, less than half of the known $nosZ_{II}$-carrying microorganisms possess genes of the "upstream" denitrification steps, and $nosZ_{II}$-type reductase was thus named as "non-denitrifier nitrous oxide reductase" [7]. As a consequence, many of the $nosZ_{II}$-carrying microbes are incapable of using nitrate (or nitrite) as an electron acceptor, and are, therefore, less affected by ambient nitrate availability. The variable prevalence of denitrifying versus non-denitrifying $nosZ_{II}$ subsets may explain the above-described differences in the correlation analyses between the inter-lake and intra-lake studies (genetic relationships versus NO_3^- levels).

Although it has been shown that denitrification is the major N_2O source in lake ecosystems [19, 27], it is likely that nitrifiers (i.e. ammonium oxidation and nitrifier-denitrification) also contribute to N_2O production in these environments. In the lake transect, where sampling sites where characterized by different hypolimnetic oxygen regimes, N_2O accumulation patterns were clearly linked to oxygen concentration. Concentration of N_2O peaked near the oxic-anoxic interface, which was located either in the sediment surface or in the water column. This could be due to O_2 availability just above the oxic-anoxic interface, which would increase N_2O production via nitrification [36, 37]. On the other hand, the presence of O_2 even at low levels likely inhibits N_2O reduction compared to other reduction steps in denitrification [37]. Therefore, truncated denitrification would also lead to observed accumulation patterns of N_2O, with concentration maxima in the vicinity of the redox transition zones. The lack of N_2O accumulation in the anoxic water layers of the lakes further supports the notion that stable anoxic conditions are conducive to full denitrification to N_2, while microaerophilic conditions would rather support truncated denitrification and/or slowed nitrous oxide reduction. In addition, dissimilatory nitrate reduction to ammonium (DNRA), in which N_2O can also be formed as a by-product [38], competes with denitrification for nitrate. The most important factor controlling competition between these two processes appears to be the C:N ratio [39, 40], where high ratios favors DNRA over denitrification. In addition, the supply of nitrate relative to nitrite and microbial generation time are identified as key environmental factors in controlling whether nitrate is reduced to nitrogen gas in denitrification, or retained in the ecosystem as ammonium in DNRA [41]. In our study lakes, the C:N ratio of sediment organic material varied between 9 and 27 (on average 17.7), and thus DNRA may have had some role on NO_3^- reduction. However, the actual contribution of N_2O production by organisms carrying out DNRA in lake ecosystems is currently unresolved.

This study provided putative evidence for the control of both denitrifier gene composition and N_2O accumulation by nitrate concentration. This suggests that N_2O emissions from denitrification would be modulated by nitrate-induced changes in the denitrifier communities. In turn, the study indicates that recent increases in the land-based and atmospheric anthropogenic nitrogen loadings from agriculture and energy production may have caused shifts in the lacustrine denitrifier communities as well as stimulated N_2O emissions from lake ecosystems.

EXPERIMENTAL PROCEDURES

Study Sites and the Sampling Procedure

The study lakes are located within the same region in southern Finland (61°01−61°52 N and 25°02−24°09 E), except Lake Jyväsjärvi which is located 150 km north of the other lakes (62°13 N and 25°44 E) (S1 Table). The lakes are located on state land with open access, thus no permits were required for collection of samples. Further, the locations are not protected in any way and the study did not involve endangered or protected species. All the study lakes were sampled in July 2011. The lakes were chosen to cover a wide variety of lake characteristics: size (surface area 25–12000 ha), maximum depth (2–85 m), and nutrient concentrations (S1 Table). All the study lakes are ice-covered from November until the beginning of May. We divided the selected lakes into two groups based on their hypolimnetic nitrate concentrations. High-NO_3^--lakes (n = 6) comprised lakes with NO_3^- concentrations between 15.7–79.4 µmol l^{-1} and low-NO_3^--lakes (n = 6) included lakes with NO_3^- concentrations between 0.6–1.5 µmol l^{-1} (S1 Table).

Depths of the sampling sites were recorded with an echo-sounder (S1 Table) and the water samples were taken with a Limnos tube sampler (height 30 cm, volume 2.1 l). Water samples for gas analyses were collected at ca. 0.5 m, 1 m, 3 m, and 5 m above the lake bottom (if the lake was deep enough) and below/under the surface (0.5 m water depth). Three replicates (30 ml) were taken from each depth for N_2O concentration measurements in 60 ml polypropylene syringes, which were closed with three-way stopcocks after removing any gas bubbles, and transported to the laboratory on ice. Nitrogen gas (N_2) samples for membrane inlet mass spectrometry (MIMS) measurements were taken in 12 ml borosilicate glass tubes (six replicates) with screw-capped butyl rubber septa (Labco Ltd.). We allowed water overflow for at least three volumes to avoid atmospheric contamination, and samples with air bubbles were discarded. Microbial processes in borosilicate glass tubes were stopped by adding 100 µl ZnCl through the septum with a needle under water. Water for nutrient analyses were collected in 1-L bottles from the near-bottom waters of the lakes and all samples were transported to the laboratory on ice. Sediment core samples for analyses of the denitrifier communities were collected in all of the lakes using a mini gravity corer with plexiglass tubes (ø = 3.5 cm).

Water column profiles of temperature and oxygen concentrations were measured *in situ* using a portable field meter (YSI model 58, Yellow Springs Instruments). Dissolved inorganic phosphorus [42], nitrate [43], and ammonium [44] were determined with a flow injection analyzer using standard methods (QuikChem 8000) from filtered (0.2 mm filter; Millipore) water samples.

Quantification of *nirS*, *nirK*, and *nosZ* Genes

Sediment samples were collected from the surface layer (0–2 cm) of the sediment cores and freeze-dried for further use (Alpha 1–4 LD plus, Christ). DNA extraction was performed from 0.03 g of dry sediment using the bead-beating and phenol-chloroform extraction protocol of Griffiths et al. [45]. Two extractions were made from each site. The DNA concentrations were measured with a Qubit 2.0 Fluorometer (Invitrogen) and the DNA concentration of each sample was adjusted to yield a concentration of 10 ng μl^{-1}.

For qPCR quantification of the *nirK*, *nirS*, $nosZ_{I}$, and $nosZ_{II}$ genes, partial 16S rRNA was used as a reference gene, and commonly used primers were selected from previous studies (S4 Table). Amplification of qPCR and fluorescent data collection was carried out with a Bio-Rad CFX96 thermal cycler (Bio-Rad Laboratorios) in a reaction mixture of 0.5 µM of each primer for the selected target gene (except for $nosZ_{II}$ 1 µM of each primer), 10 µl 2XiQ SYBR Green supermix (BioRad), 1 µl of DNA (10 ng), and PCR-grade water (Fermentas) to yield a total volume of 20 µl. Three replicate qPCR amplifications were performed for each sample.

The PCR procedure for 16S rRNA included an initial denaturation step at 95°C for 15 min and 40 cycles of amplification (95°C for 20 s, 53°C for 35 s and 72°C for 70 s). Finally, an increase of 0.5°C s^{-1} from 65 to 95°C was performed to obtain the melting curve analysis of PCR products. The thermal cycling conditions for other genes were the same as the one just described, except that the annealing temperature was 55°C for *nirS*, 60°C for *nirK* and $nosZ_{I}$, and 54°C for $nosZ_{II}$. Standard curves were constructed from PCR amplicons extracted from agarose gel with a BioRad Gel Extraction Kit (BioRad). Amplicons were re-amplified and the resulting products were purified with Agencourt AMPure XP (Beckman Coulter). A dilution series of 10^{7}–10^{2} gene copies were used as standards in each qPCR run. Gene abundances were calculated as relative abundances from the abundance of the reference gene (16S rRNA). Replicate results were averaged (n = 6) and standard errors were calculated. Inhibition was tested from the dilution series (1, 1^{-10} and 1^{-100}) and no inhibition was detected.

N_2O Gas Concentrations

N_2O samples were analyzed according to Maljanen et al. [46] with a gas chromatograph (Agilent 6890N, Agilent Technologies) equipped with an auto sampler (Gilson) and an electron capture detector (ECD). The N_2O samples were processed according to Bellido et al. [47], and two replicates from each depth were measured. N_2O equilibrium concentrations were calculated based on Henry's law (modified from IPCC Fourth Assessment Report: Climate Change 2007 and [48]). Concentration of N_2O accumulated due to microbial reactions (N_2O_{excess}) was calculated from the difference between observed N_2O concentration and the calculated equilibrium concentration. The overall amount of accumulated N_2O per square meter was estimated from integration of the N_2O_{excess} concentration profiles, and the depth-integrated N_2O_{excess} per m^3

was obtained by division through the water depth at the sampling site. All study lakes undergo complete spring mixing after ice-off (with equilibrium concentrations throughout the water column). Assuming cumulative N_2O production in the hypolimnion, with low atmospheric exchange after the mixing period, net N_2O production rates can be estimated according to Mengis et al. [25] (with slight modifications) by dividing the amount of accumulated N_2O per square meter by the number of days since ice-off (i.e. the onset of water column stratification in early May) to the sampling date (end of July). These estimates need to be considered conservative, as at least some turbulent diffusive loss to the atmosphere is indicated by the partial N_2O pressure gradient between surface water and the atmosphere (see S1 Fig.).

Natural N_2 Gas Concentrations

N_2/Ar gas concentration ratios were determined using membrane inlet mass spectrometry (MIMS) as described in Kana et al. [49]. Equilibrium concentrations were calculated according to Weiss [50]. $N_{2excess}$ was then calculated from N_2/Ar ratio in the sample divided by the N_2/Ar ratio at equilibrium for a given temperature.

STATISTICAL ANALYSES

Data analyses were conducted using PASW 18.0 (PASW Statistics 18, Release Version 18.0.0, SPSS 2009). The normality assumption was tested with the Shapiro-Wilks test. In our dataset, the effect of nitrate concentration on process parameters and denitrifier communities was specifically addressed by comparing high-NO_3^--lake and low-NO_3^--lake data either using independent samples t-test (normally distributed variables) or Mann-Whitney U-test (non-normally distributed variables). In addition, correlation analysis (Pearson or Spearmann correlation) was performed to study potential relationships among environmental parameters (NO_3^- concentration, oxygen concentration, ammonium concentration, phosphorus concentration, depth, gene abundances, and N_2O concentrations.

SUPPORTING INFORMATION

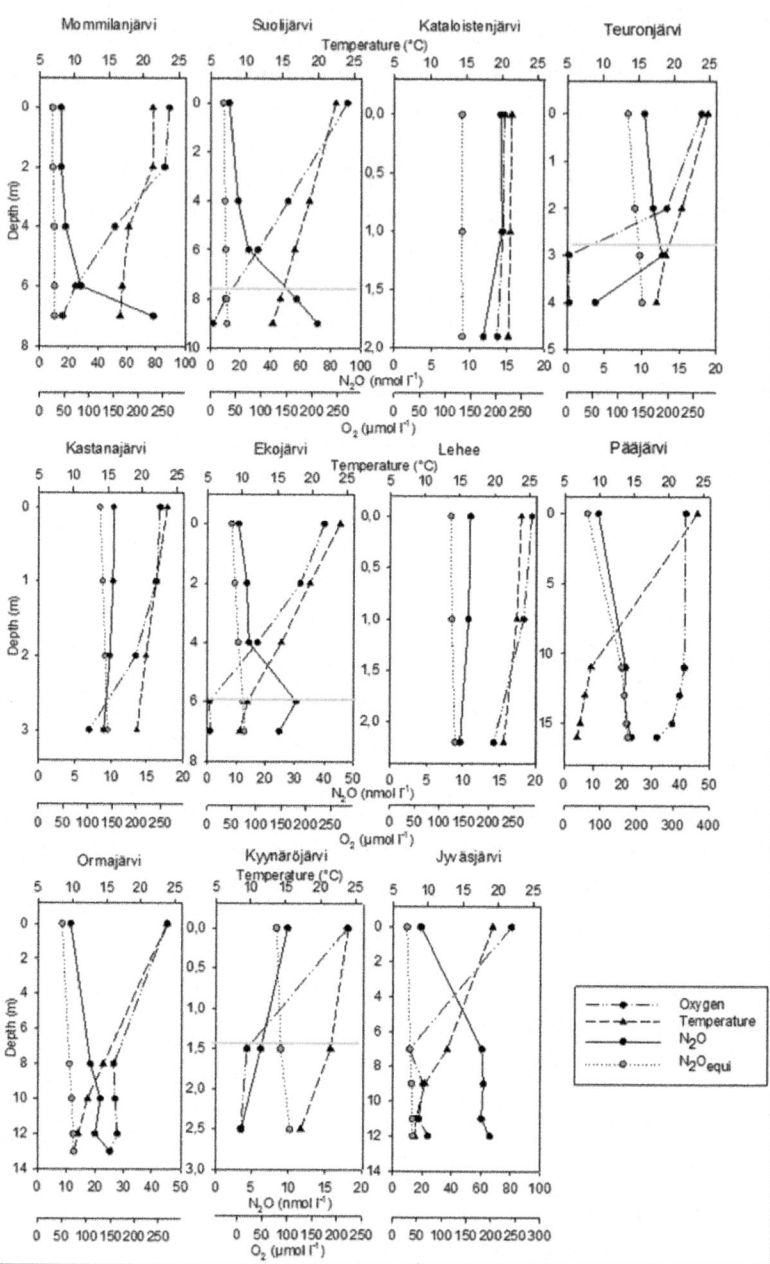

S1 Figure. Vertical profiles of measured N_2O concentrations, calculated N_2O equilibrium concentrations, oxygen concentrations, and temperatures in different lakes. The grey line indicates the respective oxic-anoxic interface.

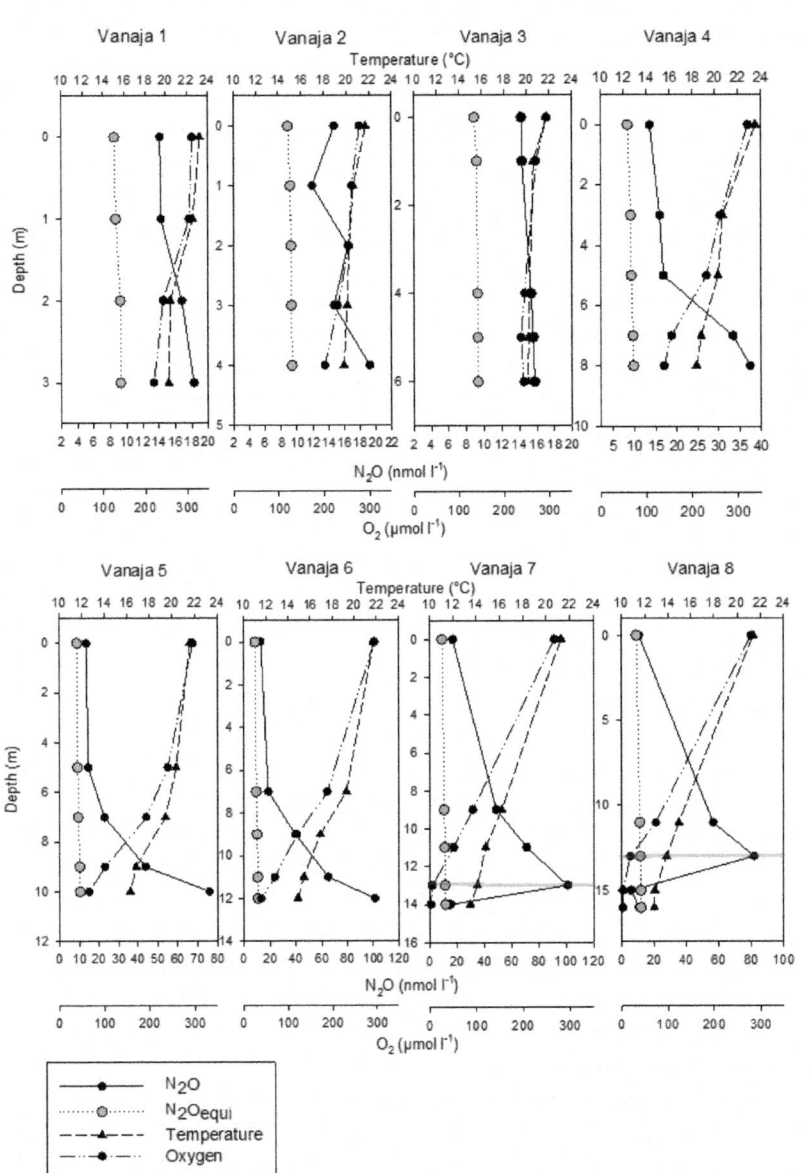

S2 Figure. Vertical profiles of measured N_2O concentrations, calculated N_2O equilibrium concentrations, oxygen concentrations, and temperatures at the different sampling sites along the depth transect of Lake Vanajavesi. The grey line indicates the oxic-anoxic interface.

S1 Table. Study site information, hypolimnetic nutrient concentrations, and oxygen status of the study lakes.

Lake	Area (ha)	Catchment field area (ha)	Max Depth (m)	Sampling depth (m)	Sampling site	Temperature °C	NO_3 (µmol l⁻¹)	PO_4 (µmol l⁻¹)	NH_4 (µmol l⁻¹)	High-nitrate/ Low-nitrate	O_2 (µmol l⁻¹)
Pääjärvi	1344	3816	85	16	N 61°03'05.8" E 25°04'43.1"	6.8	79.1	0.2	1.7	High	333.4
Mommilanjärvi	342	1585	8	7	N 61°52'49.6" E 25°2'30.2"	16.6	42.6	0.2	13.0	High	46.9
Ormajärvi	653	2067	32	13	N 61°05'43.0" E 24°59'16.5"	10.3	34.5	0.1	5.2	High	147.5
Vanajavesi	12 000	181461	24	16	N 61°09'03.2" E 24°16'11.1"	13.5	32.1	0.1	57.4	High	1.9
Jyväsjärvi	330	0*	25	13.5	N 62°13'42.2" E 25°44'05.5"	8.2	32.1	0.1	4.6	High	71.9
Suolijärvi	205	1023	10	9	N 61°7'50.5" E 24°49'13.5"	13.7	15.3	0.1	10.9	High	5.3
Ekojärvi	74	1827	8	7	N 61°11'51.3" E 24°57'23.7"	9.7	1.2	0.03	20.7	Low	5.3
Kataloistenjärvi	112	267	2	1.9	N 61°01'10.6" E 25°56'44.0"	20.9	0.6	0.03	0.9	Low	202.8
Teuronjärvi	134	528	5	4	N 61°03'41.9" E 24°51'42.3"	17.6	0.7	0.1	61.4	Low	4.7
Kyynäröjärvi	25	758	<3	2.5	N 61°07'15.2" E 24°59'32.6"	17.2	0.6	0.1	28.4	Low	6.6
Kastanajärvi	33	0	9	3	N 61°13'41.4" E 24°48'11.9"	19.4	0.6	0.03	0.6	Low	102.8
Lehee	105	110	2.5	2.2	N 61°13'41.4" E 24°48'11.9"	21.4	0.4	0.03	2.1	Low	210.0

* Jyväsjärvi is an urban lake

S2 Table. Gene copy numbers of *nirS*, *nirK*, $nosZ_I$, and $nosZ_{II}$ gene amplicons per ng of DNA (nd, no data).

Inter-lake comparison	Gene copy number (ng^{-1} of DNA)				High-nitrate/ Low-nitrate
	*nir*S	*nir*K	$nosZ_I$	$nosZ_{II}$	
Pääjärvi	1.6×10^2	2.3×10^2	1.4×10^2	6.4×10^1	High
Mommilanjärvi	2.3×10^2	1.2×10^2	1.3×10^2	1.2×10^2	High
Ormajärvi	2.6×10^2	2.8×10^2	1.7×10^2	1.5×10^2	High
Vanajavesi	8.9×10^1	1.0×10^2	7.4×10^1	4.1×10^1	High
Jyväsjärvi	9.8×10^1	1.1×10^2	8.4×10^1	1.7×10^2	High
Suolijärvi	2.8×10^1	4.4×10^1	2.9×10^1	5.1×10^1	High
Ekojärvi	4.8×10^0	9.6×10^0	8.3×10^0	1.1×10^1	Low
Kataloistenjärvi	4.8×10^2	5.8×10^2	4.3×10^2	1.4×10^2	Low
Teuronjärvi	1.9×10^2	2.3×10^2	2.2×10^2	1.1×10^2	Low
Kyynäröjärvi	4.6×10^1	6.7×10^1	5.5×10^1	4.7×10^1	Low
Kastanajärvi	3.5×10^1	2.6×10^1	1.1×10^2	2.3×10^2	Low
Lehee	4.0×10^1	7.8×10^1	4.7×10^1	8.3×10^1	Low
Intra-lake depth transect					
Vanajavesi2	5.2×10^2	2.6×10^2	2.2×10^2	1.9×10^2	
Vanajavesi3	3.3×10^2	1.9×10^2	1.8×10^2	1.3×10^2	
Vanajavesi4	2.5×10^2	1.9×10^2	1.7×10^2	8.8×10^1	
Vanajavesi5	1.0×10^2	1.3×10^2	9.1×10^1	4.3×10^1	
Vanajavesi6	8.3×10^1	1.5×10^2	8.4×10^1	2.1×10^1	
Vanajavesi7	8.9×10^1	1.0×10^2	7.4×10^1	4.2×10^1	
Vanajavesi8	1.7×10^2	1.2×10^2	6.5×10^1	nd	

S3 Table. Water temperature and pH, as well as nutrient and oxygen concentrations at various sampling sites in Lake Vanajavesi.

Site	Coordinates	Depth (m)	O_2 ($\mu mol\ l^{-1}$)	Temperature (°C)	NO_3 ($\mu mol\ l^{-1}$)	PO_4 ($\mu mol\ l^{-1}$)	NH_4 ($\mu mol\ l^{-1}$)	pH
Vanaja1	N 61°08.07.8'' E 24°17'43.8''	3	221.6	20.3	24.0	0.03	1.1	7.5
Vanaja2	N 61°07'35.9'' E 24°17'19.5''	4	201.9	19.8	33.4	0.2	8.9	7.6
Vanaja3	N 61°07'53.4'' E 24°17'18.3''	6	243.4	20.1	32.9	0.3	15.8	7.7
Vanaja4	N 61°08'17.9'' E 24°16'59.4''	8	138.1	18.4	40.2	0.6	8.4	7.1
Vanaja5	N 61°08'28.4'' E 24°16'48.4''	10	65.9	16.3	36.3	0.3	23.5	6.9
Vanaja6	N 61°08'04.12'' E 24°16'38.6''	12	38.8	14.9	44.9	0.7	13.8	6.8
Vanaja7	N 61°08'58.7'' E 24°16'38.2''	14	1.9	13.5	39.9	0.6	27.3	7.0
Vanaja8	N 61°09'03.2'' E 24°16'11.1''	16	1.9	12.8	32.1	0.1	57.4	7.1

S4 Table. Gene-specific primer pairs used in the qPCR assays.

Target gene	Primer sequence	Reference
16S rRNA 27f	5'-AGAGTTTGATCMTGGCTCAG-3'	Bacterial primer (Lane 1991)
16S rRNA 338r	5'-TGCTGCCTCCCGTAGGAGT-3'	Universal primer
nirSCd3aF	5'-AACGYSAAGGARACSGG-3'	Kandeler et al. 2006
nirSR3cd	5'-GASTTCGGRTGSGTCTTSAYGAA-3'	
nirK876	5'-ATYGGCGGVAYGGCGA-3'	Henry et al. 2004
nirK1040	5'-GCCTCGATCAGRTTRTGGTT-3'	
nosZ2F cladeI	5'-CGGRACGGCAASAAGGTSMSSGT-3'	Henry et al. 2006
nosZ2R cladeI	5'-CAKRTGCAKSGCRTGGCAGAA-3'	
nosZ-II-F cladeII	5'-CTIGGICCIYTKCAYAC-3'	Jones et al. 2013
nosZ-II-R cladeII	5'-GCIGARCARAAITCBGTRC-3'	

ACKNOWLEDGMENTS

We want to thank Saara-Maria Haapala for help in the field and Simo Jokinen for assistance in measuring N_2O samples. We thank Lammi Biological Station for helping and providing all facilities. We are grateful to M. Rollog for his laboratory assistance during MIMS analyses. Finally, we thank Jari Syväranta for valuable comments on an earlier version of the manuscript, and Sara Hallin for providing positive control samples and assistance with the $nosZ_{II}$ qPCR analyses.

AUTHOR CONTRIBUTIONS

Conceived and designed the experiments: JS LA MT. Performed the experiments: JS HN MFL. Analyzed the data: JS AJR. Contributed reagents/ materials/analysis tools: HN LA MFL MT. Wrote the paper: JS AJR LA HN MFL MT.

REFERENCES

1.   Ravishankara AR, Daniel JS, Portmann RW. Nitrous Oxide (N2O): The dominant ozone-depleting substance emitted in the 21st century. Science. 2009;326: 123–125. doi: 10.1126/science.1176985. pmid:19713491

2.   Jackson MA, Tiedje JM, Averill BA. Evidence for an NO-rebound mechanism for production of N2O from nitrite by the copper containing nitrite reductase from Achromobacter cycloclaster. FEBS Lett. 1991;291: 41–44. pmid:1936249 doi: 10.1016/0014-5793(91)81099-t

3.   Stevens RJ, Laughlin RJ, Malone JP. Soil pH affects the processes reducing nitrate to nitrous oxide and di-nitrogen. Soil Biol Biochem. 1998;30: 1119–1126. doi: 10.1016/s0038-0717(97)00227-7

4.   Zumft WG. Cell Biology and Molecular Basis of Denitrification. Microbiol Mol Biol Rev. 1997;61: 533–616. pmid:9409151

5.   Knapp C, Dodds WK, Wilson KC, O'Brien JM, Graham DW. Spatial heterogeneity of denitrification genes in a highly homogeneous urban stream. Environ Sci Technol. 2009;43: 4273–4279. pmid:19603634 doi: 10.1021/es9001407

6.   Tatariw C, Chapman EL, Sponseller RA, Mortazavi B, Edmonds JW. Denitrification in a large river: consideration of geomorphic controls on microbial activity and community structure. Ecology. 2013;94: 2249–2262 pmid:24358711 doi: 10.1890/12-1765.1

7.   Sanford RA, Wagner DD, Wu Q, Chee-Sanford JC, Thomas SH, Cruz-Carcia C, et al. Unexpected nondenitrifier nitrous oxide reductase gene diversity and abundance in soils. Proc Natl Acad Sci. 2012;109: 19709–19714. doi: 10.1073/pnas.1211238109. pmid:23150571

8.   Jones CM, Graft DRH, Bru D, Philippot L, Hallin S. The unaccounted yet abundant nitrous oxide-reducing microbial community: a potential nitrous oxide sink. ISME J. 2013;7: 417–426. doi: 10.1038/ismej.2012.125. pmid:23151640

9.   Greenberg EP, Becker GE. Nitrous oxide as end product of denitrification by strains of fluorescent pseudomonads. Can J Microbiol. 1977;23: 903–907. pmid:195699 doi: 10.1139/m77-133

10. Jones CM, Stres B, Rosenquist M, Hallin S. Phylogenetic analysis of nitrite, nitric oxide, and nitrous oxide respiratory enzymes reveal a complex evolutionary history for denitrification. Mol Biol Evol. 2008;25: 1955–1966. doi: 10.1093/molbev/msn146. pmid:18614527

11. Magalhães C, Bano N, Wiebe WJ, Bordalo AA, Hollibaugh JT. Dynamics of Nitrous Oxide Reductase Genes (nosZ) in Intertidal Rocky Biofilms and Sediments of the Douro River Estuary (Portugal), and their Relation to N-biogeochemistry. Microb Ecol. 2008;55: 259–269. pmid:17604988 doi: 10.1007/s00248-007-9273-7

12. Enwall K, Throback IN, Stenberg M, Soderstrom M, Hallin S. Soil resources influence spatial patterns of denitrifying communities at scales compatible with land management. Appl Environ Microbiol. 2010;76: 2243–2250. doi: 10.1128/AEM.02197-09. pmid:20118364

13. Henry S, Baudoin E, López-Gutiérrez JC, Martin-Laurent F, Brauman A, Philippot L. Quantification of denitrifying bacteria in soils by nirK gene targeted real-time PCR. J Microbiol Meth. 2004;59: 327–335. doi: 10.1016/j.mimet.2004.12.008

14. Henry S, Bru D, Stres B, Hallet S, Philippot L. Quantitative detection of the nosZ gene, encoding nitrous oxide reductase, and comparison of the abundance of 16S rRNA, narG, nirK, and nosZ Genes in Soils. Appl Environ Microbiol. 2006;72: 5181–5189. pmid:16885263 doi: 10.1128/aem.00231-06

15. Wallenstein MD, Vilgalys RJ. Quantitative analyses of nitrogen cycling genes in soils. Pedobiologia. 2005;49: 665–672. doi: 10.1016/j.pedobi.2005.05.005

16. Kandeler E, Deiglmayr K, Tscherko D, Bru D, Philippot L. Abundance of narG, nirS, nirK, and nosZ genes of denitrifying bacteria during primary successions of a glacier foreland. Appl Environ Microbiol. 2006;72: 5957–5962. pmid:16957216 doi: 10.1128/aem.00439-06

17. Hallin S, Jones CM, Schloter M, Philippot L. Relationship between N-cycling communities and ecosystem functioning in a 50-year-old fertilization experiment. ISME J. 2009;3: 597–605. doi: 10.1038/ismej.2008.128. pmid:19148144

18. Kortelainen P, Huttunen JT, Väisänen T, Mattsson T, Karjalainen P, Martikainen PJ. CH4, CO2 and N2O supersaturation in 12 Finnish lakes before and after ice-melt. Verh Internat Verein Limnol. 2000;27: 1410–1414.

19. McCrackin ML, Elser JJ. Atmospheric nitrogen deposition influences denitrification and nitrous oxide production in lakes. Ecology. 2010;91:

528–539. pmid:20392017 doi: 10.1890/08-2210.1

20. Weier KL, Doran JW, Power JF, Walters DT. Denitrification and the dinitrogen/nitrous oxide ratio as affected by soil water, available carbon, and nitrate. Soil Sci Soc Am J. 1993;57: 66–72. doi: 10.2136/sssaj1993.03615995005700010013x

21. Cuhel J, Simek M, Laughlin RJ, Bru D, Chéneby D, Watson CJ, et al. Insights into the effect of soil pH on N2O and N2 emissions and denitrifier community size and activity. Appl Environ Microbiol. 2010;76: 1870–1878. doi: 10.1128/AEM.02484-09. pmid:20118356

22. Philippot L, Cuhel J, Saby NPA, Chéneby D, Chronáková A, Bru D, et al. Mapping field-scale spatial patterns of size and activity of the denitrifier community. Environ Microbiol. 2009;11: 1518–1526. doi: 10.1111/j.1462-2920.2009.01879.x. pmid:19260937

23. Ferrell RT, Himmelblau DM. Diffusion coefficients of nitrogen and oxygen in water. J Chem Eng Data. 1967;12: 111–115. doi: 10.1021/je60032a036

24. Whitfield CJ, Aherne J, Baulch HM. Controls on greenhouse gas concentrations in polymictic headwater lakes in Ireland. Sci Total Environ. 2011;410–411: 217–225. doi: 10.1016/j.scitotenv.2011.09.045

25. Mengis M, Gächter R, Wehrli B. Sources and sinks of nitrous oxide (N2O) in deep lakes. Biogeochemistry. 1997;38: 281–301. doi: 10.1023/a:1005814020322

26. McCrackin ML, Elser JJ. Greenhouse gas dynamics in lakes receiving atmospheric nitrogen deposition. Global Biogeochem Cycles. 2011;25: GB4005. doi: 10.1029/2010gb003897

27. Freymond C, Wenk C, Frame CH, Lehmann MF. Year-round N2O production by benthic NOx reduction in a monomictic south-alpine lake. Biogeosciences. 2013;10: 8373–8383. doi: 10.5194/bg-10-8373-2013

28. Kim O-S, Imhoff JF, Witzel K-P, Junier P. Distribution of denitrifying bacterial communities in the stratified water column and sediment-water interface in two freshwater lakes and the Baltic Sea. Aquatic Ecology. 2011;45: 99–112. doi: 10.1007/s10452-010-9335-7

29. Rissanen AJ, Tiirola M, Ojala A. Spatial and temporal variation in denitrification and in the denitrifier community in a boreal lake. Aquat Microb Ecol. 2011;64: 27–40. doi: 10.3354/ame01506

30. Ducey TF, Shriner AD, Hunt PG. Nitrification and denitrification gene abundances in swine wastewater anaerobic lagoons. J Environ Qual. 2011;40: 610–619. pmid:21520768 doi: 10.2134/jeq2010.0387

31. Martins G, Terada A, Ribeiro DC, Corral AM, Brito AG, Smet BF, et al. Structure and activity of lacustrine sediment bacteria involved in nutrient and iron cycles. FEMS Microbiol Ecol. 2011;77: 666–679. doi: 10.1111/j.1574-6941.2011.01145.x. pmid:21635276

32. Rissanen AJ, Tiirola M, Hietanen S, Ojala A. Interlake variation and environmental controls of denitrification across different geographical scales. Aquat Microb Ecol. 2013;69: 1–16. doi: 10.3354/ame01619

33. Seitzinger SP. Denitrification in freshwater and coastal marine ecosystems: ecological and geochemical significance. Limnol Oceanogr. 1988;33: 702–724. doi: 10.4319/lo.1988.33.4_part_2.0702

34. Codispoti LA, Brandes JA, Christensen JP, Devol AH, Naqvi SWA, Paerl HW, et al. The oceanic fixed nitrogen and nitrous oxide budgets: Moving targets as we enter the anthropocene? Sci Mar. 2011;65 Suppl 2: 85–105. doi: 10.3989/scimar.2001.65s285

35. Swerts M, Merckx R, Vlassak K. Denitrification, N2 fixation and fermentation during anaerobic incubation of soils amended with glucose and nitrate. Biol Fert Soils. 1996;23: 229–235. doi: 10.1007/bf00335949

36. Goreau TJ, Kaplan WA, Wofsy SC, McElroy MB, Valois FW, Watson SW. Production of NO2- and N20 by nitrifying bacteria at reduced concentrations of oxygen. Appl Environ Microbiol. 1980;40: 526–532. pmid:16345632

37. Kampschreur MJ, Temmink H, Kleerebezem R, Jetten MSM, van Loosdrecht MCM. Nitrous oxide emission during wastewater treatment. Water Res. 2009;43: 4093–4103. doi: 10.1016/j.watres.2009.03.001. pmid:19666183

38. Bleakley BH, Tiedje JM. Nitrous oxide production by organisms other than nitrifiers or denitrifiers. Appl Environ Microbiol. 1982;44: 1342–1348. pmid:16346152

39. Kelso BHL, Smith RV, Laughlin RJ, Lennox SD. Dissimilatory nitrate reduction in anaerobic sediments leading to river nitrite accumulation. Appl Environ Microb. 1997;63: 4679–4685. pmid:16535749

40. Burgin AJ, Hamilton SK. Have we overemphasized the role of denitrification in aquatic ecosystems? A review of nitrate removal pathways. Fron Ecol Environ. 2007;5: 89–96. doi: 10.1890/1540-9295(2007)5[89:hwotro]2.0.co;2

41. Kraft B, Tegetmeyer HE, Sharma R, Klotz MG, Ferdelman TG, Hettich RL, et al. The environmental controls that govern the end product of bacterial nitrate respiration. Science. 2014;345: 676–679. doi: 10.1126/science.1254070. pmid:25104387

42. Murphy J, Riley JP. A Modified single solution method for determination of phosphate in natural waters. Analytica Chimica Acta. 1962;26: 31–36. doi: 10.1016/s0003-2670(00)88444-5

43. Wood ED, Armstron FA, Richards FA. Determination of nitrate in sea water by cadmium-copper reduction to nitrite. J Mar Biol Assoc UK. 1967;47: 23–31. doi: 10.1017/s002531540003352x

44. Solorzano L. Determination of ammonia in natural waters by the phenolhypochlorite method. Limnol Oceanogr. 1969;14: 799–801. doi: 10.4319/lo.1969.14.5.0799

45. Griffiths RI, Whiteley AS, O'Donnell AG, Bailey MJ. Rapid method for coextraction of DNA and RNA from natural environments for analysis of ribosomal DNA- and rRNA-based microbial community composition. Appl Environ Microbiol. 2000;66: 5488–5491. pmid:11097934 doi: 10.1128/aem.66.12.5488-5491.2000

46. Maljanen M, Virkajärvi P, Hytönen J, Oquist M, Sparrman T, Martikainen PJ. Nitrous oxide production in boreal soils with variable organic matter content at low temperature—snow manipulation experiment. Biogeosciences. 2009;6: 2461–2473. doi: 10.5194/bg-6-2461-2009

47. Bellido JL, Tulonen T, Kankaala P, Ojala A. CO2 and CH4 fluxes during spring and autumn mixing periods in a boreal lake (Pääjärvi, Southern Finland). J Geophy Res. 2009;114 G04007. doi: 10.1029/2009jg000923

48. Lide DR, Frederikse HPR. CRC Handbook of Chemistry and Physics, 76th Edition CRC Press, Inc, Boca Raton, FL; 1995.

49. Kana TM, Darkangelo C, Hunt M.D, Oldham JB, Bennett GE, Cornwell JC. A membrane inlet mass spectrometer for rapid high precision determination of N2, O2 and Ar in environmental water samples. Anal Chem. 1994;66: 4166–4170. doi: 10.1021/ac00095a009

50. Weiss RF. The solubility of nitrogen, oxygen and argon in water and seawater. Deep-Sea Res. 1970;17: 721–735. doi: 10.1016/0011-7471(70)90037-9

51. Henry S, Baudoin E, López-Gutiérrez JC, Martin-Laurent F, Brauman A and Philippot L (2004) Quantification of denitrifying bacteria in soils by nirK gene targeted real-time PCR. J Microbiol Meth 59: 327-335.

52. Henry S, Bru D, Stres B, Hallet S and Philippot L (2006) Quantitative Detection of the nosZ Gene, Encoding Nitrous Oxide Reductase, and Comparison of the Abundance of 16S rRNA, narG, nirK, and nosZ Genes in Soils. Appl Environ Micorbiol 72 5181-5189.

53. Jones CM, Graft DRH, Bru D, Philippot L and Hallin S (2013) The

unaccounted yet abundant nitrous oxide-reducing microbial community: a potential nitrous oxide sink. ISME J 7: 417-426.

54. Kandeler E, Deiglmayr K, Tscherko D, Bru D & Philippot L (2006) Abundance of narG, nirS, nirK, and nosZ Genes of Denitrifying Bacteria during Primary Successions of a Glacier Foreland. Appl Environ Microbiol 72: 5957–5962.

55. Lane DJ (1991) 16S/23S rRNA sequencing In: E Stackebrandt & M Goodfellow (ed) Nucleic acid techniques in bacterial systematic John Wiley & Sons pp115-175

Chapter 11

ENVIRONMENTAL PREDICTORS OF ICE SEAL PRESENCE IN THE BERING SEA

Jennifer L. Miksis-Olds and Laura E. Madden

Applied Research Laboratory, The Pennsylvania State University, State College, Pennsylvania, United States of America

ABSTRACT

Ice seals overwintering in the Bering Sea are challenged with foraging, finding mates, and maintaining breathing holes in a dark and ice covered environment. Due to the difficulty of studying these species in their natural environment, very little is known about how the seals navigate under ice. Here we identify specific environmental parameters, including components of the ambient background sound, that are predictive of ice seal presence in the Bering Sea. Multi-year mooring deployments provided synoptic time series of acoustic and oceanographic parameters from which environmental parameters predictive of species presence were identified through a series of mixed models. Ice cover and 10 kHz sound level were significant predictors of seal presence, with 40 kHz sound and prey presence (combined with ice cover) as potential predictors as well. Ice seal presence showed a strong positive correlation with ice cover and a negative association with 10 kHz environmental sound. On average, there was a 20–30 dB difference between sound levels during solid ice conditions compared to open water or melting conditions, providing a salient acoustic gradient between open water and solid ice conditions by which ice seals could orient. By constantly assessing the acoustic environment associated with the seasonal ice movement in the Bering Sea, it is possible that ice seals could utilize aspects of the soundscape to gauge their safe distance to open water or the ice edge by orienting in the direction of higher sound levels indicative of open water, especially in the frequency range above 1 kHz. In rapidly changing Arctic and sub-Arctic environments, the seasonal ice conditions and soundscapes are likely to change which may impact the ability of animals using ice presence and cues to successfully function during the winter breeding season.

INTRODUCTION

Ribbon (*Histriophoca fasciata*) and bearded seal (*Erignathus barbatus*) vocalizations are salient vocalizations recorded seasonally in the Bering Sea from January-June when sea ice is present [1]–[2]. These calls are most likely produced by males as a display to attract females and establish territory during the mating season [3]–[8]. During the winter breeding season, the Bering Sea is cold, dark, and ice covered; consequently, these aquatically mating species must locate potential mates while also maintaining positions within the ice sheets where breathing holes can be maintained or where they have access to open water at the ice edge or within polynyas. How they navigate in the low visibility conditions of the dynamic ice sheets to locate potential mates and maintain access to open water for breathing and mating is not fully known. Artificially introduced acoustic cues were shown to be extremely important for a blindfolded spotted seal (*Phoca largha*) in navigating under ice to locate breathing holes [9]; therefore, it is not unreasonable to hypothesize that ribbon and bearded seals may also be using soundscape cues to orient under the ice. The goal of this work was to determine the strongest environmental predictors of ice seal vocal presence in the Bering Sea. Multiple environmental (ice and prey) and acoustic variables were considered in predictive models of ribbon and bearded seal presence during the winter in the Bering Sea, and it was the modeling results that provided insight as to the potential role the soundscape may play in under-ice navigation of these ice seals.

It is widely known that animals use sound to navigate through the environment. Bats and dolphins actively probe the environment with echolocation [10], and non-visual communication signals from conspecifics and heterospecifics guide animals in acquiring mates, foraging, and defense [11]. Over the past decade, there have been an increasing number of studies that explore how animals use information from the overall environmental "soundscape", the combination of biologic (biophony), abiotic (i.e. wind, rain, and other geologic sounds referred to as geophony), and man-made (anthrophony) sounds, gained via passive listening for orientation and navigation [12]–[15]. The concept of using ambient or reflected sounds (as opposed to specific communication signals) to direct movement or identify appropriate habitats has recently been identified as a new field of study referred to as soundscape orientation, and the concept is also included within the broader field of soundscape ecology in the scientific literature [14], [16]–[17].

In the marine environment where visual signals do not propagate very far, animals rely on sound as their primary means of obtaining information over any significant distance. It has been speculated that large baleen whales use ambient acoustic cues or acoustic landmarks to guide their migration [18]–[22]. Laboratory and field studies have demonstrated that both invertebrates (oyster and crab) and fish use soundscape cues for orientation and localization of appropriate settlement habitat. The commonality between all the soundscape orientation experiments conducted in the marine environment, and select terrestrial habitats, is that the soundscapes identified as having an impact on animal behavior originated from areas of high species diversity in association with reefs or rainforests [12], [15], [23]–[27]. Habitats with greater biodiversity are associated with richer acoustic soundscapes compared to low diversity habitats, which in itself may be an important cue for animal orientation [16], [28]–[30]. This author is only aware of two studies that identify specific acoustic characteristics of the soundscape that are predictors of behavioral response. Stanley et al. (2011) [31] measured the sound intensity level required to elicit settlement and metamorphosis in several species of crab larvae, and Simpson et al. (2008) [32] discovered that coral reef fish responded more strongly to the higher frequency components (>570 Hz) of the reef soundscape.

It has been impossible to truly assess specific predictors, acoustic or otherwise, of behavioral response for ice seals living in conditions unhospitable to direct observation. However, advances in remote sensing technology and capabilities have provided the means to begin identifying potentially important parameters that are deserving of more in-depth study. This study used multiple, synoptically sampled time series from remotely deployed sensors to gain a better understanding of the environmental parameters most likely influencing ice seal behavior during the breeding season in the Bering Sea.

METHODS

Moorings

Active and passive acoustic sensors were incorporated into subsurface acoustic moorings deployed at two locations on the 70-m isobath of the eastern Bering Sea shelf at sites M2 (56° 51.570′N, 164° 3.801′W) and M5 (59° 54.285′N, 171° 42.285′W) in 2009 and 2007, respectively. Moorings at these locations have been deployed and maintained as part of the NOAA Ecosystems and Fisheries-Oceanography Coordinated Investigations (EcoFOCI) Program (http://www.ecofoci.noaa.gov) since 1995 and 2004, respectively for the M2 and M5 moorings [33] (Figure 1). The acoustic sensors were integrated into the NOAA-deployed, observational moorings under a NOAA Request for Blanket

Scientific Research Permit and did not require a specific permit for remote sensing. Moorings were deployed subsurface to prevent entanglement in seasonal sea ice and were serviced in Spring (April/May) and Fall (September/ October) each year depending on weather conditions. The acoustic sensors were deployed on a separate, short mooring in conjunction with oceanographic moorings at each location. The oceanographic and acoustic moorings were separated by a distance of approximately 1 km to minimize noise produced by the oceanographic mooring hardware and sensors in the acoustic recordings. The data used in this study comes from acoustic data acquired 27 Sep 2009–19 May 2011 at location M2 and 26 Sep 2008–20 May 2011 at location M5.

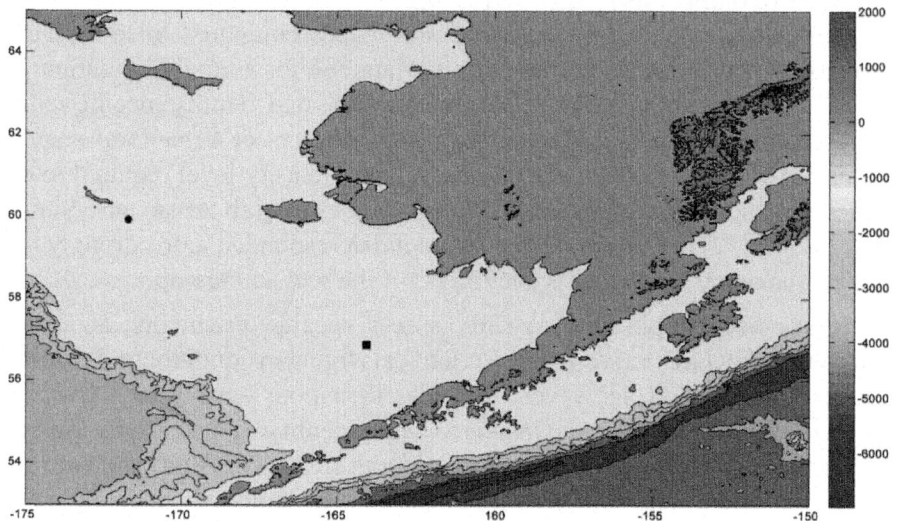

Figure 1. Mooring site locations in the Bering Sea.

The acoustic mooring consisted of a series of active and passive sensors including a 300 kHz RDI ADCP, three-frequency (125 kHz, 200 kHz, and 460 kHz) scientific echosounder system of Acoustic Water Column Profilers (AWCPs: ASL Environmental Sciences, Inc, Sidney, BC), Passive Aquatic Listener (PAL) recorder, and an AURAL (Autonomous Underwater Recorder for Acoustic Listening) (Multi-Électronique (MTE) Inc., Québec, Canada). The mooring was constructed in the following order from top (approximately 60–62 m) to bottom (approximately 70 m): 36″ floatation, 300 kHz RDI ADCP, 30″ floatation, AURAL, AWCP system, PAL, and acoustic release. The AWCP system was mounted in an upward-looking direction 15° off vertical to eliminate interference from flotation and instruments in the mooring line directly above the active acoustic system. This study utilized data from the PAL and AWCP system. AWCPs record acoustic backscatter to monitor the presence and location of acoustic scatterers such as zooplankton and fish within the water column [34]–[35]. The transducers of the three different frequencies were positioned in the mooring cage so that the beam patterns were aligned to sample the same parcel of water nearly simultaneously. The echosounders sampled the water column for 5 min. each half hour. During each 5 min. sampling period, acoustic backscatter measurements were recorded every 2 s with 20 cm range bins from approximately 0.75 m above the transducer face to the water surface. Zooplankton net tows were conducted during mooring maintenance activities and on separate research cruises in the area using either a 25-cm diameter CalVET system (CalCOFI Vertical Egg Tow; [36]) having 0.15 mm mesh nets or double-oblique tows of paired bongo frames (60-cm frame with 0.333 mm mesh and 20-cm frame with 0.150 mm mesh) [2]. Data from the net tows provided information on dominant species, species composition, and numerical density to aid in defining size classes and interpretation of the acoustic data.

The PAL is an event detector, or adaptive sub-sampling acoustic recorder, with a temporal sampling strategy designed to allow the instrument to record data for up to one year[37]–[40]. The default sampling strategy was to record a 4.5 sec acoustic time series, or soundbite, at a sampling rate of 100 kHz every 5 minutes corresponding to a 1.5% duty cycle. When sampling in the default mode, onboard processing algorithms sub-sampled the 4.5 sec soundbite eight times and generated a power spectrum for each sub-sample. A preliminary detection algorithm identified signals of interest when a temporal feature of the sub-sampled power spectra in a soundbite exceeded one of three threshold criteria: 1) the matching of spectrum characteristics to known spectra, 2) exceeding a 12 dB threshold level between sequential samples indicating a transient source, or 3) the matching of predefined peaks (e.g. 300 Hz–3 kHz) indicating possible tonal or click vocalizations from marine mammals. If no signals of interest were detected, the spectra were averaged, and a single spectrum was saved to the hard disk. The soundbite time series was discarded in the default sampling mode. During periods of increased acoustic activity where signals of interest triggered a modified sampling protocol, the sampling interval was decreased to 2 minute intervals corresponding to a 4% duty cycle. In the modified sampling mode, individual spectra and the soundbites were saved to the hard disk. The PAL continued to operate in the modified sampling mode until no signals of interest were detected. The PAL then returned to the default sampling mode. Details on the adaptive sampling algorithms of the PAL are found in Miksis-Olds et al. (2010) [39].

Field data was collected under Observational Institutional Animal Care and Use Committee (IACUC) #36003 "Characterizing Biological Scatter and Its

Implications for Marine Mammals in the Bering Sea" from The Pennsylvania State University. There was no direct interaction with any vertebrates in this study, as all data from marine mammals were obtained remotely through passive acoustic listening.

Ice Data

Daily mean ice cover (or percent cover in this specific region) and ice thickness data were obtained from the images produced by the NOAA Ice Desk at the National Weather Service in Anchorage, Alaska. The images are posted on http://pafc.arh.noaa.gov/ice.php. Ice conditions surrounding the mooring locations were estimated within an approximate 20 km^2 around the mooring.

Data Processing

Ribbon and bearded seal presence was determined from the PAL soundbites with the understanding that detection of vocalizations indicates seal presence, and the lack of acoustic detection does not imply animal absence. Soundbites were reviewed by a human classifier and verified by a second independent human classifier blind to the results of the first reviewer. Sound sources present in the soundbites were identified from spectrograms (1024 point FFT, Hamming window, 87.5% overlap) made from the original 100 kHz recordings downsampled to 48 kHz using Adobe Audition 3.0 (Adobe Systems Incorporated). These settings provided a bandwidth of 61 Hz, with a frequency resolution of 47 Hz, and a time resolution of 2.7 ms. Marine mammal vocalizations were classified aurally and visually from the spectrograms by species (bowhead (*Balaena mysticetus*), gray whale (*Eschrichtius robustus*), killer whale (*Orcinus orca*), beluga whale (*Delphinapterus leucas*), walrus (*Odobenus rosmarus*), ribbon, and bearded seals). Ribbon seal grunts, roars, and downsweeps were used to indicate presence [1], [41]–[42]. Bearded seal vocal presence was determined by the identification of trills, the most salient of the bearded seal vocalizations [6]–[7]. The adaptive sampling protocol and low sampling duty cycle of the PAL prevented calculations of the daily detection rate or overall number of seal vocalizations per day.

Analysis of PAL spectra included examination of spectral shape and levels. Temporal clusters of similarly distinctive sound spectra lasting tens of minutes to hours were manually identified and classified. Sound levels were computed from the time series of spectra. Each spectrum was computed from 1024 point samples of the 4.5 s time series. This resulted in a 513 point power spectral density with each of the bins covering 97 Hz of the 50 kHz usable bandwidth. The spectra were then reduced from 513 points to 64 points by averaging spectra levels over two bins below 3 kHz and over ten bins from 3 to 50 kHz.

The resulting power spectral density, relative to $\mu 1 Pa^2/Hz$, represents energy from the complete 50 kHz bandwidth with variable frequency resolution. To compute the sound level from these spectra, the values were converted to linear power spectral density and multiplied by the frequency resolution of the bins and then summed. The unit of the full bandwidth average is a sound pressure level, re 1 μPa. Processing of power spectral density was conducted for five frequencies over seven octaves (500 Hz, 2 kHz, 10 kHz, 20 kHz, 40 kHz) with units of dB re 1 $\mu Pa^2/Hz$.

To assess prey parameters related to zooplankton/fish abundance and community composition, the AWCP data were processed in 5 m vertical depth bins. Daily mean volume backscatter coefficient (mean S_v in units m^2/m^3) was calculated from 24 hour integrations over each 5 m depth layer using EchoView software (Myriax, Tasmania). Targets within each depth and time bin were classified as to the likely source of the scattering based on differences in scattering amplitude between the three frequencies. Analyses using this dB-difference approach [2], [43]–[45] are typically groundtruthed with information from net tows or video observations. However, given the low level of direct sampling of the water column in this study, a different approach was used and was consistent with Miksis-Olds et al. (2013) [2] summarized here. If scattering assemblages were monospecific, then the dB-difference for a single scatterer type and an aggregation of scatterers of this type would be identical, although the volume backscattering at each frequency would be different. Theoretical scattering curves for four different types of individual scatterers were generated and dB-differences between the three AWCP frequencies were calculated. Scattering amplitudes (and the subsequent dB differences) were generated using a Stochastic Distorted Wave Born Approximation model [46] for the following scatterers: 1) small scatterers such as neritic copepods (lengths of 1–5 mm) (*Pseudocalanus* spp., *Acartia longiremis, Oithona* spp. and *Calanus*), 2) medium scatterers (lengths of 5–15 mm) which includes juvenile krill, chaetognaths, and amphipods, 3) large scatterers such as adult euphausiids (lengths of 15–30 mm), 4) resonant scatterers which represents an organism with a gas-inclusion such as a swim-bladdered fish or siphonophore, and 5) unknown. The acoustic system was not able to detect the weak scattering strengths of scatterers less than approximately 5 mm in length unless they were present in extremely dense aggregations. Aggregations were classified as belonging to one of the five categories (small, medium, or large scatterer; resonant; or unknown) by determining the shortest geometric distance between the three dB differences calculated for the aggregation and that of the theoretical scatterers. If the closest geometric distance was more than 12 dB (an arbitrarily chosen value), then the aggregation was classified as unknown.

Modeling

Daily presence-absence data for ribbon and bearded seals identified in the passive acoustic recordings was the response variable in the generalized linear and generalized additive models (GLM and GAM) designed to identify predictor variables of ice seal presence (Table S1). There is a high degree of temporal overlap between ribbon and bearded seal detections in the Bering Sea [1]–[2], so daily presence-absence data for the two species was combined into a single ice seal response variable to increase statistical power. Initial models included the following predictor variables: ice thickness, % ice cover, 200 kHz Sv, % prey composition (small, medium, large, and resonant scatterers), and mean daily sound level (500 Hz, 2 kHz, 10 kHz, 20 kHz, 40 kHz) (Table S1 and Table S2). Data was first explored to identify potential outliers and evaluate distribution and collinearity among predictor variables and also with marine mammal presence using functions of the AED package in R [47]–[48]. Explanatory variables were centered to allow better model convergence and interpretation, with the exception of ice cover and thickness. Ice cover and ice thickness showed a zero-inflated distribution and transformations failed to sufficiently address the skewed distributions. Zero values are meaningful for these measurements and thus these variables were not truncated or transformed. High collinearity was found between environmental sound level variables with the correlation highest between close frequencies. Ice cover and ice thickness were also highly collinear, although one or both of these variables were removed from the models during the selection process so this collinearity did not pose a problem. Final models including multiple noise variables were checked for collinearity using the *corvif* function from the R package *AED* [47]–[48]. All variables included in final models had VIFs well below 10 (the maximum VIF was 2.32), indicating sufficiently low collinearity [49].

Generalized linear models (GLMs) and generalized additive models (GAMs) allow model fitting to describe relationships between variables without constraints of linear regression models [50]. Generalized additive mixed models (GAMMs) and generalized linear mixed models (GLMMs) extend GAMs and GLMs to include random effects and correlation structures to deal with violations of independence that are often present in observational and time series data and are becoming popular in the analysis of ecological data (for example: Friedlaender et al., 2006 [51]; Wagner & Sweka, 2011 [52]). GLMs and GLMMs with a binomial distribution and logit link function were fit using a backward stepwise approach. Variables were selected for removal using the *drop1* command from the basic*stats* package in R (version 2.14.1; [53]) to apply an analysis of deviance test following a Chi-square distribution. Variables were removed based on a significance criteria of p<0.01 until all variables in the model were considered significant. Significance tests and p-values for analysis of deviance are approximate, thus a selection criteria below the standard 95% significance level was used to avoid inclusion of unnecessary terms. GLMMs were fit using the *glmmPQL* function from the *MASS* package in R [54]. This approach allowed the inclusion of a random effect for site to allow inference beyond the two stations sampled and a temporal correlation structure to address the lack of independence due to repeated sampling at each site. Convergence problems were frequently encountered when a temporal correlation structure for date grouped by site was included. Auto-correlation in the model residuals was examined to determine whether the temporal correlation structure was needed, as including a random effect for site imposes an implicit compound symmetry correlation structure that assumes a constant correlation within data points from the same site. GAMs and GAMMs with a binomial distribution and logit link function were fit using the same procedure described above with the *gamm* function from the *mgcv* package in R[55]–[56] to explore potential non-linear relationships.

GLMM and GAMM techniques are on the "frontier of statistical research" and as such model selection and validation for generalized models on absence-presence response data is difficult [47]. Standardized residuals were extracted and plotted against predictor variables and fitted values to look for patterns. Greater variation in residuals at zero ice coverage was discovered, likely due to the large number of zero values for ice cover. A new data set, zero-truncated for ice cover, was then used to fit the final models to explore the potential for zero values to interfere with model function and selection.

RESULTS

Seasonal ice was present at both mooring locations in the Bering Sea during each winter of the study (Figure 2). The ice cover over M5 on the central shelf was thicker and present longer compared to M2 on the southeastern shelf. Bearded seals were detected on 340 days, and ribbon seals were detected on 161 days over the study period from Sep 2008-May 2011. Seals were detected on fewer days at the southern mooring (M2) most likely due to the less persistent and shorter duration of ice cover compared to M5, but the proportion of daily detections for each species was similar at both mooring locations (39 days (63%) bearded and 23 days (37%) ribbon detected at M2; 301 days (68%) bearded and 138 days (32%) ribbon at M5) (Figure 2).

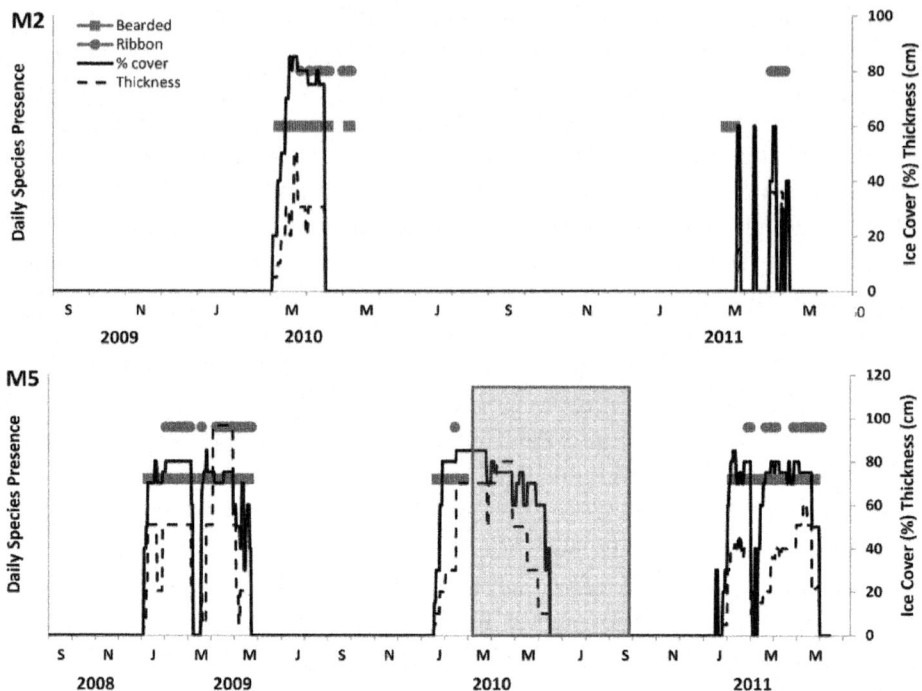

Figure 2. Time series of daily ice and ice seal presence over the M2 and M5 moorings where data exist from 2008–2011. Acoustic presence of species does not correspond to a numerical value on the y axis. The species-specific symbols reflect daily acoustic presence and are separated spatially for easy visualization. The blue box indicates a period of time where no acoustic data were available from the PAL.

The grouping of PAL spectra identified four general sea surface conditions (open water, freeze up, solid ice, seasonal melting) (Figure 3). Validation of sea ice conditions from the passive acoustic data was inferred from the satellite ice thickness and mean ice cover calculations, seasonality, and recorded soundbites of physical processes. Overall sound levels during open water conditions were generally greater than when ice was present for frequencies of 1–10 kHz, which was consistent with previous studies (Figures 3 and 4) [2]. Above 10 kHz, melting conditions produced the greatest sound intensity. For frequencies less than 1 kHz, open water and initial freeze-up conditions had the greatest sound intensity. When solid ice was present above the moorings, sound levels were observed to be the lowest across the full frequency spectrum, with extremely low level intensity (<40 dB re 1 $\mu Pa^2/Hz$) above 5 kHz. The internal noise floor of the recorder was likely a limiting factor for sound levels below approximately 32 dB.

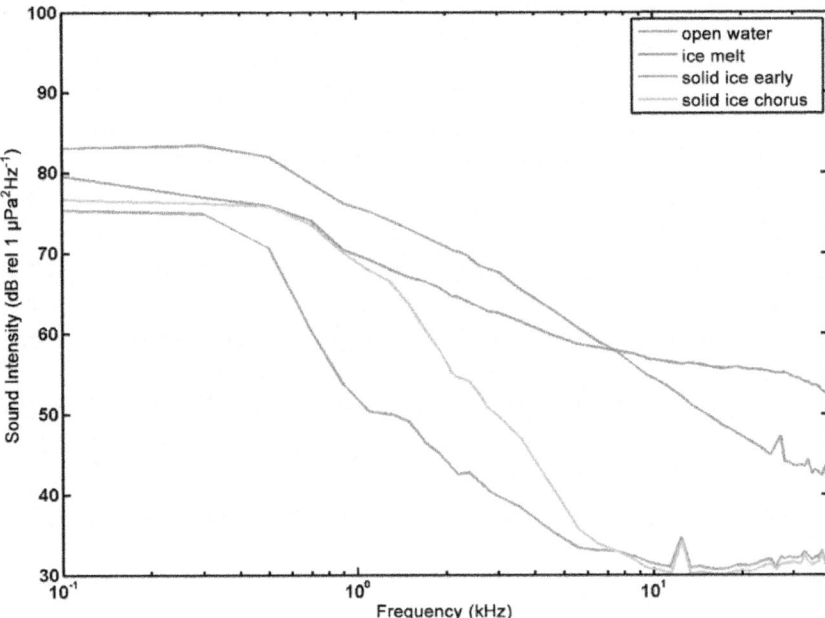

Figure 3. Representative spectra from the Bering Sea under different surface conditions. The Solid Ice Early spectrum represents the acoustic environment prior to the seasonal arrival of chorusing ice seals. The Solid Ice Chorus spectrum captures the acoustic environment when ice seals were observed to be chorusing in the acoustic record.

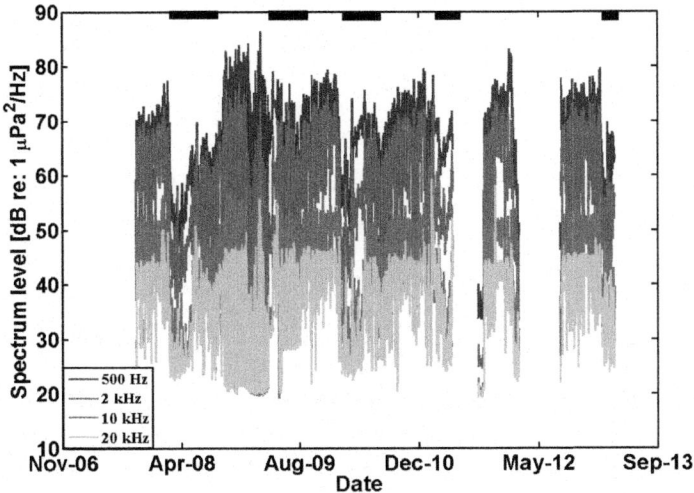

Figure 4. Sound pressure levels at four frequencies from the M5 location in the Bering

Sea over a 4 year time period. Gaps in the data are periods when no data was available from the PAL. The black bars across the top of the figure indicate presence of regional ice cover.

The initial models to determine predictors of ice seal presence included ice thickness, % ice cover, 200 kHz Sv, four size categories of % prey composition (small, medium, large, and resonant scatterers), and five sound levels (500 Hz, 2 kHz, 10 kHz, 20 kHz, 40 kHz). The final GLM and GLMM model both included ice cover, 10 kHz sound, 40 kHz sound, and an interaction between ice cover and large crustaceans as significant predictors of ice seal presence (Table 1). Ice seal presence showed a strong positive correlation with ice cover and a negative association with 10 kHz environmental sound levels (Table 1). Prey alone was not a good predictor of seal presence and an interaction between ice cover and large crustaceans indicates a negative relationship with seal presence, likely due to the somewhat non-linear relationship between ice seals and ice cover at M5 (discussed below with GAMM models).

Table 1. GLM and GLMM final model results.

Variable	Parameter Estimate	Std. Error	DF	p-value
GLM				
Intercept	−4.353	0.380	902	<0.001
Ice cover	8.253	0.865	902	<0.001
c 10 kHz sound	−0.146	0.039	902	<0.001
c 40 kHz sound	0.130	0.045	902	0.004
c Large crustacean	0.036	0.017	902	0.031
Ice cover: c Large crustacean	−0.104	0.034	902	0.002
GLMM				
Random effect: site				
Intercept	−4.353	0.332	904	<0.001
Ice cover	8.254	0.756	904	<0.001
c 10 kHz noise	−0.146	0.034	904	<0.001
c 40 kHz noise	0.130	0.039	904	0.001
c Large crustacean	0.036	0.015	904	0.014
Ice cover: c Large crustacean	−0.104	0.030	904	<0.001
GLMM				
Random effect: site				
Correlation: CAR1				
Intercept	−3.700	0.314	1315	<0.001
Ice cover	5.947	0.482	1315	<0.001
c 10 kHz noise	−0.071	0.016	1315	<0.001

The letter c denotes centered variables. Ice cover is given as a fraction of cover from 0 to 1, large crustacean represents a percent composition, and both 10 kHz and 40 kHz are given in dB re 1 µPa²/Hz. The random intercept for site in the GLMM has a standard error of 0.0001 and residual standard error of 0.871. The explanatory variable large crustacean does not meet significance selection criteria (p<0.01), however is included due to the significance of its interaction term with ice cover. doi:10.1371/journal.pone.0106998.t001

Parameters estimated by both the GLM and GLMM are nearly identical (Table 1), suggesting little difference between M2 and M5. However, the inclusion of a random site effect in the GLMM was highly effective in addressing residual autocorrelation (Figure 5B). The inclusion of a temporal correlation structure in the GLMM reduced numerical stability (increased non-convergence problems) and captured less of the residual auto-correlation in the final model (Figure 5C). The GLMM with a random site effect and no temporal correlation structure (Figure 5B) was selected as the optimal model.

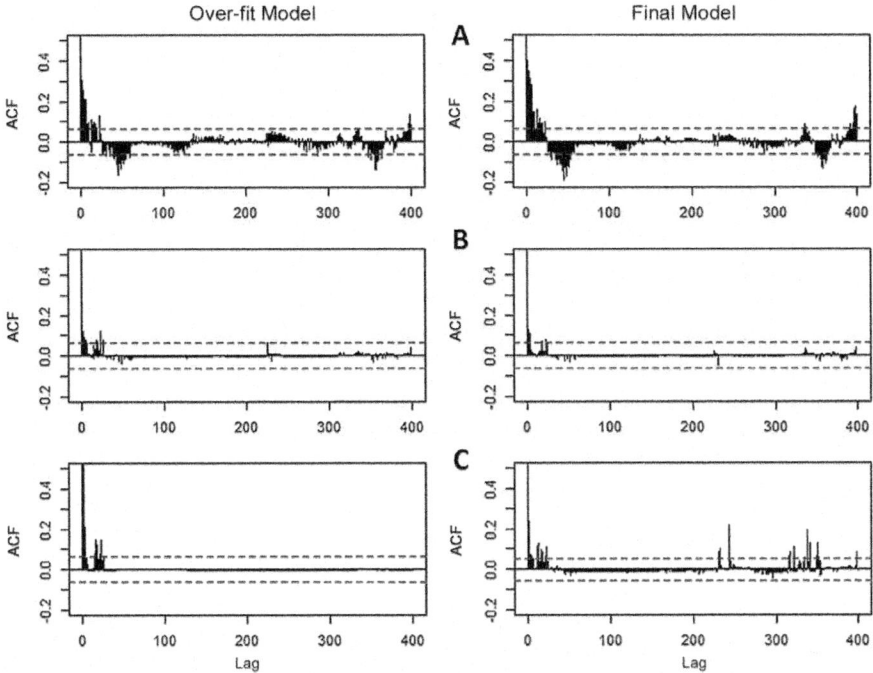

Figure 5. Auto-correlation function (ACF) of model residuals. Plots show auto-correlation of model residuals to 400 lags (400 days) for A) GLM with no random effects or temporal correlation structure, B) GLMM with random site effect and no temporal correlation structure and C) GLMM with random site effect and continuous AR-1 correlation structure. Over-fit models include all explanatory variables and interactions under consideration. Final models include only significant predictor variables after model selection.

GAMM models showed primarily linear relationships between seal presence and ice cover and 10 kHz sound, with the exception of ice cover at M5 (Figure 6B, Table 2). The plateau in the seal-ice cover smoother seen at M5 may explain the negative slope of the ice cover: large crustacean interaction terms in Table 1. The final GAMM model included smooth terms for ice cover and 10 kHz sound with a random smoother for ice cover (Table 2). Including random smoothers for ice cover and 10 kHz sound resulted in neither 10 kHz smoother being significant. All GAMM models with 40 kHz sound failed to converge.

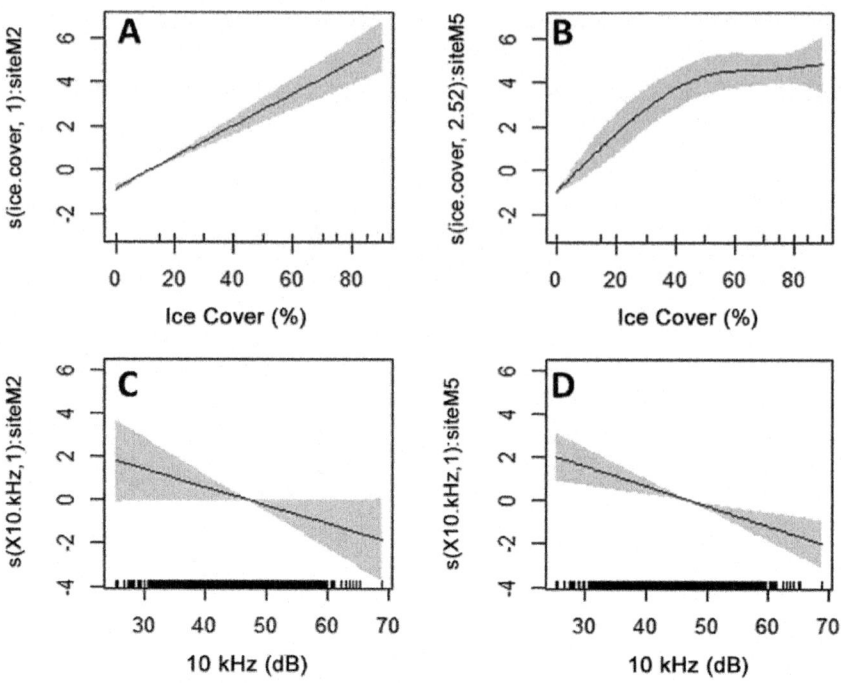

Figure 6. GAMM comparison of smooth functions by site for (A–B) percent ice cover and (C–D) 10 kHz sound. Shaded areas denote 95% confidence intervals. The smooth for M2 on 10 kHz sound was not significant (C). Increase in ice seal presence slows beyond 50% ice cover at M5 (B), although the relationship is still generally linear.

Table 2. Final GAMM model parameters including a random smoother by site for ice cover.

Variable	edf	Std. Error	p-value
GAMM			
Intercept	–3.394	0.225	<0.001
s(ice cover): M2	1.000	–	0.003
s(ice cover): M5	2.522	–	<0.001
s(10 kHz)	1.000	–	<0.001

The estimated degrees of freedom (edf) indicate the "curviness" of the smooth terms with 1.00 representing a straight line. A linear relationship is indicated for 10 kHz sound (both sites) and ice cover at M2.
doi:10.1371/journal.pone.0106998.t002

Figure 6B suggests a non-linear effect of ice cover on seal presence as the increasing trend in the smoother levels near 50 percent cover. However, 2.52 degrees of freedom alone is not strong evidence against a GLMM [47]. This relationship must also be regarded with caution as non-convergence issues disallowed inclusion of all predictor variables of interest in the fitting of this model. As a result, the GLMM with a random site effect was selected as the optimal model for this data. Ice cover and 10 kHz sound level appear to be significant predictors of seal presence, with 40 kHz sound and prey presence (combined with ice cover) as potential predictors as well.

DISCUSSION

Sea ice and 10 kHz sound levels were the strongest predictors of ice seal vocal presence during the winter breeding season in the Bering Sea. The results indicate that as 10 kHz (and to a lesser extend 40 kHz) sound levels increased, the detection of ice seal vocalizations decreased. Neither ribbon seals nor bearded seals have a significant amount of energy in their vocalizations above 10 kHz [1], [6], but if the underwater hearing capabilities of ribbon and bearded seals are comparable to other phylogenetically related, ice-dependent species (e.g. spotted seal (*Phoca largha*), ring seal (*Pusa hispida*), harbor seal (*Phoca vitulina*) [57]) then they are capable of hearing sound above 70 kHz [58]–[61]. The 10 kHz frequency falls directly within the frequency range of best hearing for related phocid species [58]–[61], so it is appropriate to conclude that ribbon and bearded seals can both detect and respond to sound signals in the 10–40 kHz range. Although it is known that ice seals hear and vocalize underwater, there is little direct evidence about how they use or rely on sound to direct their movements and behavior.

The modeling results directed a more detailed examination of the acoustic environment that ice seals encounter on an annual basis. Open water conditions are the loudest up to approximately 8 kHz (Figure 3). Above 8 kHz, conditions associated with the process of ice melting were loudest. Conversely, acoustic conditions associated with solid ice cover were the quietest over the entire spectrum from <5 Hz to 50 kHz. On average, there was a 20–30 dB difference between sound levels during solid ice conditions compared to open water or melting conditions. This difference provides a salient acoustic gradient between open water and solid ice conditions by which ice seals could orient to maintain their horizontal position within the ice sheet or proximity to the ice edge so that access to open water for breathing is preserved. By constantly assessing the acoustic environment to navigate along with the seasonal ice movement in the Bering Sea, it is possible that ice seals can gauge their safe distance to open water or the ice edge through the soundscape in dark, ice covered surroundings by orienting in the direction of higher sound levels, especially in the frequency range above 1 kHz.

This observational study was not able to establish a cause-effect relationship or identify a specific threshold or optimal sound level range that ribbon and bearded seals may employ to navigate under ice. Long-term tagging studies with acoustic dosimeters and GPS location capabilities will be needed to confirm this theory and provide direct evidence of the mechanisms of under-ice navigation in ice seals. It will also be useful to investigate this relationship across locations and regions in both the Arctic and Antarctic to assess whether this concept can be generalized to all ice-dependent species required to navigate under ice. This work presents a particularly timely observation, as the sea ice and acoustic conditions of the oceans, the strongest predictors of ice seal vocal presence during the winter breeding season, are changing due to climate change and industrialization related to shipping and energy exploration/production [62]–[63]. In order to fully access the risk this poses to ice seals, it is critical to gain a better understanding of how the seals use and rely on ice presence and its associated sound to survive in their extreme environments.

SUPPORTING INFORMATION

Table S1. Response and predictor variables used in the GLM and GAM modeling at the M2 (A) and M5 (B) locations. Bearded and ribbon seal acoustic presence is a binary response: 0 for absent, 1 for present. The 200 kHz Sv are daily mean values, and the scatterer percent composition is reflective of the daily mean values within each size category. Ice cover % and ice thickness are daily assessment values.

Date	Beard-ed seal pres-ence	Rib-bon seal pres-ence	Ice cover (%)	Ice thick-ness (cm)	200 kHz Sv (dB)	Small scat-ter % comp	Me-dium scat-ter % comp	Large scat-ter % comp	Weak reso-nator % comp	Strong reso-nator % comp	Un-class % comp
9/27/2009	0	0	0	0	-60.4	0	25	75	0	0	0
9/28/2009	0	0	0	0	-57.3	0	21	79	0	0	0
9/29/2009	0	0	0	0	-68.1	0	9	91	0	0	0
9/30/2009	0	0	0	0	-68.1	0	6	94	0	0	0
10/1/2009	0	0	0	0	-64.7	0	18	82	0	0	0
10/2/2009	0	0	0	0	-62.8	0	13	87	0	0	0
10/3/2009	0	0	0	0	-66.3	0	15	85	0	0	0
10/4/2009	0	0	0	0	-69.3	0	12	88	0	0	0
10/5/2009	0	0	0	0	-66.6	0	15	85	0	0	0
10/6/2009	0	0	0	0	-62.9	0	18	82	0	0	0
10/7/2009	0	0	0	0	-65.4	0	14	86	0	0	0
10/8/2009	0	0	0	0	-69.3	0	13	87	0	0	0

Date										
10/9/2009	0	0	0	-66.3	0	19	81	0	0	0
10/10/2009	0	0	0	-65.5	0	12	88	0	0	0
10/11/2009	0	0	0	-68.0	0	13	87	0	0	0
10/12/2009	0	0	0	-68.5	0	15	85	0	0	0
10/13/2009	0	0	0	-70.0	0	13	87	0	0	0
10/14/2009	0	0	0	-65.9	0	15	85	0	0	0
10/15/2009	0	0	0	-58.7	0	28	72	0	0	0
10/16/2009	0	0	0	-59.2	0	27	73	0	0	0
10/17/2009	0	0	0	-61.9	0	26	74	0	0	0
10/18/2009	0	0	0	-63.0	0	18	82	0	0	0
10/19/2009	0	0	0	-64.6	0	23	77	0	0	0
10/20/2009	0	0	0	-62.0	0	25	75	0	0	0
10/21/2009	0	0	0	-62.1	0	20	80	0	0	0
10/22/2009	0	0	0	-65.6	0	13	87	0	0	0
10/23/2009	0	0	0	-64.0	0	11	89	0	0	0
10/24/2009	0	0	0	-66.9	0	8	92	0	0	0
10/25/2009	0	0	0	-65.7	0	11	89	0	0	0
10/26/2009	0	0	0	-63.7	0	17	83	0	0	0
10/27/2009	0	0	0	-63.0	0	21	79	0	0	0
10/28/2009	0	0	0	-60.9	0	27	73	0	0	0
10/29/2009	0	0	0	-61.5	0	14	86	0	0	1

Date											
10/30/2009	0	0	0	86	14	0	-59.1	0	0	0	0
10/31/2009	1	0	0	75	24	0	-56.6	0	0	0	0
11/1/2009	0	0	0	88	12	0	-65.0	0	0	0	0
11/2/2009	0	0	0	85	14	0	-64.8	0	0	0	0
11/3/2009	1	0	0	83	17	0	-61.5	0	0	0	0
11/4/2009	1	0	0	80	19	0	-60.7	0	0	0	0
11/5/2009	4	0	0	79	18	0	-58.2	0	0	0	0
11/6/2009	5	0	0	84	11	0	-60.5	0	0	0	0
11/7/2009	0	0	0	83	17	0	-54.1	0	0	0	0
11/8/2009	0	0	0	82	18	0	-56.3	0	0	0	0
11/9/2009	0	0	0	80	20	0	-61.4	0	0	0	0
11/10/2009	0	0	0	80	20	0	-60.9	0	0	0	0
11/11/2009	2	0	0	80	18	0	-59.0	0	0	0	0
11/12/2009	7	0	0	82	11	0	-67.8	0	0	0	0
11/13/2009	3	0	0	83	14	0	-64.6	0	0	0	0
11/14/2009	7	0	0	78	15	0	-61.4	0	0	0	0
11/15/2009	3	0	0	79	19	0	-58.1	0	0	0	0
11/16/2009	0	0	0	79	21	0	-58.2	0	0	0	0
11/17/2009	0	0	0	79	21	0	-60.2	0	0	0	0
11/18/2009	0	0	0	84	16	0	-55.2	0	0	0	0
11/19/2009	0	0	0	82	18	0	-57.8	0	0	0	0

Date										
11/20/2009	0	0	0	-60.5	0	20	80	0	0	0
11/21/2009	0	0	0	-65.8	0	15	79	0	0	5
11/22/2009	0	0	0	-61.3	0	21	79	0	0	0
11/23/2009	0	0	0	-61.8	0	17	81	0	0	2
11/24/2009	0	0	0	-60.8	0	18	80	0	0	2
11/25/2009	0	0	0	-64.6	0	19	78	0	0	3
11/26/2009	0	0	0	-65.6	0	17	80	0	0	3
11/27/2009	0	0	0	-67.3	0	13	82	0	0	5
11/28/2009	0	0	0	-63.9	0	18	82	0	0	0
11/29/2009	0	0	0	-63.8	0	20	78	0	0	2
11/30/2009	0	0	0	-62.1	0	16	81	0	0	3
12/1/2009	0	0	0	-63.2	0	27	73	0	0	0
12/2/2009	0	0	0	-63.2	0	29	71	0	0	1
12/3/2009	0	0	0	-64.2	0	28	72	0	0	0
12/4/2009	0	0	0	-59.7	0	24	76	0	0	0
12/5/2009	0	0	0	-64.5	0	18	82	0	0	0
12/6/2009	0	0	0	-58.5	0	29	71	0	0	0
12/7/2009	0	0	0	-58.1	0	34	66	0	0	1
12/8/2009	0	0	0	-59.0	0	28	72	0	0	0
12/9/2009	0	0	0	-60.0	0	21	79	0	0	0
12/10/2009	0	0	0	-60.1	0	20	80	0	0	0

Date										
12/11/2009	0	0	0	76	24	0	-61.2	0	0	0
12/12/2009	0	0	0	78	22	0	-61.4	0	0	0
12/13/2009	0	0	0	86	14	0	-64.1	0	0	0
12/14/2009	0	0	0	80	20	0	-62.6	0	0	0
12/15/2009	0	0	0	82	18	0	-57.8	0	0	0
12/16/2009	0	0	0	83	16	0	-59.3	0	0	0
12/17/2009	0	0	0	76	24	0	-58.1	0	0	0
12/18/2009	0	0	0	80	20	0	-60.3	0	0	0
12/19/2009	0	0	0	78	22	0	-61.6	0	0	0
12/20/2009	1	0	0	77	22	0	-61.0	0	0	0
12/21/2009	0	0	0	84	16	0	-63.9	0	0	0
12/22/2009	0	0	0	78	22	0	-62.6	0	0	0
12/23/2009	0	0	0	83	17	0	-65.7	0	0	0
12/24/2009	0	0	0	81	19	0	-64.2	0	0	0
12/25/2009	0	0	0	89	11	0	-69.7	0	0	0
12/26/2009	0	0	0	80	20	0	-65.2	0	0	0
12/27/2009	0	0	0	81	19	0	-66.4	0	0	0
12/28/2009	0	0	0	87	13	0	-65.6	0	0	0
12/29/2009	0	0	0	85	15	0	-65.2	0	0	0
12/30/2009	0	0	0	79	21	0	-65.4	0	0	0
12/31/2009	0	0	0	90	10	0	-69.3	0	0	0

Date										
1/1/2010	0	0	0	90	10	0	-69.1	0	0	0
1/2/2010	0	0	0	90	10	0	-68.0	0	0	0
1/3/2010	0	0	0	77	23	0	-62.1	0	0	0
1/4/2010	0	0	0	74	26	0	-60.6	0	0	0
1/5/2010	0	0	0	75	25	0	-60.2	0	0	0
1/6/2010	0	0	0	80	20	0	-60.4	0	0	0
1/7/2010	0	0	0	78	22	0	-61.3	0	0	0
1/8/2010	0	0	0	82	18	0	-62.5	0	0	0
1/9/2010	0	0	0	80	20	0	-61.3	0	0	0
1/10/2010	0	0	0	76	24	0	-60.4	0	0	0
1/11/2010	0	0	0	82	18	0	-61.4	0	0	0
1/12/2010	0	0	0	80	20	0	-59.3	0	0	0
1/13/2010	0	0	0	77	23	0	-60.1	0	0	0
1/14/2010	0	0	0	83	17	0	-62.1	0	0	0
1/15/2010	0	0	0	81	19	0	-62.4	0	0	0
1/16/2010	0	0	0	79	21	0	-64.0	0	0	0
1/17/2010	0	0	0	79	21	0	-61.3	0	0	0
1/18/2010	1	0	0	81	18	0	-59.7	0	0	0
1/19/2010	1	0	0	76	23	0	-57.7	0	0	0
1/20/2010	0	0	0	88	12	0	-64.0	0	0	0
1/21/2010	0	0	0	87	13	0	-63.4	0	0	0

Date											
1/22/2010	0	0	0	0	-58.9	0	24	75	0	0	0
1/23/2010	0	0	0	0	-67.6	0	14	86	0	0	0
1/24/2010	0	0	0	0	-64.4	0	20	80	0	0	0
1/25/2010	0	0	0	0	-62.0	0	24	76	0	0	0
1/26/2010	0	0	0	0	-59.8	0	20	80	0	0	0
1/27/2010	0	0	0	0	-57.7	0	22	78	0	0	0
1/28/2010	0	0	0	0	-57.5	0	23	75	0	0	2
1/29/2010	0	0	0	0	-67.8	0	15	85	0	0	0
1/30/2010	0	0	0	0	-65.6	0	23	77	0	0	0
1/31/2010	0	0	0	0	-64.0	0	22	78	0	0	0
2/1/2010	0	0	0	0	-58.1	0	15	85	0	0	0
2/2/2010	0	0	0	0	-60.6	0	17	83	0	0	0
2/3/2010	0	0	0	0	-59.0	0	19	81	0	0	0
2/4/2010	0	0	0	0	-55.8	0	17	82	0	0	1
2/5/2010	0	0	0	0	-50.5	0	20	80	0	0	0
2/6/2010	0	0	0	0	-58.3	0	12	87	0	0	1
2/7/2010	0	0	0	0	-59.5	0	19	81	0	0	0
2/8/2010	0	0	0	0	-57.0	0	17	82	0	0	1
2/9/2010	0	0	0	0	-65.6	0	9	85	0	0	6
2/10/2010	0	0	0	0	-61.3	0	12	84	0	0	4
2/11/2010	0	0	0	0	-62.2	0	14	85	0	0	0

Date											
2/12/2010	5	0	0	85	11	0	-67.0	0	0	0	0
2/13/2010	2	0	0	89	9	0	-69.6	0	0	0	0
2/14/2010	0	0	0	83	17	0	-62.6	0	0	0	0
2/15/2010	0	0	0	80	20	0	-60.7	0	0	0	0
2/16/2010	0	0	0	83	17	0	-62.5	0	0	0	0
2/17/2010	1	0	0	83	16	0	-63.3	0	0	0	0
2/18/2010	0	0	0	82	18	0	-61.7	0	0	0	0
2/19/2010	2	0	0	81	18	0	-62.6	0	0	0	0
2/20/2010	3	0	0	83	14	0	-64.8	0	0	0	0
2/21/2010	1	0	0	78	22	0	-61.2	0	0	0	0
2/22/2010	0	0	0	76	24	0	-58.5	0	0	0	0
2/23/2010	0	0	0	77	23	0	-58.6	0	0	0	0
2/24/2010	0	0	0	76	24	0	-58.4	0	0	0	0
2/25/2010	0	0	0	79	21	0	-58.5	0	0	0	0
2/26/2010	0	0	0	80	20	0	-56.7	0	0	0	0
2/27/2010	0	0	0	85	15	0	-52.8	0	0	0	0
2/28/2010	0	0	0	83	15	0	-58.0	0	0	0	0
3/1/2010	NA	NA	NA	NA	NA	NA	-60.1	5	20	0	0
3/2/2010	NA	NA	NA	NA	NA	NA	-61.4	5	20	0	0
3/3/2010	NA	NA	NA	NA	NA	NA	-61.0	5	20	0	0
3/4/2010	NA	NA	NA	NA	NA	NA	-58.2	5	20	0	0

Date												
3/5/2010	0	0	40	10	-56.8	NA	NA	NA	NA	NA	NA	NA
3/6/2010	0	0	40	10	-58.8	NA	NA	NA	NA	NA	NA	NA
3/7/2010	1	0	40	10	-60.7	NA	NA	NA	NA	NA	NA	NA
3/8/2010	1	0	50	20	-60.5	NA	NA	NA	NA	NA	NA	NA
3/9/2010	1	0	50	20	-61.3	NA	NA	NA	NA	NA	NA	NA
3/10/2010	1	0	50	20	-63.0	NA	NA	NA	NA	NA	NA	NA
3/11/2010	1	0	50	20	-63.1	NA	NA	NA	NA	NA	NA	NA
3/12/2010	1	1	70	30	-62.7	NA	NA	NA	NA	NA	NA	NA
3/13/2010	1	0	70	30	-62.0	NA	NA	NA	NA	NA	NA	NA
3/14/2010	1	0	70	30	-63.7	NA	NA	NA	NA	NA	NA	NA
3/15/2010	1	1	85	20	-65.2	NA	NA	NA	NA	NA	NA	NA
3/16/2010	1	0	85	20	-64.8	NA	NA	NA	NA	NA	NA	NA
3/17/2010	1	0	80	30	-67.8	NA	NA	NA	NA	NA	NA	NA
3/18/2010	1	0	85	30	-62.9	NA	NA	NA	NA	NA	NA	NA
3/19/2010	1	0	85	51	-69.6	NA	NA	NA	NA	NA	NA	NA
3/20/2010	1	0	85	51	-65.9	NA	NA	NA	NA	NA	NA	NA
3/21/2010	1	0	85	51	-63.9	NA	NA	NA	NA	NA	NA	NA
3/22/2010	0	0	80	30	-67.1	NA	NA	NA	NA	NA	NA	NA
3/23/2010	1	0	80	30	-63.4	NA	NA	NA	NA	NA	NA	NA
3/24/2010	1	1	80	30	-64.3	NA	NA	NA	NA	NA	NA	NA
3/25/2010	1	0	80	30	-66.9	NA	NA	NA	NA	NA	NA	NA

Date												
3/26/2010	0	0	80	30	-68.1	NA	NA	NA	NA	NA	NA	NA
3/27/2010	0	0	80	30	-64.8	NA	NA	NA	NA	NA	NA	NA
3/28/2010	1	0	80	30	-72.3	NA	NA	NA	NA	NA	NA	NA
3/29/2010	1	0	80	20	-72.6	NA	NA	NA	NA	NA	NA	NA
3/30/2010	1	0	80	20	-77.3	NA	NA	NA	NA	NA	NA	NA
3/31/2010	1	0	75	30	-72.6	NA	NA	NA	NA	NA	NA	NA
4/1/2010	1	1	75	30	-75.0	NA	NA	NA	NA	NA	NA	NA
4/2/2010	1	0	75	30	-77.1	NA	NA	NA	NA	NA	NA	NA
4/3/2010	1	0	75	30	-70.7	NA	NA	NA	NA	NA	NA	NA
4/4/2010	1	0	75	30	-73.2	NA	NA	NA	NA	NA	NA	NA
4/5/2010	1	0	75	30	-71.6	NA	NA	NA	NA	NA	NA	NA
4/6/2010	1	0	75	30	-71.7	NA	NA	NA	NA	NA	NA	NA
4/7/2010	1	1	80	30	-66.6	NA	NA	NA	NA	NA	NA	NA
4/8/2010	1	0	80	30	-65.5	NA	NA	NA	NA	NA	NA	NA
4/9/2010	1	1	75	30	-66.8	NA	NA	NA	NA	NA	NA	NA
4/10/2010	1	1	75	30	-73.1	NA	NA	NA	NA	NA	NA	NA
4/11/2010	0	0	75	30	-62.1	NA	NA	NA	NA	NA	NA	NA
4/12/2010	0	0	75	30	-55.9	NA	NA	NA	NA	NA	NA	NA
4/13/2010	0	0	75	30	-62.1	NA	NA	NA	NA	NA	NA	NA
4/14/2010	0	0	0	0	-63.0	NA	NA	NA	NA	NA	NA	NA
4/15/2010	0	1	0	0	-63.0	NA	NA	NA	NA	NA	NA	NA

Date											
4/16/2010	1	0	0	0	-62.5	NA	NA	NA	NA	NA	NA
4/17/2010	0	1	0	0	-69.0	NA	NA	NA	NA	NA	NA
4/18/2010	0	0	0	0	-64.5	NA	NA	NA	NA	NA	NA
4/19/2010	0	0	0	0	-59.3	NA	NA	NA	NA	NA	NA
4/20/2010	0	0	0	0	-62.1	NA	NA	NA	NA	NA	NA
4/21/2010	0	0	0	0	-65.5	NA	NA	NA	NA	NA	NA
4/22/2010	0	0	0	0	-62.8	NA	NA	NA	NA	NA	NA
4/23/2010	0	0	0	0	-64.2	NA	NA	NA	NA	NA	NA
4/24/2010	0	0	0	0	-62.0	NA	NA	NA	NA	NA	NA
4/25/2010	0	0	0	0	-62.0	NA	NA	NA	NA	NA	NA
4/26/2010	0	0	0	0	-61.9	NA	NA	NA	NA	NA	NA
4/27/2010	0	0	0	0	-59.2	NA	NA	NA	NA	NA	NA
4/28/2010	0	1	0	0	-52.7	NA	NA	NA	NA	NA	NA
4/29/2010	0	0	0	0	-74.0	NA	NA	NA	NA	NA	NA
4/30/2010	0	0	40	20	-72.9	NA	NA	NA	NA	NA	NA
5/1/2010	0	0	40	20	-73.8	NA	NA	NA	NA	NA	NA
5/2/2010	0	1	40	20	-72.0	NA	NA	NA	NA	NA	NA
5/3/2010	1	1	60	30	-72.5	NA	NA	NA	NA	NA	NA
5/4/2010	1	1	60	30	-70.8	NA	NA	NA	NA	NA	NA
5/5/2010	0	1	60	30	-71.0	NA	NA	NA	NA	NA	NA
5/6/2010	0	0	60	30	-70.9	NA	NA	NA	NA	NA	NA

Date												
5/7/2010	0	0	0	0	0	-70.7	NA	NA	NA	NA	NA	NA
5/8/2010	0	0	0	0	0	-70.6	NA	NA	NA	NA	NA	NA
5/9/2010	0	0	0	0	0	-73.7	NA	NA	NA	NA	NA	NA
5/10/2010	0	0	0	0	0	-72.4	NA	NA	NA	NA	NA	NA
5/11/2010	0	0	0	0	0	-72.0	NA	NA	NA	NA	NA	NA
5/12/2010	0	0	0	0	0	-74.0	NA	NA	NA	NA	NA	NA
5/13/2010	0	0	0	0	0	-75.0	NA	NA	NA	NA	NA	NA
5/14/2010	0	0	0	0	0	-73.2	NA	NA	NA	NA	NA	NA
5/15/2010	0	0	0	0	0	-68.9	NA	NA	NA	NA	NA	NA
5/16/2010	0	0	0	0	0	-73.3	NA	NA	NA	NA	NA	NA
5/17/2010	0	0	0	0	0	-75.9	NA	NA	NA	NA	NA	NA
5/18/2010	0	0	0	0	0	-76.3	NA	NA	NA	NA	NA	NA
5/19/2010	0	0	0	0	0	-77.7	NA	NA	NA	NA	NA	NA
5/20/2010	0	0	0	0	0	-74.4	NA	NA	NA	NA	NA	NA
5/21/2010	0	0	0	0	0	-79.1	NA	NA	NA	NA	NA	NA
5/22/2010	0	0	0	0	0	-79.7	NA	NA	NA	NA	NA	NA
5/23/2010	0	0	0	0	0	-79.1	NA	NA	NA	NA	NA	NA
5/24/2010	0	0	0	0	0	-80.3	NA	NA	NA	NA	NA	NA
5/25/2010	0	0	0	0	0	-80.2	NA	NA	NA	NA	NA	NA
5/26/2010	0	0	0	0	0	-77.1	NA	NA	NA	NA	NA	NA
5/27/2010	0	0	0	0	0	-72.0	NA	NA	NA	NA	NA	NA

5/28/2010	0	0	0	0	-73.3	NA	NA	NA	NA	NA	NA
5/29/2010	0	0	0	0	-78.8	NA	NA	NA	NA	NA	NA
5/30/2010	0	0	0	0	-79.9	NA	NA	NA	NA	NA	NA
5/31/2010	0	0	0	0	-79.8	NA	NA	NA	NA	NA	NA
6/1/2010	0	0	0	0	-77.2	NA	NA	NA	NA	NA	NA
6/2/2010	0	0	0	0	-75.0	NA	NA	NA	NA	NA	NA
6/3/2010	0	0	0	0	-76.4	NA	NA	NA	NA	NA	NA
6/4/2010	0	0	0	0	-75.5	NA	NA	NA	NA	NA	NA
6/5/2010	0	0	0	0	-76.2	NA	NA	NA	NA	NA	NA
6/6/2010	0	0	0	0	-75.3	NA	NA	NA	NA	NA	NA
6/7/2010	0	0	0	0	-74.7	NA	NA	NA	NA	NA	NA
6/8/2010	0	0	0	0	-77.3	NA	NA	NA	NA	NA	NA
6/9/2010	0	0	0	0	-75.4	NA	NA	NA	NA	NA	NA
6/10/2010	0	0	0	0	-68.5	NA	NA	NA	NA	NA	NA
6/11/2010	0	0	0	0	-72.1	NA	NA	NA	NA	NA	NA
6/12/2010	0	0	0	0	-75.0	NA	NA	NA	NA	NA	NA
6/13/2010	0	0	0	0	-74.3	NA	NA	NA	NA	NA	NA
6/14/2010	0	0	0	0	-70.4	NA	NA	NA	NA	NA	NA
6/15/2010	0	0	0	0	-67.3	NA	NA	NA	NA	NA	NA
6/16/2010	0	0	0	0	-70.7	NA	NA	NA	NA	NA	NA
6/17/2010	0	0	0	0	-68.0	NA	NA	NA	NA	NA	NA

6/18/2010	0	0	0	0	-69.5	NA	NA	NA	NA	NA	NA
6/19/2010	0	0	0	0	-69.4	NA	NA	NA	NA	NA	NA
6/20/2010	0	0	0	0	-70.5	NA	NA	NA	NA	NA	NA
6/21/2010	0	0	0	0	-72.1	NA	NA	NA	NA	NA	NA
6/22/2010	0	0	0	0	-71.6	NA	NA	NA	NA	NA	NA
6/23/2010	0	0	0	0	-71.1	NA	NA	NA	NA	NA	NA
6/24/2010	0	0	0	0	-72.2	NA	NA	NA	NA	NA	NA
6/25/2010	0	0	0	0	-72.8	NA	NA	NA	NA	NA	NA
6/26/2010	0	0	0	0	-72.6	NA	NA	NA	NA	NA	NA
6/27/2010	0	0	0	0	-72.1	NA	NA	NA	NA	NA	NA
6/28/2010	0	0	0	0	-70.4	NA	NA	NA	NA	NA	NA
6/29/2010	0	0	0	0	-68.1	NA	NA	NA	NA	NA	NA
6/30/2010	0	0	0	0	-70.5	NA	NA	NA	NA	NA	NA
7/1/2010	0	0	0	0	-68.8	NA	NA	NA	NA	NA	NA
7/2/2010	0	0	0	0	-68.8	NA	NA	NA	NA	NA	NA
7/3/2010	0	0	0	0	-68.8	NA	NA	NA	NA	NA	NA
7/4/2010	0	0	0	0	-68.8	NA	NA	NA	NA	NA	NA
7/5/2010	0	0	0	0	-68.8	NA	NA	NA	NA	NA	NA
7/6/2010	0	0	0	0	-68.8	NA	NA	NA	NA	NA	NA
7/7/2010	0	0	0	0	-68.8	NA	NA	NA	NA	NA	NA
7/8/2010	0	0	0	0	-68.8	NA	NA	NA	NA	NA	NA

Date											
7/9/2010	0	0	0	0	-68.1	NA	NA	NA	NA	NA	NA
7/10/2010	0	0	0	0	-68.7	NA	NA	NA	NA	NA	NA
7/11/2010	0	0	0	0	-68.8	NA	NA	NA	NA	NA	NA
7/12/2010	0	0	0	0	-68.8	NA	NA	NA	NA	NA	NA
7/13/2010	0	0	0	0	-68.8	NA	NA	NA	NA	NA	NA
7/14/2010	0	0	0	0	-68.8	NA	NA	NA	NA	NA	NA
7/15/2010	0	0	0	0	-68.8	NA	NA	NA	NA	NA	NA
7/16/2010	0	0	0	0	-68.7	NA	NA	NA	NA	NA	NA
7/17/2010	0	0	0	0	-68.8	NA	NA	NA	NA	NA	NA
7/18/2010	0	0	0	0	-68.8	NA	NA	NA	NA	NA	NA
7/19/2010	0	0	0	0	-68.7	NA	NA	NA	NA	NA	NA
7/20/2010	0	0	0	0	-68.7	NA	NA	NA	NA	NA	NA
7/21/2010	0	0	0	0	-68.6	NA	NA	NA	NA	NA	NA
7/22/2010	0	0	0	0	-68.7	NA	NA	NA	NA	NA	NA
7/23/2010	0	0	0	0	-68.5	NA	NA	NA	NA	NA	NA
7/24/2010	0	0	0	0	-68.6	NA	NA	NA	NA	NA	NA
7/25/2010	0	0	0	0	-68.7	NA	NA	NA	NA	NA	NA
7/26/2010	0	0	0	0	-68.7	NA	NA	NA	NA	NA	NA
7/27/2010	0	0	0	0	-68.8	NA	NA	NA	NA	NA	NA
7/28/2010	0	0	0	0	-68.8	NA	NA	NA	NA	NA	NA
7/29/2010	0	0	0	0	-68.8	NA	NA	NA	NA	NA	NA

Date										
7/30/2010	0	0	0	-68.7	NA	NA	NA	NA	NA	NA
7/31/2010	0	0	0	-68.7	NA	NA	NA	NA	NA	NA
8/1/2010	0	0	0	-75.4	NA	NA	NA	NA	NA	NA
8/2/2010	0	0	0	-73.8	NA	NA	NA	NA	NA	NA
8/3/2010	0	0	0	-74.4	NA	NA	NA	NA	NA	NA
8/4/2010	0	0	0	-75.4	NA	NA	NA	NA	NA	NA
8/5/2010	0	0	0	-75.0	NA	NA	NA	NA	NA	NA
8/6/2010	0	0	0	-72.9	NA	NA	NA	NA	NA	NA
8/7/2010	0	0	0	-71.7	NA	NA	NA	NA	NA	NA
8/8/2010	0	0	0	-72.8	NA	NA	NA	NA	NA	NA
8/9/2010	0	0	0	-73.9	NA	NA	NA	NA	NA	NA
8/10/2010	0	0	0	-75.6	NA	NA	NA	NA	NA	NA
8/11/2010	0	0	0	-76.0	NA	NA	NA	NA	NA	NA
8/12/2010	0	0	0	-74.4	NA	NA	NA	NA	NA	NA
8/13/2010	0	0	0	-74.4	NA	NA	NA	NA	NA	NA
8/14/2010	0	0	0	-74.4	NA	NA	NA	NA	NA	NA
8/15/2010	0	0	0	-75.6	NA	NA	NA	NA	NA	NA
8/16/2010	0	0	0	-74.3	NA	NA	NA	NA	NA	NA
8/17/2010	0	0	0	-73.4	NA	NA	NA	NA	NA	NA
8/18/2010	0	0	0	-74.9	NA	NA	NA	NA	NA	NA
8/19/2010	0	0	0	-74.8	NA	NA	NA	NA	NA	NA

Date											
8/20/2010	0	0	0	0	-74.5	NA	NA	NA	NA	NA	NA
8/21/2010	0	0	0	0	-73.3	NA	NA	NA	NA	NA	NA
8/22/2010	0	0	0	0	-73.3	NA	NA	NA	NA	NA	NA
8/23/2010	0	0	0	0	-74.2	NA	NA	NA	NA	NA	NA
8/24/2010	0	0	0	0	-74.8	NA	NA	NA	NA	NA	NA
8/25/2010	0	0	0	0	-75.1	NA	NA	NA	NA	NA	NA
8/26/2010	0	0	0	0	-75.0	NA	NA	NA	NA	NA	NA
8/27/2010	0	0	0	0	-74.6	NA	NA	NA	NA	NA	NA
8/28/2010	0	0	0	0	-75.1	NA	NA	NA	NA	NA	NA
8/29/2010	0	0	0	0	-75.4	NA	NA	NA	NA	NA	NA
8/30/2010	0	0	0	0	-75.7	NA	NA	NA	NA	NA	NA
8/31/2010	0	0	0	0	-75.6	NA	NA	NA	NA	NA	NA
9/1/2010	0	0	0	0	-68.7	NA	NA	NA	NA	NA	NA
9/2/2010	0	0	0	0	-68.7	NA	NA	NA	NA	NA	NA
9/3/2010	0	0	0	0	-68.5	NA	NA	NA	NA	NA	NA
9/4/2010	0	0	0	0	-68.6	NA	NA	NA	NA	NA	NA
9/5/2010	0	0	0	0	-68.7	NA	NA	NA	NA	NA	NA
9/6/2010	0	0	0	0	-68.6	NA	NA	NA	NA	NA	NA
9/7/2010	0	0	0	0	-68.3	NA	NA	NA	NA	NA	NA
9/8/2010	0	0	0	0	-67.9	NA	NA	NA	NA	NA	NA
9/9/2010	0	0	0	0	-68.6	NA	NA	NA	NA	NA	NA

Date										
9/10/2010	0	0	0	-68.7	NA	NA	NA	NA	NA	NA
9/11/2010	0	0	0	-68.7	NA	NA	NA	NA	NA	NA
9/12/2010	0	0	0	-68.7	NA	NA	NA	NA	NA	NA
9/13/2010	0	0	0	-68.5	NA	NA	NA	NA	NA	NA
9/14/2010	0	0	0	-68.6	NA	NA	NA	NA	NA	NA
9/15/2010	0	0	0	-68.7	NA	NA	NA	NA	NA	NA
9/16/2010	0	0	0	-68.7	NA	NA	NA	NA	NA	NA
9/17/2010	0	0	0	-68.6	NA	NA	NA	NA	NA	NA
9/18/2010	0	0	0	-68.6	NA	NA	NA	NA	NA	NA
9/19/2010	0	0	0	-68.7	NA	NA	NA	NA	NA	NA
9/20/2010	0	0	0	-68.7	NA	NA	NA	NA	NA	NA
9/21/2010	0	0	0	-68.6	NA	NA	NA	NA	NA	NA
9/22/2010	0	0	0	-68.7	NA	NA	NA	NA	NA	NA
9/23/2010	0	0	0	-68.2	NA	NA	NA	NA	NA	NA
9/24/2010	0	0	0	-68.3	NA	NA	NA	NA	NA	NA
9/25/2010	0	0	0	-67.6	NA	NA	NA	NA	NA	NA
9/26/2010	0	0	0	-67.5	NA	NA	NA	NA	NA	NA
9/27/2010	0	0	0	-68.5	NA	NA	NA	NA	NA	NA
9/28/2010	0	0	0	-68.3	NA	NA	NA	NA	NA	NA
9/29/2010	0	0	0	-66.8	NA	NA	NA	NA	NA	NA
9/30/2010	0	0	0	-67.2	NA	NA	NA	NA	NA	NA

Date											
10/1/2010	0	0	0	0	NA	NA	NA	NA	NA	NA	NA
10/2/2010	0	0	0	0	NA	NA	NA	NA	NA	NA	NA
10/3/2010	0	0	0	0	NA	NA	NA	NA	NA	NA	NA
10/4/2010	0	0	0	0	NA	NA	NA	NA	NA	NA	NA
10/5/2010	0	0	0	0	-61.8	0	37	32	30	0	0
10/6/2010	0	0	0	0	-63.0	0	32	40	28	0	0
10/7/2010	0	0	0	0	-62.6	0	29	41	30	0	0
10/8/2010	0	0	0	0	-62.2	0	31	35	34	0	0
10/9/2010	0	0	0	0	-61.9	1	32	35	32	0	0
10/10/2010	0	0	0	0	-61.5	1	22	38	39	0	0
10/11/2010	0	0	0	0	-61.6	0	18	34	48	0	0
10/12/2010	0	0	0	0	-59.7	1	21	26	53	0	0
10/13/2010	0	0	0	0	-59.8	0	26	28	45	0	0
10/14/2010	0	0	0	0	-58.7	0	22	27	50	0	0
10/15/2010	0	0	0	0	-50.6	1	26	10	63	0	0
10/16/2010	0	0	0	0	-51.2	2	34	13	51	0	0
10/17/2010	0	0	0	0	-57.5	0	35	12	53	1	0
10/18/2010	0	0	0	0	-58.1	0	54	8	38	0	0
10/19/2010	0	0	0	0	-58.7	1	43	22	34	0	0
10/20/2010	0	0	0	0	-57.6	8	32	22	38	0	0
10/21/2010	0	0	0	0	-56.8	1	41	17	41	0	0

10/22/2010	0	0	0	0	-56.5	0	40	21	39	0	0
10/23/2010	0	0	0	0	-56.5	2	44	18	36	0	0
10/24/2010	0	0	0	0	-57.9	1	48	19	32	0	0
10/25/2010	0	0	0	0	-57.3	0	45	23	31	0	0
10/26/2010	0	0	0	0	-57.9	1	42	17	40	0	0
10/27/2010	0	0	0	0	-57.3	0	41	19	40	0	0
10/28/2010	0	0	0	0	-57.4	0	40	15	45	0	0
10/29/2010	0	0	0	0	-57.1	0	46	10	43	0	0
10/30/2010	0	0	0	0	-55.6	0	50	18	32	0	0
10/31/2010	0	0	0	0	-55.8	0	51	22	26	0	0
11/1/2010	0	0	0	0	-56.7	0	51	15	33	0	0
11/2/2010	0	0	0	0	-57.7	0	52	11	37	0	0
11/3/2010	0	0	0	0	-57.2	0	52	15	33	0	0
11/4/2010	0	0	0	0	-58.0	1	56	14	29	0	0
11/5/2010	0	0	0	0	-57.7	0	54	15	30	0	0
11/6/2010	0	0	0	0	-57.5	0	51	13	36	0	0
11/7/2010	0	0	0	0	-57.3	0	45	21	34	0	0
11/8/2010	0	0	0	0	-57.0	0	39	23	37	0	0
11/9/2010	0	0	0	0	-57.0	0	28	22	50	0	0
11/10/2010	0	0	0	0	-55.9	0	28	11	61	0	0
11/11/2010	0	0	0	0	-55.0	0	32	14	53	0	0

Date											
11/12/2010	0	0	0	0	-54.2	0	34	14	52	0	0
11/13/2010	0	0	0	0	-55.3	0	42	13	44	0	0
11/14/2010	0	0	0	0	-55.7	0	39	14	46	0	0
11/15/2010	0	0	0	0	-56.7	0	23	28	49	0	0
11/16/2010	0	0	0	0	-56.1	0	12	30	58	0	0
11/17/2010	0	0	0	0	-55.9	0	22	25	53	0	0
11/18/2010	0	0	0	0	-55.5	0	18	19	62	0	0
11/19/2010	0	0	0	0	-55.4	0	22	20	57	0	0
11/20/2010	0	0	0	0	-56.5	0	30	23	47	0	0
11/21/2010	0	0	0	0	-55.5	0	29	18	52	1	0
11/22/2010	0	0	0	0	-55.6	0	34	17	48	0	0
11/23/2010	0	0	0	0	-55.9	0	28	17	55	0	0
11/24/2010	0	0	0	0	-55.4	0	37	16	47	0	0
11/25/2010	0	0	0	0	-56.6	0	24	18	58	0	0
11/26/2010	0	0	0	0	-58.0	0	23	13	63	0	0
11/27/2010	0	0	0	0	-56.8	0	26	17	57	0	0
11/28/2010	0	0	0	0	-59.6	1	25	21	53	1	0
11/29/2010	0	0	0	0	-58.4	2	27	23	48	0	0
11/30/2010	0	0	0	0	-58.2	0	33	23	43	0	0
12/1/2010	0	0	0	0	-58.5	0	30	22	48	0	0
12/2/2010	0	0	0	0	-58.2	0	26	24	50	0	0

Date											
12/3/2010	0	0	0	0	-56.8	0	47	8	44	0	0
12/4/2010	0	0	0	0	-55.7	4	72	1	23	0	0
12/5/2010	0	0	0	0	-54.9	16	68	0	17	0	0
12/6/2010	0	0	0	0	-56.9	5	81	0	13	0	0
12/7/2010	0	0	0	0	-57.6	3	78	1	19	0	0
12/8/2010	0	0	0	0	-55.3	1	73	1	24	0	0
12/9/2010	0	0	0	0	-55.6	1	69	1	29	0	0
12/10/2010	0	0	0	0	-55.9	0	58	1	41	0	0
12/11/2010	0	0	0	0	-57.4	0	37	9	53	0	0
12/12/2010	0	0	0	0	-57.4	0	27	10	63	0	0
12/13/2010	0	0	0	0	-60.9	0	15	25	60	0	0
12/14/2010	0	0	0	0	-61.6	0	18	22	60	0	0
12/15/2010	0	0	0	0	-59.6	0	16	21	63	0	0
12/16/2010	0	0	0	0	-57.7	1	20	13	67	0	0
12/17/2010	0	0	0	0	-61.2	3	15	26	56	0	0
12/18/2010	0	0	0	0	-61.3	1	20	25	54	0	0
12/19/2010	0	0	0	0	-62.0	0	13	20	67	0	0
12/20/2010	0	0	0	0	-61.3	0	13	21	66	0	0
12/21/2010	0	0	0	0	-58.8	1	14	14	71	0	0
12/22/2010	0	0	0	0	-55.4	0	30	10	60	0	0
12/23/2010	0	0	0	0	-61.7	0	30	22	47	0	0

Date											
12/24/2010	0	0	0	0	-62.5	0	16	20	64	0	0
12/25/2010	0	0	0	0	-60.2	0	19	19	61	1	0
12/26/2010	0	0	0	0	-60.8	0	20	21	58	0	0
12/27/2010	0	0	0	0	-63.9	0	18	29	53	0	0
12/28/2010	0	0	0	0	-64.5	0	12	33	54	0	0
12/29/2010	0	0	0	0	-63.9	0	10	35	55	0	0
12/30/2010	0	0	0	0	-65.0	0	10	41	49	0	0
12/31/2010	0	0	0	0	-63.7	0	14	33	53	0	0
1/1/2011	0	0	0	0	-63.0	0	11	30	59	0	0
1/2/2011	0	0	0	0	-64.6	0	13	40	47	0	0
1/3/2011	0	0	0	0	-65.2	1	10	42	47	0	0
1/4/2011	0	0	0	0	-48.9	8	10	29	53	0	0
1/5/2011	0	0	0	0	-48.7	6	35	27	32	0	0
1/6/2011	0	0	0	0	-64.8	0	14	38	48	0	0
1/7/2011	0	0	0	0	-64.0	2	13	31	53	1	0
1/8/2011	0	0	0	0	-62.3	0	19	26	55	0	0
1/9/2011	0	0	0	0	-61.9	0	18	28	54	0	0
1/10/2011	0	0	0	0	-62.7	0	13	30	57	0	1
1/11/2011	0	0	0	0	-63.9	0	17	31	52	0	0
1/12/2011	0	0	0	0	-62.1	0	17	28	54	0	1
1/13/2011	0	0	0	0	-61.5	0	11	33	56	0	0

Date										
1/14/2011	0	0	53	31	16	0	-63.1	0	0	0
1/15/2011	0	0	54	29	17	0	-60.4	0	0	0
1/16/2011	0	0	49	24	26	0	-58.4	0	0	0
1/17/2011	0	0	46	35	19	0	-58.1	0	0	0
1/18/2011	0	2	59	19	14	6	-56.5	0	0	0
1/19/2011	0	1	47	27	20	6	-48.2	0	0	0
1/20/2011	0	0	42	38	20	0	-62.4	0	0	0
1/21/2011	0	0	52	37	12	0	-62.7	0	0	0
1/22/2011	1	2	53	30	13	1	-60.1	0	0	0
1/23/2011	0	1	55	21	14	9	-51.3	0	0	0
1/24/2011	0	1	36	31	33	0	-59.1	0	0	0
1/25/2011	0	0	40	39	20	1	-61.2	0	0	0
1/26/2011	0	1	35	38	26	0	-62.0	0	0	0
1/27/2011	0	1	54	22	16	8	-56.5	0	0	0
1/28/2011	0	1	45	30	22	2	-53.9	0	0	0
1/29/2011	0	0	39	33	25	2	-62.9	0	0	0
1/30/2011	0	0	40	37	23	0	-63.9	0	0	0
1/31/2011	0	0	41	39	19	0	-63.6	0	0	0
2/1/2011	0	3	51	27	18	1	-56.1	0	0	0
2/2/2011	0	3	38	40	20	0	-53.8	0	0	0
2/3/2011	0	2	37	47	14	0	-55.5	0	0	0

Date											
2/4/2011	0	0	0	0	-62.1	0	13	42	45	1	0
2/5/2011	0	0	0	0	-61.0	6	7	40	46	0	0
2/6/2011	0	0	0	0	-64.8	1	9	56	33	0	0
2/7/2011	0	0	0	0	-64.1	0	17	39	44	0	0
2/8/2011	0	0	0	0	-64.2	0	12	46	42	0	0
2/9/2011	0	0	0	0	-61.9	3	8	44	46	0	0
2/10/2011	0	0	0	0	-57.0	4	6	40	47	3	0
2/11/2011	0	0	0	0	-57.4	0	10	52	33	5	0
2/12/2011	0	0	0	0	-61.0	0	8	55	36	1	0
2/13/2011	0	0	0	0	-61.6	6	6	39	47	2	0
2/14/2011	0	0	0	0	-63.2	1	9	48	42	1	0
2/15/2011	0	0	0	0	-64.0	0	7	54	38	1	0
2/16/2011	0	0	0	0	-63.0	1	11	44	43	2	0
2/17/2011	0	0	0	0	-63.9	1	14	40	45	0	0
2/18/2011	0	0	0	0	-64.0	0	16	53	31	0	0
2/19/2011	0	0	0	0	-63.7	1	12	51	36	1	0
2/20/2011	0	0	0	0	-63.4	3	10	46	40	0	0
2/21/2011	0	0	0	0	-64.4	0	20	49	31	0	0
2/22/2011	0	0	0	0	-62.2	0	12	50	37	2	0
2/23/2011	0	0	0	0	-61.0	1	13	41	43	2	0
2/24/2011	0	0	0	0	-57.3	8	17	31	43	1	0

Date											
2/25/2011	0	0	0	0	-61.5	0	11	45	44	0	0
2/26/2011	0	0	0	0	-65.3	1	17	43	39	0	0
2/27/2011	0	0	0	0	-63.1	0	7	52	41	1	0
2/28/2011	0	0	0	0	-60.8	0	9	55	33	2	0
3/1/2011	0	0	0	0	-62.1	0	11	59	30	0	0
3/2/2011	0	0	0	0	-61.3	0	5	49	44	2	0
3/3/2011	0	0	0	0	-61.0	0	6	47	45	1	1
3/4/2011	0	0	0	0	-64.1	0	2	54	43	1	0
3/5/2011	0	0	0	0	-64.0	0	10	56	34	1	0
3/6/2011	0	0	0	0	-64.5	0	3	58	38	1	0
3/7/2011	0	0	0	0	-66.6	0	1	61	37	0	0
3/8/2011	0	0	0	0	-65.5	0	13	65	21	1	0
3/9/2011	1	0	0	0	-65.4	0	9	66	24	0	0
3/10/2011	0	0	0	0	-66.6	0	11	69	20	0	0
3/11/2011	1	0	0	0	-66.0	0	11	71	18	0	0
3/12/2011	0	0	0	0	-65.9	0	3	66	30	0	0
3/13/2011	0	0	0	0	-65.5	0	11	66	23	0	0
3/14/2011	0	0	0	0	-65.4	0	10	68	22	0	0
3/15/2011	0	0	0	0	-64.9	0	8	70	22	0	0
3/16/2011	1	0	0	0	-65.6	0	13	69	19	0	0
3/17/2011	0	0	NA	NA	-67.4	0	10	74	16	0	0

Date											
3/18/2011	0	0	30	66	4	0	-64.6	15	60	0	0
3/19/2011	0	1	33	58	8	0	-63.1	NA	NA	0	1
3/20/2011	0	1	41	46	11	0	-56.0	NA	NA	0	0
3/21/2011	0	1	38	47	14	0	-64.1	0	0	0	0
3/22/2011	0	0	35	50	15	0	-64.6	NA	NA	0	0
3/23/2011	0	2	24	55	18	0	-64.7	0	0	0	0
3/24/2011	0	1	27	59	13	0	-63.0	NA	NA	0	0
3/25/2011	0	0	25	60	14	0	-64.5	0	0	0	0
3/26/2011	0	0	34	51	15	0	-64.1	NA	NA	0	0
3/27/2011	0	0	32	61	7	0	-64.1	NA	NA	0	0
3/28/2011	0	1	27	69	4	0	-64.4	0	0	0	0
3/29/2011	0	2	34	56	8	0	-62.3	NA	NA	0	0
3/30/2011	0	2	42	47	9	0	-60.7	NA	NA	0	0
3/31/2011	0	0	37	56	7	0	-60.9	0	0	0	0
4/1/2011	0	1	27	64	7	0	-63.0	36	60	0	0
4/2/2011	0	0	29	60	11	0	-63.0	NA	NA	0	0
4/3/2011	0	1	34	47	17	0	-61.8	NA	NA	0	0
4/4/2011	0	1	36	57	6	0	-64.2	0	0	0	0
4/5/2011	0	1	40	47	13	0	-63.1	NA	NA	0	0
4/6/2011	0	0	41	44	14	0	-63.2	0	0	0	0
4/7/2011	0	1	24	7	61	7	-58.0	NA	NA	0	0

Date											
4/8/2011	0	0	11	1	77	11	-60.3	0	0	0	0
4/9/2011	0	0	14	3	75	8	-59.1	NA	NA	0	0
4/10/2011	0	0	28	30	42	0	-64.6	NA	NA	0	0
4/11/2011	0	0	44	42	13	0	-64.3	0	0	0	0
4/12/2011	1	0	32	58	9	0	-63.7	NA	NA	0	0
4/13/2011	0	1	26	47	26	0	-64.6	0	0	0	0
4/14/2011	0	0	41	45	14	0	-63.1	NA	NA	0	0
4/15/2011	0	0	44	41	15	0	-64.3	36	40	1	0
4/16/2011	0	0	45	37	18	0	-65.2	NA	NA	1	0
4/17/2011	1	1	44	41	14	0	-64.8	NA	NA	1	0
4/18/2011	0	0	42	39	18	0	-64.4	36	60	0	0
4/19/2011	0	0	32	41	27	0	-65.3	NA	NA	1	0
4/20/2011	0	0	23	53	24	0	-66.2	0	0	1	0
4/21/2011	0	0	33	57	10	0	-65.9	NA	NA	1	0
4/22/2011	0	0	35	58	7	0	-65.6	0	0	1	0
4/23/2011	0	0	38	54	8	0	-64.0	NA	NA	0	0
4/24/2011	0	1	38	46	15	0	-61.1	NA	NA	0	0
4/25/2011	0	0	23	51	26	0	-64.4	36	30	1	0
4/26/2011	0	0	27	64	9	0	-66.3	NA	NA	0	0
4/27/2011	0	1	32	56	11	0	-65.1	0	0	1	0
4/28/2011	0	0	26	65	9	0	-66.8	NA	NA	0	0

Date											
4/29/2011	0	0	0	40	30	-67.4	0	13	58	28	0	0
4/30/2011	0	0	0	NA	NA	-67.5	0	11	62	27	0	0
5/1/2011	0	0	0	NA	NA	-66.8	0	11	59	30	1	0
5/2/2011	0	0	0	0	0	-68.6	0	4	72	24	0	0
5/3/2011	0	0	0	NA	NA	-68.8	0	7	66	28	0	0
5/4/2011	0	0	0	0	0	-67.5	0	14	61	26	0	0
5/5/2011	0	0	0	0	0	-64.4	0	15	53	31	0	0
5/6/2011	0	0	0	0	0	-64.1	0	16	50	34	1	0
5/7/2011	0	0	0	0	0	-61.6	0	20	51	27	1	0
5/8/2011	0	0	0	0	0	-62.0	0	21	50	29	1	0
5/9/2011	0	0	0	0	0	-63.9	0	16	58	26	0	0
5/10/2011	0	0	0	0	0	-60.2	0	12	49	38	0	0
5/11/2011	0	0	0	0	0	-60.9	0	14	48	36	2	0
5/12/2011	0	0	0	0	0	-53.7	0	8	44	46	2	0
5/13/2011	0	0	0	0	0	-56.4	0	9	41	47	3	0
5/14/2011	0	0	0	0	0	-56.4	0	14	32	51	3	0
5/15/2011	0	0	0	0	0	-58.3	0	17	31	50	2	0
5/16/2011	0	0	0	0	0	-58.5	0	19	31	49	1	0
5/17/2011	0	0	0	0	0	-58.3	0	24	34	40	1	0
5/18/2011	0	0	0	0	0	-57.3	0	17	32	35	2	0
5/19/2011	0	0	0	0	0	NA	NA	NA	NA	NA	NA	NA

Table S2. Daily mean sound levels at M2 (A) and M5 (B) used in the GLM and GAM modeling. Sound level units are dB re 1 $\mu Pa^2/Hz$.

Date	500 Hz	2 kHz	10 kHz	20 kHz	40 kHz
9/27/2009	66.7	62.9	47.9	42.8	37.0
9/28/2009	67.8	63.0	49.9	44.5	38.4
9/29/2009	68.2	61.7	48.1	42.7	37.2
9/30/2009	69.4	65.1	50.9	44.5	37.5
10/1/2009	71.6	65.8	47.5	42.8	36.5
10/2/2009	75.7	69.3	50.7	44.6	38.7
10/3/2009	72.4	64.1	45.7	41.2	37.2
10/4/2009	67.0	58.1	40.9	36.5	33.4
10/5/2009	62.0	57.1	44.6	42.4	37.3
10/6/2009	69.0	65.3	51.1	43.4	37.0
10/7/2009	67.4	63.1	49.3	43.2	36.8
10/8/2009	65.9	59.6	45.6	40.7	36.1
10/9/2009	67.9	62.8	49.2	44.0	37.8
10/10/2009	68.3	64.0	50.3	44.3	37.8
10/11/2009	65.1	59.6	46.1	40.9	35.9
10/12/2009	64.8	55.1	28.3	27.9	30.5
10/13/2009	67.8	58.8	39.2	34.8	32.4
10/14/2009	69.3	62.6	47.7	42.4	37.2
10/15/2009	68.6	63.8	50.2	44.4	38.1
10/16/2009	65.3	60.2	46.9	41.8	36.8
10/17/2009	62.5	56.9	43.8	39.3	35.5
10/18/2009	61.0	54.4	39.3	35.2	33.0
10/19/2009	61.8	55.5	42.3	39.0	35.1
10/20/2009	66.9	61.3	48.1	42.9	37.4
10/21/2009	69.0	64.1	50.2	44.1	37.7
10/22/2009	70.3	66.1	51.6	44.8	37.3
10/23/2009	73.6	68.3	52.3	44.2	36.9
10/24/2009	70.8	65.3	50.3	44.4	38.2
10/25/2009	64.4	58.5	44.0	39.1	35.4
10/26/2009	65.0	58.1	44.3	41.6	36.8

10/27/2009	58.4	52.2	39.9	41.0	36.3
10/28/2009	63.3	57.2	44.2	39.5	35.5
10/29/2009	66.9	62.1	48.7	43.3	38.1
10/30/2009	76.5	69.0	53.0	49.5	40.2
10/31/2009	71.7	66.4	49.5	43.9	39.8
11/1/2009	65.8	59.0	45.0	40.5	35.8
11/2/2009	74.9	64.6	41.9	38.8	33.1
11/3/2009	69.1	63.7	49.5	44.4	37.9
11/4/2009	76.8	69.5	50.1	41.9	36.7
11/5/2009	77.6	69.8	52.4	43.7	36.4
11/6/2009	70.0	66.2	51.4	44.7	37.7
11/7/2009	70.9	66.7	51.8	44.8	38.1
11/8/2009	69.1	64.0	49.6	43.5	37.8
11/9/2009	68.9	61.7	45.7	40.9	36.1
11/10/2009	74.8	66.8	47.7	42.1	37.2
11/11/2009	73.0	67.3	51.1	44.7	38.3
11/12/2009	69.4	64.2	49.7	44.6	39.1
11/13/2009	67.4	60.1	45.5	41.7	37.0
11/14/2009	81.9	73.3	58.2	48.4	42.6
11/15/2009	79.0	70.3	53.1	45.3	41.1
11/16/2009	83.1	76.2	57.3	51.5	49.7
11/17/2009	79.9	71.8	49.4	41.0	36.7
11/18/2009	76.3	68.8	51.8	44.9	38.5
11/19/2009	68.8	64.7	50.0	43.8	38.4
11/20/2009	68.6	64.2	50.1	44.1	38.4
11/21/2009	79.0	72.5	52.6	43.0	37.3
11/22/2009	67.0	60.8	46.5	41.3	36.2
11/23/2009	81.0	72.5	52.2	44.3	40.2
11/24/2009	74.2	67.7	51.4	44.2	38.0
11/25/2009	72.5	65.2	49.5	43.1	37.8
11/26/2009	69.8	59.7	44.7	37.5	33.9
11/27/2009	71.2	60.9	43.8	37.0	33.9
11/28/2009	72.2	65.3	50.2	44.0	39.2
11/29/2009	67.5	62.8	48.3	41.8	36.9
11/30/2009	71.0	67.6	50.0	43.5	37.8
12/1/2009	65.0	60.4	46.9	41.8	37.4

12/2/2009	70.6	66.7	51.2	44.5	37.8
12/3/2009	64.3	59.7	45.7	40.7	37.0
12/4/2009	72.2	68.3	51.7	43.8	36.8
12/5/2009	72.0	68.6	51.1	43.2	36.8
12/6/2009	70.9	66.6	51.1	44.3	38.2
12/7/2009	72.6	69.3	51.4	43.1	36.5
12/8/2009	68.1	64.2	49.2	43.4	38.0
12/9/2009	66.3	62.2	47.8	42.5	37.9
12/10/2009	67.3	63.0	48.5	43.0	38.3
12/11/2009	65.7	60.4	46.6	41.6	37.6
12/12/2009	60.0	52.8	38.1	34.7	31.5
12/13/2009	64.6	59.0	45.1	40.5	36.5
12/14/2009	64.1	57.6	44.0	39.3	36.2
12/15/2009	70.7	66.6	50.8	44.1	37.7
12/16/2009	70.3	66.1	50.7	44.1	37.7
12/17/2009	70.1	65.9	50.6	44.3	38.3
12/18/2009	66.0	61.6	47.0	41.9	37.2
12/19/2009	67.5	63.1	48.8	44.1	38.5
12/20/2009	69.5	65.7	50.5	44.1	37.9
12/21/2009	69.6	65.6	51.6	45.9	40.3
12/22/2009	66.9	62.6	48.4	43.3	38.0
12/23/2009	62.1	57.6	43.7	39.3	34.9
12/24/2009	69.3	64.9	50.1	43.9	37.8
12/25/2009	67.9	63.6	49.1	43.8	38.2
12/26/2009	65.5	61.1	47.2	42.2	37.3
12/27/2009	63.3	58.2	44.6	39.9	35.6
12/28/2009	68.4	63.9	49.2	43.5	37.5
12/29/2009	68.1	63.8	48.9	43.2	37.2
12/30/2009	66.4	61.6	47.8	42.7	37.5
12/31/2009	59.4	53.5	40.3	35.9	32.1
1/1/2010	63.7	58.1	44.4	39.5	35.5
1/2/2010	66.0	61.7	47.3	42.2	37.9
1/3/2010	70.6	65.9	50.1	43.7	38.0
1/4/2010	71.5	67.5	51.4	44.2	36.8
1/5/2010	66.4	61.9	48.0	42.5	37.3
1/6/2010	69.9	65.6	50.5	44.3	37.6

1/7/2010	68.4	64.0	49.5	44.0	38.2
1/8/2010	67.2	62.5	48.5	43.0	38.0
1/9/2010	67.4	62.5	48.5	43.1	38.1
1/10/2010	68.1	63.7	49.5	43.8	38.4
1/11/2010	65.6	61.0	47.0	41.7	37.3
1/12/2010	67.5	63.1	48.1	42.2	37.4
1/13/2010	66.6	62.2	47.8	42.5	38.0
1/14/2010	64.9	59.5	45.6	40.3	36.3
1/15/2010	68.8	64.5	49.6	43.8	38.1
1/16/2010	68.5	64.1	49.5	43.5	37.9
1/17/2010	67.9	63.3	49.1	43.5	38.5
1/18/2010	69.7	65.4	50.6	44.7	38.5
1/19/2010	71.9	68.0	52.0	44.6	36.9
1/20/2010	71.3	67.4	51.6	44.6	37.2
1/21/2010	71.6	64.8	49.4	43.7	38.9
1/22/2010	69.3	62.6	48.0	42.7	38.4
1/23/2010	64.9	60.9	45.4	40.4	36.6
1/24/2010	64.7	57.6	44.1	39.3	35.4
1/25/2010	65.9	59.6	45.8	40.6	36.9
1/26/2010	69.0	64.0	49.5	43.7	38.4
1/27/2010	69.9	65.5	50.7	44.6	38.7
1/28/2010	69.4	64.6	50.0	44.2	38.6
1/29/2010	77.4	68.9	46.5	40.2	35.9
1/30/2010	80.8	71.8	46.8	38.7	36.6
1/31/2010	81.5	71.9	47.7	42.3	38.1
2/1/2010	79.8	70.0	49.3	43.4	38.1
2/2/2010	69.7	61.1	37.3	31.8	29.8
2/3/2010	70.8	62.2	46.4	41.3	36.8
2/4/2010	70.8	65.7	50.3	43.9	37.8
2/5/2010	72.7	68.5	52.4	44.5	36.4
2/6/2010	72.2	67.7	52.2	44.9	37.2
2/7/2010	68.3	61.9	47.9	43.3	39.7
2/8/2010	72.3	68.1	48.7	40.6	36.7
2/9/2010	69.4	65.3	48.1	41.0	35.4
2/10/2010	67.6	61.2	47.0	41.4	36.4
2/11/2010	67.4	61.2	47.1	41.4	36.5

2/12/2010	66.0	57.7	43.9	38.5	34.7
2/13/2010	65.5	57.6	43.7	38.2	34.6
2/14/2010	66.9	56.9	40.5	34.6	31.9
2/15/2010	66.1	59.0	45.5	40.1	37.1
2/16/2010	64.7	57.7	44.3	39.0	36.1
2/17/2010	62.9	56.1	42.7	37.2	34.1
2/18/2010	61.8	48.4	37.3	33.6	34.1
2/19/2010	65.1	57.6	46.2	44.4	43.1
2/20/2010	64.2	54.2	40.1	34.6	32.4
2/21/2010	65.6	55.5	43.4	40.8	38.8
2/22/2010	62.9	51.3	45.5	44.9	43.6
2/23/2010	64.1	57.3	44.7	41.6	39.3
2/24/2010	67.4	61.9	48.0	42.4	37.8
2/25/2010	67.5	62.1	48.0	42.5	37.9
2/26/2010	68.4	62.9	49.0	43.3	38.2
2/27/2010	73.2	68.3	54.3	47.9	43.0
2/28/2010	72.5	66.9	53.7	48.8	46.8
3/1/2010	72.8	66.2	51.0	45.3	41.7
3/2/2010	68.9	59.9	44.2	40.7	39.9
3/3/2010	69.9	63.7	51.6	47.7	45.8
3/4/2010	69.2	63.0	50.6	46.7	44.7
3/5/2010	73.9	67.0	50.6	43.8	39.0
3/6/2010	66.1	58.9	47.0	44.2	43.4
3/7/2010	64.9	57.2	46.9	44.2	42.5
3/8/2010	62.5	54.2	38.1	34.0	34.3
3/9/2010	67.3	58.9	42.9	36.9	36.7
3/10/2010	67.8	59.3	43.6	39.4	39.3
3/11/2010	71.7	62.3	39.0	33.5	32.6
3/12/2010	69.9	59.4	40.0	36.9	35.1
3/13/2010	70.9	56.6	38.7	36.4	37.7
3/14/2010	65.7	55.3	38.3	33.5	31.3
3/15/2010	69.3	55.2	32.5	29.6	30.9
3/16/2010	71.3	61.1	38.2	35.6	42.7
3/17/2010	69.5	58.8	43.0	41.7	40.5
3/18/2010	67.5	55.6	41.4	39.0	38.0
3/19/2010	65.7	55.9	46.2	42.7	41.5

3/20/2010	62.0	53.0	44.3	42.3	41.1
3/21/2010	61.6	52.2	43.7	44.0	41.6
3/22/2010	57.5	44.3	33.1	30.7	31.7
3/23/2010	56.0	46.9	36.9	34.5	35.0
3/24/2010	59.6	47.5	38.5	41.1	40.6
3/25/2010	61.2	54.8	46.2	45.2	44.6
3/26/2010	57.2	48.8	41.1	40.0	39.6
3/27/2010	60.4	50.2	38.7	36.8	36.2
3/28/2010	57.6	47.1	35.2	32.7	34.2
3/29/2010	62.5	52.5	40.0	36.6	36.0
3/30/2010	67.1	51.3	34.9	30.8	31.7
3/31/2010	68.5	55.2	38.2	33.6	31.0
4/1/2010	69.2	57.7	42.5	38.7	37.9
4/2/2010	66.7	58.2	44.7	40.9	38.9
4/3/2010	64.4	53.9	40.8	38.1	36.8
4/4/2010	63.1	53.9	43.8	40.8	39.2
4/5/2010	66.2	54.5	45.8	42.8	41.3
4/6/2010	63.2	54.3	42.7	39.6	38.0
4/7/2010	65.1	55.9	40.5	37.9	36.3
4/8/2010	65.5	55.3	36.9	35.1	35.2
4/9/2010	63.8	54.4	40.5	37.9	36.8
4/10/2010	69.6	64.2	55.7	53.2	50.5
4/11/2010	71.2	65.8	56.9	54.4	51.5
4/12/2010	72.5	67.5	56.4	51.9	47.8
4/13/2010	66.1	60.8	46.7	41.2	36.1
4/14/2010	63.9	57.0	43.2	38.1	34.9
4/15/2010	67.2	61.3	47.0	41.2	35.8
4/16/2010	66.3	60.7	46.3	39.9	34.5
4/17/2010	59.8	51.4	37.8	32.5	31.3
4/18/2010	69.4	64.6	49.8	43.4	37.2
4/19/2010	66.6	61.0	47.0	41.0	35.7
4/20/2010	66.8	60.2	46.4	40.8	36.1
4/21/2010	65.5	57.7	43.7	38.4	34.7
4/22/2010	64.3	56.3	42.8	37.8	35.2
4/23/2010	63.0	53.3	39.6	34.8	32.6
4/24/2010	66.9	60.3	46.1	40.4	35.6

4/25/2010	66.0	59.6	45.9	40.7	36.4
4/26/2010	70.2	64.7	51.1	46.1	41.2
4/27/2010	74.0	68.5	59.6	58.1	55.8
4/28/2010	81.4	73.6	59.3	55.7	52.0
4/29/2010	70.0	64.3	55.6	53.6	50.8
4/30/2010	69.8	64.2	50.2	43.2	36.8
5/1/2010	78.2	72.0	53.3	48.9	47.4
5/2/2010	67.0	61.2	48.4	44.3	41.0
5/3/2010	61.3	52.1	39.6	36.7	34.1
5/4/2010	60.2	44.1	32.4	33.2	32.6
5/5/2010	63.6	56.6	44.4	41.8	39.8
5/6/2010	61.5	53.2	40.1	36.1	33.7
5/7/2010	66.7	60.6	47.2	41.1	36.1
5/8/2010	63.9	55.6	42.7	37.5	34.2
5/9/2010	71.8	63.4	44.1	39.6	34.9
5/10/2010	64.1	52.9	39.0	35.2	31.8
5/11/2010	67.1	61.6	48.2	41.5	35.6
5/12/2010	61.5	54.0	41.4	36.1	33.8
5/13/2010	64.4	58.4	45.3	39.8	41.2
5/14/2010	57.8	47.2	33.4	33.6	47.1
5/15/2010	61.1	54.9	42.2	38.0	41.4
5/16/2010	70.8	63.4	49.0	42.5	37.0
5/17/2010	67.5	62.0	49.6	42.4	39.4
5/18/2010	66.8	62.0	48.8	42.3	40.8
5/19/2010	58.1	52.7	39.8	35.4	45.3
5/20/2010	67.2	61.9	48.7	42.0	39.9
5/21/2010	70.7	66.0	51.4	43.0	40.3
5/22/2010	63.1	53.8	39.6	34.5	32.4
5/23/2010	62.7	57.6	44.8	38.8	47.3
5/24/2010	72.5	63.8	47.0	40.3	41.6
5/25/2010	61.6	56.8	44.0	38.4	44.2
5/26/2010	59.4	52.5	39.9	35.0	43.2
5/27/2010	67.5	63.3	50.4	42.9	44.6
5/28/2010	68.7	64.4	51.0	43.6	41.3
5/29/2010	67.3	62.8	51.0	43.1	38.2
5/30/2010	62.8	57.8	45.2	39.5	44.4

5/31/2010	64.6	58.5	45.6	40.2	45.2
6/1/2010	58.0	52.1	39.9	36.7	38.7
6/2/2010	68.0	63.8	51.2	43.2	42.4
6/3/2010	68.2	64.0	51.5	42.9	39.5
6/4/2010	60.9	56.9	44.0	38.2	42.4
6/5/2010	60.8	56.5	44.2	38.3	43.0
6/6/2010	70.6	63.3	43.1	37.3	45.3
6/7/2010	62.0	57.2	44.7	38.8	36.3
6/8/2010	57.7	52.8	41.2	41.9	48.1
6/9/2010	55.1	50.4	38.6	33.5	46.2
6/10/2010	58.3	53.7	41.5	36.3	49.0
6/11/2010	63.2	58.6	46.2	40.3	36.1
6/12/2010	62.0	57.1	45.0	39.1	35.1
6/13/2010	61.0	56.1	46.1	38.4	34.7
6/14/2010	66.5	61.9	52.1	45.9	37.8
6/15/2010	64.4	60.3	51.3	41.8	37.0
6/16/2010	64.3	60.2	48.5	42.8	37.4
6/17/2010	66.5	62.5	52.8	43.3	37.8
6/18/2010	63.8	59.3	51.1	41.7	37.0
6/19/2010	64.4	57.8	43.2	42.0	37.3
6/20/2010	57.6	51.9	39.0	39.0	34.3
6/21/2010	68.1	60.5	44.2	38.6	35.8
6/22/2010	62.7	58.1	46.4	40.1	36.1
6/23/2010	56.9	53.2	59.7	43.2	37.8
6/24/2010	54.0	48.7	37.3	32.8	33.0
6/25/2010	62.1	57.3	45.4	39.6	35.6
6/26/2010	65.1	60.3	48.1	41.7	36.9
6/27/2010	64.4	60.1	48.2	41.9	37.8
6/28/2010	60.3	55.8	46.4	38.7	36.6
6/29/2010	56.0	51.3	40.9	43.1	37.2
6/30/2010	58.6	53.2	40.7	38.0	33.4
7/1/2010	55.7	50.7	40.6	43.4	37.2
7/2/2010	60.0	51.3	36.9	37.0	34.3
7/3/2010	51.0	42.3	31.7	30.9	30.4
7/4/2010	55.2	50.2	39.3	35.9	33.4
7/5/2010	64.8	53.8	35.7	41.0	34.8

7/6/2010	60.9	54.8	42.9	37.5	34.4
7/7/2010	59.5	54.7	42.9	38.1	37.1
7/8/2010	51.3	45.9	35.2	32.5	37.1
7/9/2010	63.4	58.5	46.8	40.5	36.4
7/10/2010	64.6	60.2	49.5	41.8	37.3
7/11/2010	55.3	49.7	38.4	34.2	32.5
7/12/2010	60.9	50.4	39.8	35.4	38.5
7/13/2010	60.7	49.3	41.6	43.0	36.1
7/14/2010	56.3	51.5	40.0	34.6	32.3
7/15/2010	54.0	49.1	38.3	33.8	33.5
7/16/2010	49.7	44.4	32.9	29.1	29.9
7/17/2010	48.4	41.2	30.6	28.8	30.4
7/18/2010	60.1	50.2	34.6	35.9	31.8
7/19/2010	59.6	55.1	43.2	38.0	34.3
7/20/2010	60.1	54.2	42.5	39.9	35.4
7/21/2010	64.0	59.0	48.8	41.8	37.3
7/22/2010	62.3	56.8	44.8	39.8	35.6
7/23/2010	64.1	59.7	50.8	41.7	37.2
7/24/2010	66.0	61.1	53.7	43.3	38.9
7/25/2010	60.8	56.3	45.2	38.9	40.1
7/26/2010	60.2	55.8	44.2	38.1	42.0
7/27/2010	55.6	50.9	39.1	33.7	38.3
7/28/2010	58.9	53.7	42.0	36.4	38.3
7/29/2010	61.1	56.9	48.1	46.2	43.5
7/30/2010	56.4	51.8	41.9	43.4	40.4
7/31/2010	65.4	60.1	49.8	41.5	40.4
8/1/2010	55.1	49.8	38.3	33.6	39.9
8/2/2010	56.2	50.8	39.8	39.0	43.4
8/3/2010	61.0	56.2	44.7	42.8	38.4
8/4/2010	59.7	53.7	46.9	37.6	41.8
8/5/2010	62.7	48.8	33.6	30.6	43.9
8/6/2010	63.0	58.8	47.4	42.8	40.4
8/7/2010	66.4	61.6	53.2	44.0	37.9
8/8/2010	64.3	58.1	49.4	44.2	42.9
8/9/2010	65.8	62.0	54.1	43.4	37.8
8/10/2010	69.8	63.5	56.5	43.7	38.8

8/11/2010	65.1	60.3	49.3	42.5	38.5
8/12/2010	71.7	55.1	41.3	38.7	37.3
8/13/2010	72.0	61.1	43.0	34.9	40.8
8/14/2010	67.6	62.3	54.7	44.5	44.6
8/15/2010	65.5	61.6	56.4	45.5	44.3
8/16/2010	63.9	59.1	49.6	42.6	43.9
8/17/2010	66.4	61.8	53.6	48.4	46.5
8/18/2010	65.2	61.0	58.6	43.2	44.0
8/19/2010	61.2	56.6	49.6	40.4	45.4
8/20/2010	57.4	52.1	45.2	35.5	40.6
8/21/2010	61.7	55.7	44.0	40.1	43.2
8/22/2010	64.0	58.1	46.1	40.3	42.0
8/23/2010	63.8	59.2	47.8	41.3	42.1
8/24/2010	61.8	56.6	46.7	39.1	36.5
8/25/2010	64.7	53.8	33.3	30.4	44.1
8/26/2010	64.7	55.2	41.3	40.3	36.6
8/27/2010	62.8	56.4	44.0	39.2	35.7
8/28/2010	67.9	59.9	53.1	43.2	44.3
8/29/2010	59.1	53.4	41.3	36.3	45.6
8/30/2010	60.9	53.7	40.6	36.0	48.3
8/31/2010	62.6	57.7	45.3	39.5	45.9
9/1/2010	64.3	59.9	47.7	41.6	46.5
9/2/2010	65.4	57.8	45.3	39.7	47.9
9/3/2010	63.1	57.9	48.5	40.1	46.1
9/4/2010	66.7	61.2	51.6	43.3	41.7
9/5/2010	66.2	59.5	47.1	41.2	44.5
9/6/2010	64.0	58.4	54.8	42.9	44.9
9/7/2010	64.3	57.8	48.5	45.3	45.2
9/8/2010	68.1	64.2	55.9	45.1	40.5
9/9/2010	75.3	66.2	52.3	43.3	44.5
9/10/2010	71.7	59.9	40.3	33.1	47.6
9/11/2010	63.4	53.9	34.4	32.6	48.5
9/12/2010	60.6	49.0	36.2	31.8	50.7
9/13/2010	64.4	59.5	47.9	41.5	45.1
9/14/2010	64.8	60.6	50.5	42.4	45.6
9/15/2010	71.6	60.2	42.8	37.4	48.2

9/16/2010	61.4	52.3	38.3	33.7	48.1
9/17/2010	61.5	56.6	44.7	39.3	48.0
9/18/2010	65.4	59.0	46.7	41.0	46.4
9/19/2010	66.0	60.8	53.8	43.1	43.6
9/20/2010	63.1	57.9	46.5	40.5	46.4
9/21/2010	61.0	55.9	44.1	38.8	45.3
9/22/2010	61.6	54.8	42.8	37.4	47.8
9/23/2010	68.3	59.9	46.2	39.0	46.2
9/24/2010	68.6	63.3	63.6	47.4	44.6
9/25/2010	67.9	62.3	60.7	44.5	44.2
9/26/2010	70.1	66.4	63.1	47.9	44.0
9/27/2010	72.2	64.2	61.3	46.4	47.7
9/28/2010	75.6	69.3	60.6	47.6	48.0
9/29/2010	68.9	65.2	62.4	50.7	44.2
9/30/2010	71.1	68.1	65.4	48.2	42.9
10/1/2010	70.5	67.1	57.7	47.8	42.9
10/2/2010	70.7	67.1	57.5	45.8	41.0
10/3/2010	71.4	66.1	58.9	45.6	41.5
10/4/2010	67.0	61.2	56.4	43.4	50.8
10/5/2010	75.4	66.9	58.5	54.6	48.2
10/6/2010	70.6	61.7	42.1	38.4	34.6
10/7/2010	69.2	63.9	42.1	38.1	34.6
10/8/2010	72.8	67.1	44.3	37.8	34.8
10/9/2010	70.3	62.7	41.7	37.0	34.6
10/10/2010	69.1	59.4	39.1	35.1	33.3
10/11/2010	70.9	63.5	48.3	42.9	37.3
10/12/2010	69.8	64.0	50.1	44.6	38.0
10/13/2010	69.7	62.0	46.0	41.2	36.3
10/14/2010	64.9	57.7	44.8	41.4	36.2
10/15/2010	70.3	60.4	47.1	44.2	37.7
10/16/2010	72.9	65.6	52.8	46.1	37.7
10/17/2010	69.0	64.3	50.9	45.0	37.7
10/18/2010	70.2	65.3	50.9	44.7	37.6
10/19/2010	76.2	70.3	51.9	46.0	39.9
10/20/2010	79.8	71.6	50.4	42.0	37.1
10/21/2010	68.5	62.8	48.5	43.5	37.6

10/22/2010	67.4	61.6	47.7	42.5	36.5
10/23/2010	68.8	62.5	48.7	43.8	37.6
10/24/2010	67.8	59.2	40.7	41.0	35.0
10/25/2010	70.6	62.8	41.3	34.0	32.8
10/26/2010	68.8	62.3	48.5	43.5	37.6
10/27/2010	69.3	63.4	49.6	45.7	38.7
10/28/2010	69.0	63.8	49.4	44.2	38.1
10/29/2010	70.8	67.5	53.1	47.0	38.4
10/30/2010	66.7	61.1	47.1	44.4	37.9
10/31/2010	70.0	61.6	45.0	39.5	35.7
11/1/2010	70.6	64.0	48.5	43.8	38.1
11/2/2010	70.3	65.0	50.4	45.1	38.5
11/3/2010	69.7	63.6	49.1	44.1	38.6
11/4/2010	67.1	60.7	46.5	41.8	37.1
11/5/2010	66.4	59.4	45.1	40.7	36.4
11/6/2010	70.1	64.9	50.2	44.8	38.6
11/7/2010	69.7	62.4	47.3	42.2	37.1
11/8/2010	70.1	60.6	46.0	41.7	37.1
11/9/2010	71.7	61.7	46.0	42.2	37.2
11/10/2010	72.0	64.3	49.8	44.3	37.9
11/11/2010	68.8	62.9	49.0	44.1	38.2
11/12/2010	66.4	52.4	35.3	31.3	32.3
11/13/2010	66.3	53.4	37.6	34.2	33.7
11/14/2010	71.8	63.7	45.1	40.0	36.0
11/15/2010	66.5	57.9	43.6	39.1	35.3
11/16/2010	70.6	64.5	50.1	45.1	39.0
11/17/2010	68.3	62.0	48.0	43.1	37.8
11/18/2010	70.0	62.3	48.3	43.6	38.2
11/19/2010	70.5	65.2	50.9	45.4	38.4
11/20/2010	70.6	64.2	49.9	44.3	37.7
11/21/2010	69.4	63.2	49.5	44.6	38.1
11/22/2010	67.3	59.7	47.3	50.9	42.5
11/23/2010	68.9	62.0	48.8	46.2	39.8
11/24/2010	69.7	65.1	49.8	45.3	39.0
11/25/2010	73.5	65.8	50.5	44.7	38.3
11/26/2010	76.7	66.6	48.8	43.9	38.4

11/27/2010	78.0	68.4	46.7	42.4	37.6
11/28/2010	77.8	68.4	51.2	45.2	38.8
11/29/2010	85.2	75.1	52.2	45.3	41.0
11/30/2010	70.2	62.5	45.5	40.8	36.6
12/1/2010	63.4	57.8	43.3	39.8	34.1
12/2/2010	73.6	65.4	49.2	43.6	37.5
12/3/2010	75.4	68.6	52.1	45.4	38.3
12/4/2010	71.6	67.8	52.1	45.0	38.1
12/5/2010	72.8	68.9	52.1	44.4	36.9
12/6/2010	71.6	67.1	51.8	45.5	38.2
12/7/2010	70.5	65.7	51.0	45.7	38.7
12/8/2010	72.8	68.8	52.0	44.4	37.0
12/9/2010	72.7	69.0	52.0	44.0	36.5
12/10/2010	72.6	68.9	51.9	44.1	36.4
12/11/2010	72.3	68.2	52.0	44.4	37.4
12/12/2010	72.5	68.5	51.6	44.1	38.3
12/13/2010	69.0	64.4	49.9	44.2	37.9
12/14/2010	69.9	65.7	50.6	44.9	38.2
12/15/2010	72.6	68.4	52.0	44.4	36.9
12/16/2010	72.9	68.9	52.0	43.6	36.2
12/17/2010	66.9	62.0	47.7	42.1	36.5
12/18/2010	64.9	55.7	41.9	37.2	34.2
12/19/2010	68.0	63.1	49.2	44.1	38.0
12/20/2010	66.2	61.0	47.0	42.0	37.1
12/21/2010	71.2	66.9	50.9	44.0	37.3
12/22/2010	74.5	71.1	51.6	41.1	34.8
12/23/2010	69.0	64.6	49.4	43.4	37.7
12/24/2010	70.7	66.4	51.1	44.7	37.6
12/25/2010	72.8	68.4	51.8	43.9	36.4
12/26/2010	71.9	67.7	51.6	44.4	36.7
12/27/2010	68.1	63.7	49.1	43.7	37.7
12/28/2010	65.3	60.3	46.7	41.8	37.2
12/29/2010	71.0	66.3	51.4	45.0	37.8
12/30/2010	70.9	65.4	52.2	48.6	44.1
12/31/2010	69.2	63.0	49.3	45.1	40.7
1/1/2011	70.7	65.1	50.8	46.3	40.3

1/2/2011	67.2	62.3	47.9	42.0	36.2
1/3/2011	67.0	62.0	48.2	42.3	36.6
1/4/2011	71.2	67.1	50.6	43.1	36.6
1/5/2011	72.6	68.4	51.4	43.6	36.6
1/6/2011	64.6	59.5	46.1	40.4	35.7
1/7/2011	69.3	64.6	50.4	44.2	37.5
1/8/2011	67.7	62.9	49.2	43.4	37.4
1/9/2011	66.9	61.9	48.4	42.8	37.7
1/10/2011	66.5	61.7	48.3	43.9	37.6
1/11/2011	62.8	57.7	45.1	39.3	35.1
1/12/2011	64.0	58.8	46.0	40.2	35.9
1/13/2011	64.3	59.1	46.3	40.5	36.3
1/14/2011	67.2	62.3	48.5	42.8	38.4
1/15/2011	67.9	62.8	49.3	43.7	38.1
1/16/2011	63.1	57.6	45.2	39.2	35.5
1/17/2011	60.3	54.6	43.5	36.7	34.0
1/18/2011	71.8	67.5	51.6	44.0	37.4
1/19/2011	73.5	70.1	52.0	43.4	37.9
1/20/2011	64.5	59.1	46.5	40.4	35.6
1/21/2011	66.9	61.3	48.4	42.5	37.2
1/22/2011	67.3	61.8	48.4	42.2	37.7
1/23/2011	74.5	70.9	51.3	41.8	35.1
1/24/2011	66.3	60.3	47.1	42.2	36.6
1/25/2011	68.7	63.3	49.4	46.3	38.5
1/26/2011	68.8	63.8	49.8	43.8	37.9
1/27/2011	72.8	68.8	51.9	44.4	36.8
1/28/2011	72.1	68.1	51.8	44.5	36.8
1/29/2011	69.2	64.3	50.3	44.4	38.2
1/30/2011	66.4	61.4	48.1	42.5	37.2
1/31/2011	68.0	63.1	49.2	43.6	38.0
2/1/2011	69.8	65.0	50.5	44.6	38.1
2/2/2011	68.5	62.3	48.5	43.4	37.8
2/3/2011	67.3	58.1	45.1	39.3	35.4
2/4/2011	67.3	59.9	46.4	40.3	35.5
2/5/2011	72.1	67.6	52.2	45.0	37.2
2/6/2011	68.7	63.5	49.5	43.8	37.5

2/7/2011	66.6	59.0	46.2	40.8	36.0
2/8/2011	67.9	62.3	49.2	45.4	38.7
2/9/2011	70.4	65.8	51.0	44.8	37.8
2/10/2011	71.1	66.5	51.2	44.8	37.9
2/11/2011	69.4	64.2	49.6	43.3	37.4
2/12/2011	66.9	60.3	46.9	41.4	36.8
2/13/2011	72.2	67.9	52.1	44.9	37.4
2/14/2011	71.2	66.1	51.6	46.6	42.3
2/15/2011	69.6	61.9	49.7	46.1	42.1
2/16/2011	70.4	63.0	48.9	43.1	37.3
2/17/2011	70.4	64.6	50.7	45.0	38.1
2/18/2011	67.8	58.4	44.9	39.5	35.2
2/19/2011	71.1	62.9	48.9	43.3	37.5
2/20/2011	70.6	64.9	50.8	44.5	37.8
2/21/2011	67.3	57.0	44.1	39.0	35.1
2/22/2011	70.9	63.6	49.6	43.8	37.5
2/23/2011	70.4	64.2	50.1	44.2	37.8
2/24/2011	73.1	69.0	52.9	45.4	37.8
2/25/2011	66.8	61.8	48.0	42.3	36.8
2/26/2011	69.9	64.4	50.0	44.2	38.0
2/27/2011	70.1	63.9	49.7	44.2	38.4
2/28/2011	70.3	58.4	44.1	39.0	35.1
3/1/2011	71.0	57.8	42.9	37.7	34.4
3/2/2011	70.9	62.4	48.3	42.8	37.2
3/3/2011	73.3	67.3	52.7	47.6	43.2
3/4/2011	70.5	65.5	54.7	52.3	49.7
3/5/2011	67.5	57.4	45.0	41.7	38.8
3/6/2011	70.2	63.8	51.6	48.4	44.5
3/7/2011	70.9	64.4	53.3	51.6	48.9
3/8/2011	69.1	61.2	53.5	53.0	50.3
3/9/2011	66.1	51.0	40.3	38.6	38.4
3/10/2011	68.2	58.9	51.0	50.3	47.6
3/11/2011	69.4	54.6	43.4	41.4	39.3
3/12/2011	70.5	62.4	49.0	45.1	39.5
3/13/2011	67.1	56.4	47.9	46.8	44.2
3/14/2011	68.2	57.7	48.8	47.4	44.9

3/15/2011	68.0	59.8	50.0	48.1	45.2
3/16/2011	65.1	55.2	47.3	46.4	43.8
3/17/2011	65.2	58.0	46.6	44.1	42.6
3/18/2011	73.6	68.0	58.6	56.1	50.7
3/19/2011	67.1	61.0	47.0	40.7	35.3
3/20/2011	74.0	70.1	52.6	43.9	36.2
3/21/2011	66.9	61.7	48.2	42.5	36.9
3/22/2011	66.9	61.5	48.0	43.4	37.7
3/23/2011	66.1	61.3	46.7	40.9	36.0
3/24/2011	65.0	57.1	44.5	38.9	35.2
3/25/2011	63.6	53.3	41.1	34.7	32.6
3/26/2011	66.8	60.6	47.0	41.6	36.8
3/27/2011	67.2	61.1	47.5	42.1	37.0
3/28/2011	67.3	61.2	47.7	42.2	37.0
3/29/2011	68.9	63.6	49.6	44.1	38.2
3/30/2011	72.5	67.5	54.1	50.1	43.7
3/31/2011	71.9	66.9	56.4	53.6	49.4
4/1/2011	64.0	56.8	45.9	43.2	40.6
4/2/2011	70.2	65.2	56.0	55.0	49.9
4/3/2011	72.7	68.2	53.1	46.9	39.8
4/4/2011	72.2	67.3	57.9	56.7	52.5
4/5/2011	69.9	64.8	50.8	44.7	37.9
4/6/2011	70.9	66.5	51.8	45.4	37.4
4/7/2011	73.9	69.9	52.9	43.4	35.9
4/8/2011	72.8	68.8	52.1	43.6	35.9
4/9/2011	72.7	68.8	51.8	43.7	36.3
4/10/2011	66.9	61.3	47.6	41.8	36.5
4/11/2011	70.3	65.8	51.3	45.1	38.1
4/12/2011	71.7	67.3	52.1	44.8	37.1
4/13/2011	71.4	67.0	53.9	49.6	44.8
4/14/2011	66.7	61.0	51.2	48.3	43.6
4/15/2011	63.3	55.9	43.4	39.2	36.4
4/16/2011	59.5	46.9	39.2	34.5	33.7
4/17/2011	61.4	52.6	43.7	42.0	40.1
4/18/2011	65.4	59.6	50.7	48.8	45.1
4/19/2011	67.8	56.5	43.8	42.1	40.3

4/20/2011	63.5	54.1	43.5	41.5	40.6
4/21/2011	65.9	59.6	49.6	48.8	46.8
4/22/2011	69.6	64.3	54.6	53.8	51.3
4/23/2011	72.1	66.9	55.5	53.2	47.9
4/24/2011	74.1	69.5	60.0	58.3	52.7
4/25/2011	70.8	65.6	58.0	55.3	50.4
4/26/2011	71.2	64.0	54.8	52.8	49.9
4/27/2011	71.4	64.1	49.2	42.4	35.6
4/28/2011	69.0	60.7	46.6	40.9	35.4
4/29/2011	64.4	53.5	41.0	35.0	32.8
4/30/2011	65.4	53.5	40.7	34.2	32.5
5/1/2011	68.5	57.0	40.8	34.1	32.5
5/2/2011	79.9	68.4	52.8	46.7	45.3
5/3/2011	68.7	60.5	46.4	40.9	35.7
5/4/2011	62.4	56.7	44.2	38.5	34.1
5/5/2011	64.4	59.1	46.1	40.6	35.0
5/6/2011	64.7	59.7	46.5	40.7	34.7
5/7/2011	64.1	59.4	46.0	43.0	35.7
5/8/2011	62.9	48.0	37.0	34.4	31.5
5/9/2011	67.0	60.3	46.7	40.7	34.5
5/10/2011	72.9	66.0	51.0	44.6	37.1
5/11/2011	66.0	57.9	42.9	37.2	33.6
5/12/2011	69.9	62.1	44.2	38.6	34.4
5/13/2011	65.9	59.2	45.5	41.4	35.9
5/14/2011	69.7	64.9	51.1	45.2	37.1
5/15/2011	73.2	64.8	50.3	44.0	36.6
5/16/2011	64.3	57.8	44.7	39.1	34.0
5/17/2011	52.1	42.2	35.4	29.3	29.9
5/18/2011	57.6	50.0	38.1	33.7	31.6
5/19/2011	NA	NA	NA	NA	NA

ACKNOWLEDGMENTS

Thanks are extended to the captains and crews of the R/V Miller Freeman and R/V Oscar Dyson for their efforts in deploying and recovering the mooring instruments. Support from Phyllis Stabeno, Bill Floering and Carol Dewitt (NOAA PMEL) made it possible to include the passive acoustic recorders into the moorings. Jeffrey Nystuen (APL UW) graciously provided the soundscape spectra image.

AUTHOR CONTRIBUTIONS

Conceived and designed the experiments: JLMO LEM. Performed the experiments: JLMO. Analyzed the data: JLMO LEM. Contributed reagents/ materials/analysis tools: JLMO LEM. Contributed to the writing of the manuscript: JLMO LEM.

REFERENCES

1. Miksis-Olds JL, Parks SE (2011) Seasonal trends in acoustic detection of ribbon seals (*Histriophoca fasciata*) in the Bering Sea. Aquatic Mammals 37: 464–471. doi: 10.1578/am.37.4.2011.464

2. Miksis-Olds JL, Stabeno PJ, Napp JM, Pinchuk AI, Nystuen JA, et al. (2013) Ecosystem response to a temporary sea ice retreat in the Bering Sea. Prog Oceanography 111: 38–51. doi: 10.1016/j.pocean.2012.10.010

3. Poulter TC (1968) Marine mammals. p. 405–465. *in* Seboek T, ed. Animal Communication. Indiana University Press.

4. Ray C, Watkins WA, Burns J (1969) The underwater song of *Erignathus* (bearded seal). Zoologica 54: 79–83.

5. Burns JJ (1981) Bearded seal, *Erignathus barbatus* (Erxleben, 1777). P. 145–170.*in* Ridgway SS, Harrison RJ, eds. Handbook of Marine Mammals. Academic Press.

6. Stirling I, Calvert W, Cleator H (1983) Underwater vocalizations as a tool for studying the distribution and relative abundance of wintering pinnipeds in the High Arctic. Arctic 36: 262–274. doi: 10.14430/ arctic2275

7. Cleator H, Stirling I, Smith TG (1989) Underwater vocalizations of the bearded seal (*Erignathus barbatus*). Can J Zool 67: 1900–1910. doi: 10.1139/z89-272

8. Cleator HJ, Stirling I (1990) Winter distribution of bearded seals (*Erignathus barbatus*) in the Penny Strait Area, Northwest Territories, as determined by underwater vocalisations. Can J Fish Aquat Sci 47: 1071–1109. doi: 10.1139/f90-123

9. Sonafrank N, Elsner R, Wartzok D (1983) Under-ice navigation by the spotted seal,*Phoca largha*. Abstract. Fifth Biennial Conf. on the Biol. of Mar. Mammals, Boston, November 1983.

10. Madsen PT, Surlykke A (2013) Functional convergence in bat and

toothed whale biosonars. Physiology 28: 276–283. doi: 10.1152/physiol.00008.2013

11. Bradbury JW, Vehrencamp SL (1998) Principles of Animal Communication. Sinauer Associates, Inc. 882 p.

12. Simpson SD, Meekan M, Montgomery J, McCauley R, Jeffs A (2005) Homeward sound. Science 308: 221. doi: 10.1126/science.1107406

13. Slabbekoorn H, Bouton N (2008) Soundscape orientation: a new field in need of sound investigation. Anim Behav 76: e5–e8. doi: 10.1016/j.anbehav.2008.06.010

14. van Opzeeland I, Slabbekoorn H (2012) Importance of underwater sounds for migration of fish and aquatic mammals. 357–359. *in* Popper AN, Hawkins A, eds. Effects of Noise on Aquatic Life. Springer Science+Business Media, LLC.

15. Lillis A, Eggleston DB, Bohnenstiehl DR (2013) Oyster larvae settle in response to habitat-associated underwater sounds. PLoS ONE 8: e79337. doi: 10.1371/journal.pone.0079337

16. Pijanowski BC, Villanueva-Rivera LJ, Dumyahn SL, Farina A, Krause BL, et al. (2011) Soundscape ecology the science of sound in the landscape. BioScience 61 (3): 203–216. doi: 10.1525/bio.2011.61.3.6

17. Bormpoudakis D, Sueur J, Pantis JD (2013) Spatial heterogeneity of ambient sound at the habitat level: ecological implications and applications. Landscape Ecol 28: 495–506. doi: 10.1007/s10980-013-9849-1

18. Norris KS (1967) Some observations on the migration and orientation of marine mammals. 101–102. *in* Storm RM, ed. Animal Orientation and Navigation. Proceedings of the 27th Annual Biology Colloquium. Oregon State University Press. 134p.

19. Able KP (1980) Mechanisms of orientation, navigation, and homing. 283–373. *in*Gauthreaux Jr SA, ed. Animal Migration, Orientation, and Navigation. Academic Press.

20. Kenney RD, Mayo CA, Winn HE (2001) Migration and foraging strategies at varying spatial scales in western North Atlantic right whales: a review of hypothesis. J Cet Res Manag (Special Issue) 2: 251–260.

21. Gordon JCD, Tyack P (2002) Acoustic techniques for studying cetaceans. 293–324.*in* Evans PGH, Raga JA, eds. Marine Mammals: Biology and Conservation. Kluwer Academic.

22. Mate BR, Urban-Ramirez J (2003) A note on the route and speed of a gray whale in its northern migration from Mexico to central California,

tracked by satellite-monitored radio tag. J Cet Res Manag 5(2): 155–157.

23.   Jeffs A, Tolimieri N, Montomery JC (2003) Crabs on cue for the coast: the use of underwater sound for orientation by pelagic crab stages. Mar Freshwat Res 54: 841–845.

24.   Tolimieri N, Haine O, Jeffs A, McCauley R, Montgomery J (2004) Directional orientation of pomacentrid larvae to ambient reef sound. Coral Reefs 23: 184–191. doi: 10.1007/s00338-004-0383-0

25.   Radford CA, Stanley JA, Simpson SD, Jeffs AG (2011) Juvenile coral reef fish use sound to locate habitats. Coral Reefs 30: 295–305. doi: 10.1007/s00338-010-0710-6

26.   Stanley JA, Radford CA, Jeffs AG (2012) Location, location, location: finding a suitable home among the noise. Proc R Soc B 270: 3622–3631. doi: 10.1098/rspb.2012.0697

27.   Rodriguez A, Gasc A, Pavoine S, Grandcolas P, Gaucher P, et al. (2014) Temporal and spatial variability of animal sounds within a Neotropical forest. Ecological Informatics 21: 133–143. doi: 10.1016/j.ecoinf.2013.12.006

28.   Sueur J, Pavoine S, Hamerlynck O, Duvail S (2008) Rapid acoustic survey for biodiversity appraisal. PLoS ONE 3: e4065. doi: 10.1371/journal.pone.0004065

29.   Radford CA, Stanley JA, Tindle CT, Montgomery JC, Jeffs AG (2010) Localised coastal habitats have distinct underwater sound signatures. Mar Ecol Prog Ser 401: 21–29. doi: 10.3354/meps08451

30.   McWilliam JN, Hawkins AD (2013) A comparison of inshore marine soundscapes. J Exp Mar Biol Ecol 446: 166–176. doi: 10.1016/j.jembe.2013.05.012

31.   Stanley JA, Radford CA, Jeffs AG (2011) Behavioural response thresholds in New Zealand crab megalopae to ambient underwater sound. PLoS ONE 6: e28572. doi: 10.1371/journal.pone.0028572

32.   Simpson SD, Meekan MG, Jeffs A, Montogomery JC, McCauley RD (2008) Settlement-stage coral reef fish prefer the higher-frequency invertebrate-generated audible component of reef noise. Anim Behav 75: 1861–1868. doi: 10.1016/j.anbehav.2007.11.004

33.   Stabeno PJ, Napp J, Mordy C, Whitledge T (2010) Factors influencing physical structure and lower trophic levels of the eastern Bering Sea shelf in 2005: Sea ice, tides and winds. Prog Oceanography 85(3–4): 180–196. doi: 10.1016/j.pocean.2010.02.010

34.   Brierley AS, Saunders RA, Bone DG, Murphy EJ, Enderlein P, et al. (2006)

Use of moored acoustic instruments to measure short-term variability in abundance of Antarctic krill. Limnology and Oceanography: Methods 4: 18–29. doi: 10.4319/lom.2006.4.18

35. Kunze E, Dower JF, Beveridge I, Dewey R, Bartlett KP (2006) Observations of Biologically Generated Turbulence in a Coastal Inlet. Science 22: 1768–1770. doi: 10.1126/science.1129378

36. Smith PE, Flerx W, Hewitt RP (1985) The CalCOFI vertical egg tow (CalVET) net. p. 23–33. *in* Lasker R, ed. An Egg Production Method for Estimating Spawning Biomass of Pelagic Fish: Application to the northern anchovy *Engraulis mordax*. NOAA Technical Report NMFS 36, U.S. Department of Commerce.

37. Nystuen JA (1998) Temporal sampling requirements for Autonomous Rain Gauges, J. Atmos. Ocean. Technol. 15: 1254–1261. doi: 10.1175/1520-0426(1998)015<1253:tsrfar>2.0.co;2

38. Nystuen JA, Amitai E, Anagnostou EN, Anagnostou MN (2008) Spatial averaging of oceanic rainfall variability using underwater sound: Ionian Sea Rainfall Experiment 2004, J Acoust Soc Am. 123: 1952–1962. doi: 10.1121/1.2871485

39. Miksis-Olds JL, Nystuen JA, Parks SE (2010) Detecting marine mammals with an adaptive sub-sampling recorder in the Bering Sea. J App Acoust 71: 1087–1092. doi: 10.1016/j.apacoust.2010.05.010

40. Sousa-Lima RS, Norris TF, Oswald JN, Fernandes DP (2013) A review and inventory of fixed autonomous recorders for passive acoustic monitoring of marine mammals. Aquatic Mammals 39: 23–53. doi: 10.1578/am.39.1.2013.23

41. Watkins WA, Ray GC (1977) Underwater sounds from ribbon seal, *Phoca (Histriophoca) fasciata.*. Fish Bulletin 75: 450–453.

42. Jones JM, Thayre BJ, Roth EH, Mahoney M, Sia I, et al. (2014) Ringed, bearded, and ribbon seal vocalizations north of Barrow, Alaska: Seasonal presence and relationship with sea ice. Arctic 67: 203–222. doi: 10.14430/arctic4388

43. Watkins JL, Brierley AS (2002) Verification of acoustic techniques used to identify and size Antarctic krill. ICES J Mar Sci 59: 1326–1336. doi: 10.1006/jmsc.2002.1309

44. Reiss CR, Cossio AM, Loeb VL, Demer DA (2008) Variations in the biomass of Antarctic krill (*Euphausia superba*) around the South Shetland Islands, 1996–2006. ICES J Mar Sci 65 (4): 497–508. doi: 10.1093/icesjms/fsn033

45. De Robertis A, McKelvey DR, Ressler PH (2010) Development and application of an empirical multifrequency method for backscatter classification. Can J Fish Aquat Sci 67: 1459–1474. doi: 10.1139/f10-075

46. Demer DA, Conti SG (2003) Validation of the stochastic distorted-wave Born approximation model with broad bandwidth total target strength measurements of Antarctic krill. ICES J Mar Sci 60: 625–635. doi: 10.1016/s1054-3139(03)00063-8

47. Zuur AF, Ieno EN, Walker NJ, Saveliev AA, Smith GM (2009) Mixed effects models and extensions in ecology with R: Springer Science+Buisness Media.

48. Zuur A (2010) AED: Data files used in Mixed effects models and extensions in ecology with R (in Zuur et al. 2009). R package version 1.0.

49. Chatterjee S, Hadi AS (2006) Regression Analysis by Example. John Wiley and Sons.

50. McCullagh P, Nelder JA (1989) Generalized Linear Models, Chapman & Hall/CRC.

51. Friedlaender AS, Halpin PN, Qian SS, Lawson GL, Wiebe PH, et al. (2006) Whale distribution in relation to prey abundance and oceanographic processes in shelf waters of the Western Antarctic Peninsula. Mar Ecol Prog Ser 317: 297–310. doi: 10.3354/meps317297

52. Wagner T, Sweka JA (2011) Evaluation of Hypotheses for Describing Temporal Trends in Atlantic Salmon Parr Densities in Northeast US Rivers. N Am J Fish Manag 31(2): 340–351. doi: 10.1080/02755947.2011.574081

53. R Development Core Team (2011) R: A Language and Environment for Statistical Computing. R Foundation for Statistical Computing, Vienna, Austria. Available:http://www.R-project.org/.

54. Venables WN, Ripley BD (2002) Modern Applied Statistics with S, Springer Verlag.

55. Wood SN (2006) Generalized Additive Models: An Introduction with R, CRC Press.

56. Wood SN (2011) Fast stable restricted maximum likelihood and marginal likelihood estimation of semiparametric generalized linear models. J Roy Stat Soc: Ser B (Statistical Methodology).

57. Berta A, Churchill M (2012) Pinniped taxonomy: review of currently recognized species and subspecies, and evidence used for their description. Mamm Rev 42: 207–234. doi: 10.1111/j.1365-2907.2011.00193.x

58. Terhune JM, Ronald K (1975) Underwater hearing sensitivity of two ringed seals (*Pusa hispida*). Can J Zool 53: 227–231. doi: 10.1139/z75-

028

59. Terhune JM (1988) Detection thresholds of a harbour seal to repeated underwater high-frequency, short-duration sinusoidal pulses. Can J Zool 66: 1578–1582. doi: 10.1139/z88-230

60. Reichmuth C, Holt MM, Mulsow J, Sills JM, Southall BL (2013) Comparative assessment of amphibious hearing in pinnipeds. Journal of Comparative Physiology A 199: 491–507. doi: 10.1007/s00359-013-0813-y

61. Sills JM, Southall BL, Reichmuth C (2014) Amphibious hearing in spotted seals (*Phoca largha*): underwater audiograms, aerial audiograms, and critical ratio measurements. J Exp Biol 217: 726–734. doi: 10.1242/jeb.097469

62. Joseph JE, Chiu CS (2010) A computational assessment of the sensitivity of ambient noise level to ocean acidification. J Acoust Soc Am 128(3): EL144–EL149. doi: 10.1121/1.3425738

63. Boyd IL, Frisk G, Urban E, Tyack P, Ausubel J, et al. (2011) An International Quiet Ocean Experiment. Oceanography 24(2): 174–181. doi: 10.5670/oceanog.2011.37

CHAPTER 1

Evgenii Sharkov and Valentina Svalova (2011). Geological-Geomechanical Simulation of the Late Cenozoic Geodynamics in the Alpine-Mediterranean Mobile Belt, New Frontiers in Tectonic Research - General Problems, Sedimentary Basins and Island Arcs, Prof. Evgenii Sharkov (Ed.), ISBN: 978-953-307-595-2, InTech, DOI: 10.5772/25250.

CHAPTER 2

B.A. Dyachkov, M.A. Mizernaya, Nina Maiorova, Zinaida Chernenko, Victor Maiorov and O.N. Kuzmina (2011). Geotectonic Position and Metallogeny of the Greater Altai Geological Structures in the System of the Central-Asian Mobile Belt, New Frontiers in Tectonic Research - General Problems, Sedimentary Basins and Island Arcs, Prof. Evgenii Sharkov (Ed.), ISBN: 978-953-307-595-2, InTech, DOI: 10.5772/21607.

CHAPTER 3

Vsevolod Yutsis, Antonio Tamez-Ponce and Konstantin Krivosheya (2011). Geophysical Modeling of the Surroundings of La Popa Basin, NE Mexico, with Gravity and Magnetic Data, New Frontiers in Tectonic Research - General Problems, Sedimentary Basins and Island Arcs, Prof. Evgenii Sharkov (Ed.), ISBN: 978-953-307-595-2, InTech, DOI: 10.5772/18760.

CHAPTER 4

Feng Gui, Khalil Rahman, Daniel Moos, George Vassilellis, Chao Li, Qing Liu, Fuxiang Zhang, Jianxin Peng, Xuefang Yuan and Guoqing Zou (2013). Optimizing Hydraulic Fracturing Treatment Integrating Geomechanical Analysis and Reservoir Simulation for a Fractured Tight Gas Reservoir, Tarim Basin, China, Effective and Sustainable Hydraulic Fracturing, Dr. Rob Jeffrey (Ed.), ISBN: 978-953-51-1137-5, InTech, DOI: 10.5772/56384.

CHAPTER 5

Nima Gholizadeh Doonechaly, Sheik S. Rahman and Andrei Kotousov (2013). A New Approach to Hydraulic Stimulation of Geothermal Reservoirs by Roughness Induced Fracture Opening, Effective and Sustainable Hydraulic Fracturing, Dr. Rob Jeffrey (Ed.), ISBN: 978-953-51-1137-5, InTech, DOI: 10.5772/56447.

CHAPTER 6

Arellano, V., Barragán, R., Ramírez, M., López, S., Aragón, A., Paredes, A., Casimiro, E. and Reyes, L. (2015) Reservoir Processes Related to Exploitation in Los Azufres (México) Geothermal Field Indicated by Geochemical and Production Monitoring Data. International Journal of Geosciences, 6, 1048-1059. doi: 10.4236/ijg.2015.69083.

CHAPTER 7

Evgenii Sharkov (2011). Does the Tethys Begin to Open Again? Late Cenozoic Tectonomagmatic Activization of the Eurasia from Petrological and Geomechanical Points of View, New Frontiers in Tectonic Research - General Problems, Sedimentary Basins and Island Arcs, Prof. Evgenii Sharkov (Ed.), ISBN: 978-953-307-595-2, InTech, DOI: 10.5772/18839.

CHAPTER 8

Moss, C. and Schmitz, A. (2014) Environmental Flows from Alternate Land Uses in the Delta, Pacific, and the Southeastern States: 1947-2007. Journal of Environmental Protection, 5, 1531-1540. doi:10.4236/jep.2014.516145.

CHAPTER 9

Yugui Yang, Feng Gao, and Yuanming Lai, "Compressive Mechanical Properties and Micromechanical Characteristics of Warm and Ice-Rich Frozen Silt," Advances in Materials Science and Engineering, vol. 2015, Article ID 379560, 7 pages, 2015. doi:10.1155/2015/379560

CHAPTER 10

Saarenheimo J, Rissanen AJ, Arvola L, Nykänen H, Lehmann MF, Tiirola M (2015) Genetic and Environmental Controls on Nitrous Oxide Accumulation in Lakes. PLoS ONE 10(3): e0121201. doi:10.1371/journal.pone.0121201

CHAPTER 11

Miksis-Olds JL, Madden LE (2014) Environmental Predictors of Ice Seal Presence in the Bering Sea. PLoS ONE 9(9): e106998. doi:10.1371/journal.pone.0106998

INDEX